Electrical Designing Workbook
By Tom Henry and Tim Henry

Based on the 1996 NEC® Second Printing August 1996

HENRY PUBLICATIONS SINCE 1985

ISBN 0 - 945495 - 44 - 7

Preface

Electrical designing is basically a matter of providing an arrangement of electrical conductors and equipment to safely and effectively transfer electrical energy from the source to the load.

Perhaps electrical design is not why you got into this business, but it has been said that 70% of all electrical designing is done by the electrician. While the electrical designers are busy designing the large buildings, the electrician is called upon to wire the new piece of equipment just purchased, correct Code violations, or to give a bid for an add-on, or install a larger service to an existing building, etc.

The three major elements of a wiring system are the branch circuit, feeder and service. The electrician is required to be able to select the proper conductors and other equipment which make up the system.

Remember, conductors are not intelligent. They don't know where they are going to be installed (in the basement, attic, soil, free air). They don't know the environment where they have been placed (wet, dry, hot, cold). They do not know how crowded it will be in the conduit or how many conductors beside them will be carrying current. They do not know how far they will be run or how long and hard they will have to work. They do not know if they will be properly protected and insulated from heat and moisture. Conductors aren't very smart.

Electrical terms and definitions must be fully understood, as the rules must be followed for each part of the system.

The electrician must understand the definition of a demand factor and where it applies in designing. The Code permits a reduction in the computed load in some instances as all of the connected load will not be operating at the same time. Connected load x demand factor = maximum demand.

Lower demand loads result in a smaller service size which is a savings when bidding the job. But always remember the Code is not for future expansion of electrical use. Allowance for load growth is probably the most neglected consideration in electrical design. In this book you will learn to design both the Code minimum and for future growth and cost savings.

Installing *larger* conductors initially provides a savings in energy costs. All conductors waste energy in the form of heat; I^2R. By using larger conductors than the Code required minimum can actually result in paying for the larger conductors within a year with the savings on the electric bill and will continue to generate significant energy and dollar savings year after year.

This is a very neglected area in the electrical industry. These kind of savings need to be explained to the customer. By upsizing wire the contractor can demonstrate the real savings to the customer as well as insurance against future needs. By using a conductor only one size larger than the Code required minimum can provide a quick payback in savings.

This book will help the electrical person understand about proper designing and how to explain to the customer proper designing for future expansion and with a cost savings.

We are the ones that should be doing the selling of GFCI receptacles, larger size wire, etc. Today the plumber is the one doing the selling. He is the one installing the gold fixtures in the bathroom and installing the whirlpool instead of the tub!

This book is not written for the first year apprentice. This is an excellent book for the final year apprentice, contractor, designer and inspector.When this book is used for classroom instructions, the instructor must first teach the formulas and variables from this book.

The world works and lives at the end of a wire. The world needs people to install the wire *correctly*.

CONTENTS

CONTENTS

PAGE

CONTENTS

THE SYSTEM

THE SYSTEM

The distribution system begins at the generating station or the substation of the utility company. The high voltage transmission lines are transformed down until the service voltage is at a safe level.

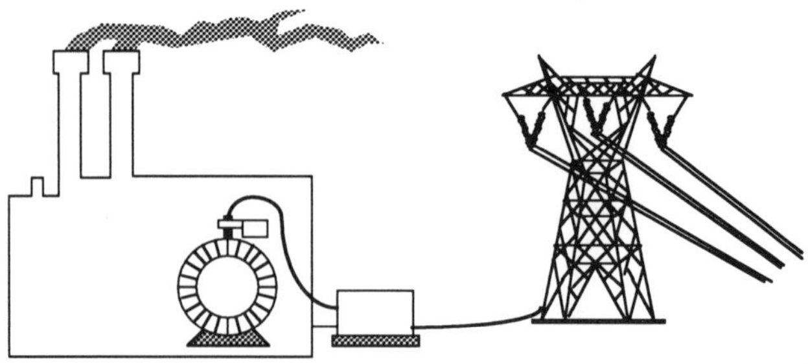

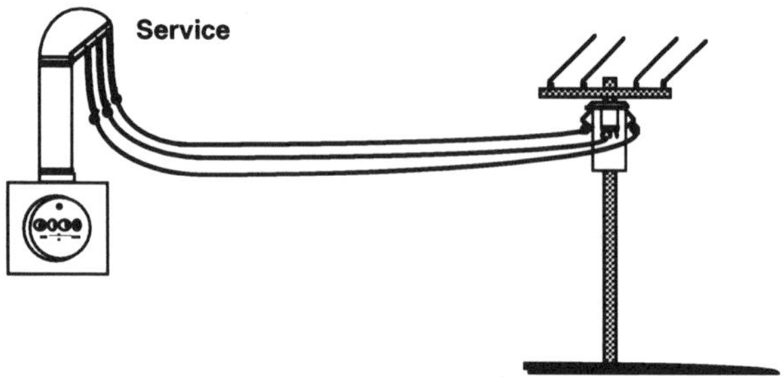

The premises wiring system begins at the service and extends to all the equipment.

The load is the first consideration, then the branch circuit is required to supply the load. Now comes the selection of conductors, equipment, etc.

In designing it is very important that you understand the definition of branch circuit, feeder, service, etc. as the design rules change on each part of the system.

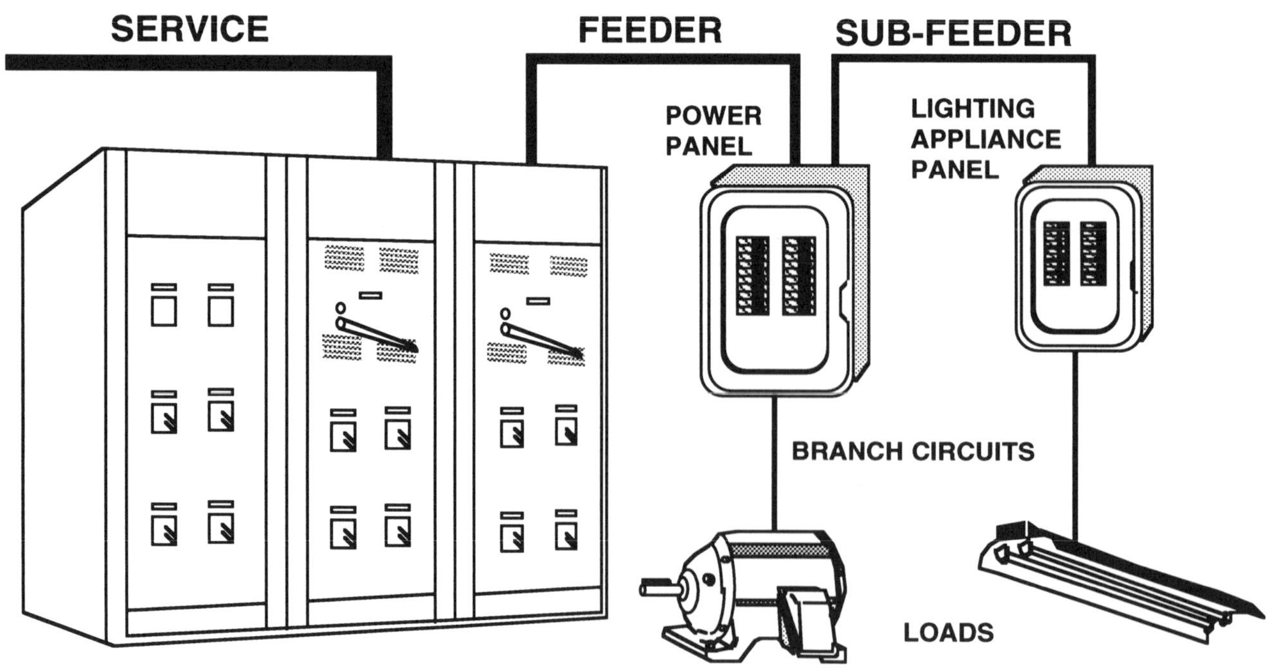

A very common example is when an electrician is called to install a branch circuit in an existing building.

Whether the branch circuit is a lighting circuit, receptacle circuit, or a circuit to a piece of equipment, the first thing that must be determined is the *load* of this circuit.

Once the load is determined, the type of branch circuit must be considered. Is the branch circuit a general purpose, appliance, individual, multiwire, motor, etc.? Each type of branch circuit has different rules in designing the circuit.

The *variables* such as continuous load, distance, temperature, number of conductors, etc. enter into the designing.

Electrical designing is *not* an automatic procedure that merely requires the use of a formula and adding the loads together.

Although the following pages may seem to be an involved and almost laborious evaluation of circuit designing, there is really no alternative if all the *related* factors are to be effectively included. Any other design method quickly becomes a guess and includes the risk of being inadequate, dangerous, and costly.

Too often designing is to the *minimum* standards of the Code.The Code permits 3% voltage drop in a branch circuit. Voltage drop is wasted electricity. In good designing, voltage drop is held to no more than 1%.

The Code is not for the future. Good designing would provide one *spare* circuit for each five circuits in use.

Homeruns should be limited to a maximum of 100 feet. Careful layout of panelboard locations and use of a sufficient number of panelboards will avoid the problem of homeruns over 100 feet.

Larger size conductors are installed to provide energy savings from the *power loss* generated by heating the smaller conductors.

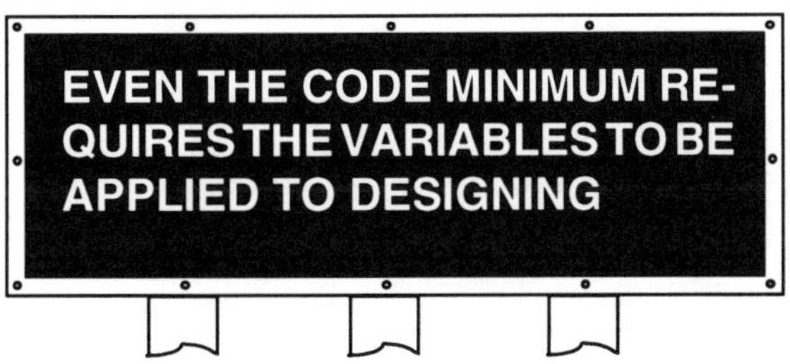

THE STEPS

We must first know the *load* in order to design a circuit. Even receptacle circuits where the actual load may be unknown as it is not known exactly what loads will be plugged into the receptacles, the designer should size the circuit to the rating of the circuit. If a 20 amp breaker is installed, the circuit is rated for 20 amps and must be designed to this load.

Once the load is determined and the overcurrent device (fuse or breaker) is selected, the wire is selected and must have the ampacity to be protected by the overcurrent device.

Too often the wire is selected before the overcurrent device resulting in the wire being not properly protected as required.

The *type* of load has an affect on the sizing of the fuse or breaker.

STEP 1

?

LOAD

noncontinuous
single
motor
receptacles
hoist
welder
continuous
multiple
a/c
lights

STEP 2

?

FUSE or CB

next higher size
dual element
inverse time
nontime delay
instantaneous

STEP 3

?

WIRE SIZE

voltage drop
note 8
correction factor
copper
60°C insulation
75°C insulation
90°C insulation
aluminum
equipment rating 60°C or 75°C

OVERCURRENT PROTECTION

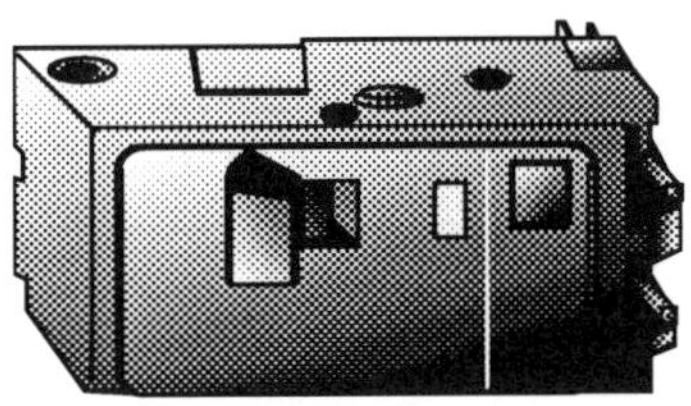

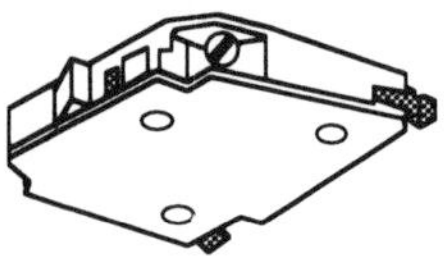

OVERCURRENT PROTECTION

The fuse or breaker for conductors or equipment must automatically open the circuit it protects if the current flowing in that circuit reaches a value which will cause excessive or dangerous temperatures in the conductors.

Overcurrent protection must provide protection for the conductors and the device from ground fault and short circuit currents. Specifically, this requires that a short-circuit overcurrent device have a proven interrupting rating at least equal to the current which the electrical system can deliver into a short on its load terminals.

The general rule is, conductors shall be protected against overcurrent in accordance with their ampacities, *unless otherwise permitted.*

GENERAL RULE

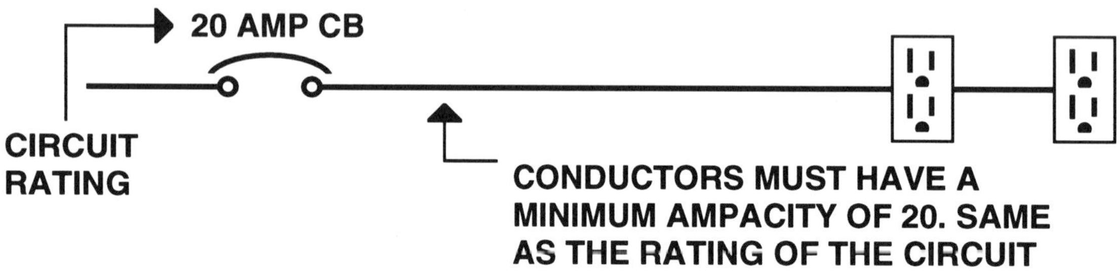

Often we choose to go up to the next higher standard size. This is permitted in some instances, but the general rule is you don't go up, *unless otherwise permitted* by the Code.

This question often arises: What size breaker is required in the condition shown below?

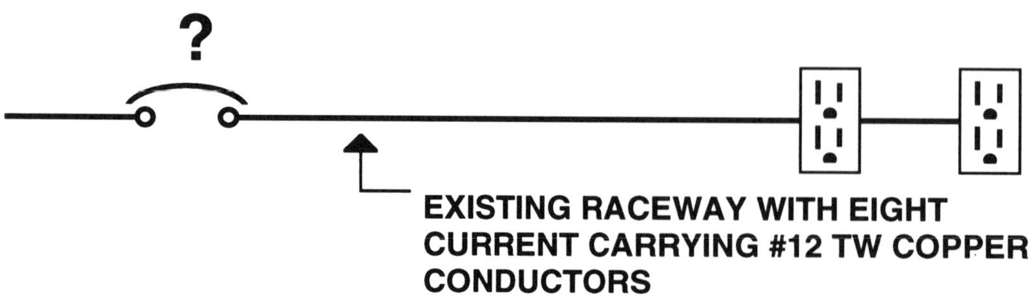

Table 310-16: A #12 TW has an ampacity of 25. The Note 8 factor requires the 25 ampacity to be reduced by 70%. 25 x 70% = 17.5 ampacity and maximum load permitted on this conductor.

Section 240-3b states the next *higher* size overcurrent protective device can be used where the ampacity of the conductor does not correspond with the standard size overcurrent device.

The reduced ampacity is 17.5 amps which is not a standard size fuse or breaker. The next higher standard size from Section 240-6 is 20 amps. If the conductor is *not* part of a *multioutlet branch circuit supplying receptacles for cord and plug connected portable loads*, then 20 amps is permitted per Section 240-3b.

Since this example shows the branch circuit supplying receptacles, the maximum size overcurrent device would be 15 amps. A 15 amp overcurrent device would protect the conductors at 15 amps. *If* a 20 amp overcurrent device was installed, loads could be plugged into the receptacles loading the circuit to the maximum rating of 20 amps. This would be in *excess* of the 17.5 ampacity or maximum loading which would result in exceeding the insulation rating of the conductor.

If the loads on this branch circuit were *hard wired* instead of plug and cord connected receptacles, then the next higher overcurrent device, 20 amp, is permitted.

The electrical designer or electrician designing a lighting branch circuit could encounter this very same example:

A raceway containing eight #12 TW current carrying conductors to light fixtures. The reduced ampacity and maximum load permitted on these conductors is 17.5 amps. If each light fixture has a load of 2 amps, 17.5/2 = 8.75 or a maximum of 8 lights at 2 amps each = 16 amps which would not exceed the maximum load permitted of 17.5.

If eight light fixtures at 2 amps each (known loads) are *hard wired* to the branch circuit, the next higher standard size, 20 amp overcurrent device is permitted.

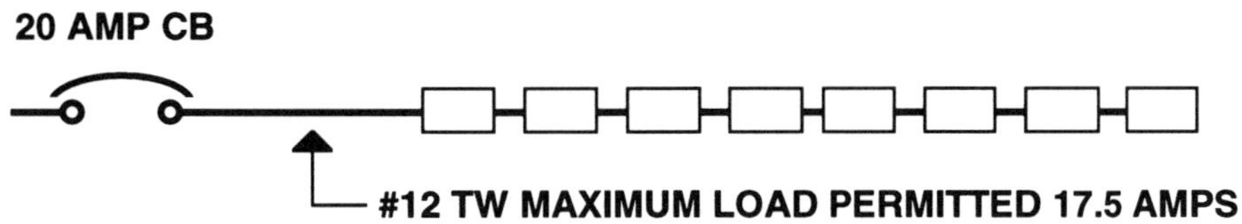

The designer of this circuit must limit the connected load to 17.5 amps or less.

There are cases where the ampacity of a conductor does *not* correspond to a standard rating of an overcurrent device from 240-6. The Code permits going up to the next higher *standard* size up to 800 amps.

A #6 TW with an ampacity of 55 can be protected by a 60 amp breaker.

#6 TW AMPACITY = 55

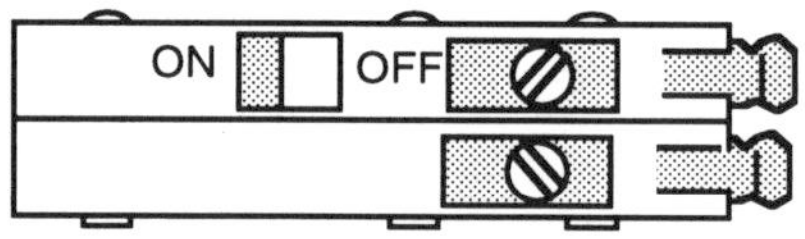

60 amp BREAKER

The example below is a #500 kcmil THW with a normal ampacity of 380 amps. Standard size breakers are 350 and 400. The Code allows a 400 amp to be used.

380 AMPACITY

400 AMP CB

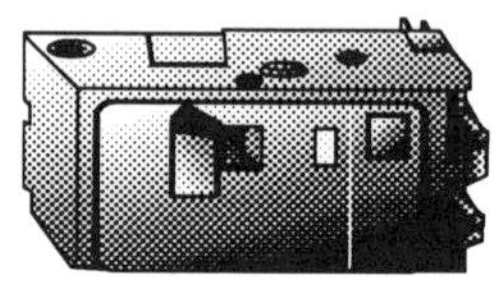

A #700 kcmil THHN with an ampacity of 520 is considered protected by the next higher standard size from section 240-6, which is a 600 amp.

520 AMPACITY

600 AMP CB

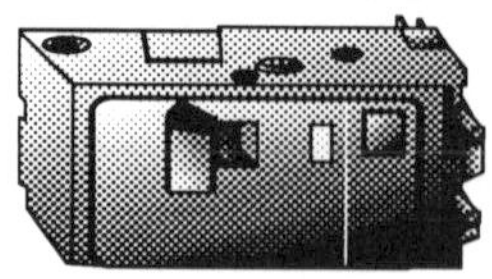

A #750 kcmil THW with an ampacity of 475 amps is considered protected by the next higher standard size; a 500 amp breaker.

By selecting THHN insulation rather than THW, the THHN will assure the conductor insulation will never be overloaded above its ampacity.

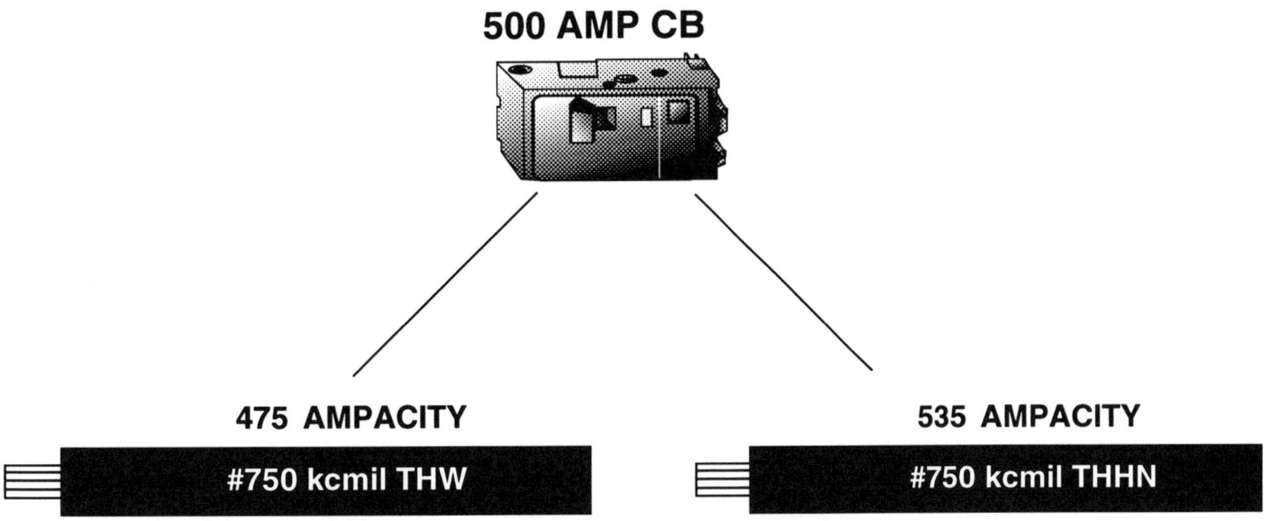

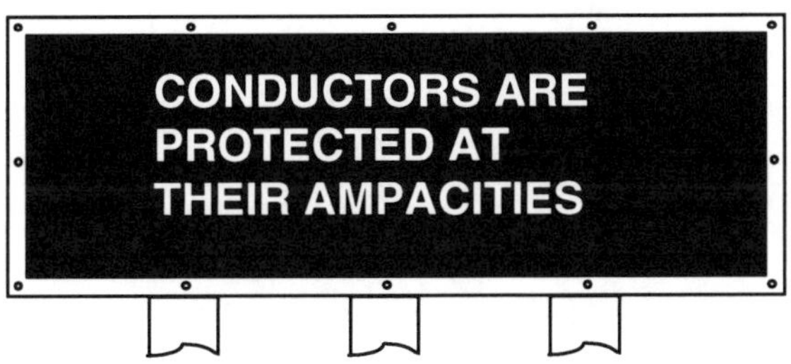

SELECTING THE WIRE

SELECTING THE PROPER CONDUCTOR

As we know, a conductor can be round, solid, stranded, rectangular, copper, aluminum, etc.

SOLID

STRANDED

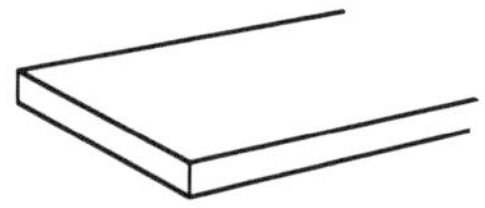

RECTANGULAR

It is very important, in designing, to get started off on the right foot by completely understanding the function of a conductor, the insulation and how temperature effects the conductor.

A conductor is a component of the electrical circuit and is used to carry the current to and from the load in the circuit. The current flowing through the conductor produces heat.

The capacity of current the conductor can carry is referred to as the *ampacity* (amp-capacity).

Ampacity is defined as the current in amperes that a conductor can carry *continuously* under the conditions of use without exceeding its temperature rating.

The current a conductor can safely carry is determined by several factors. The size of the conductor, the alloy (copper or aluminum), the type of insulation, the environment, etc.

Different types of insulations have different ampacities and different temperature ratings for the same size conductor.

#10 TW 30 ampacity
60°C - 140°F temperature rating

#10 THW 35 ampacity
75°C - 167°F temperature rating

#10 THHN 40 ampacity
90°C - 194°F temperature rating

A #10 wire is the same size with a circular mil area of 10380, but as you can see, it has three different ampacities and different temperature ratings.

Insulation around a conductor is like a pipe carrying water. With the water pressure (voltage) and the flow (amperes) of water, the pipe (insulation) is used to contain the water. If the pipe were to "spring a leak" you would lose some of the water pressure.

"Good" insulation means a high resistance to current to keep it from "leaking".

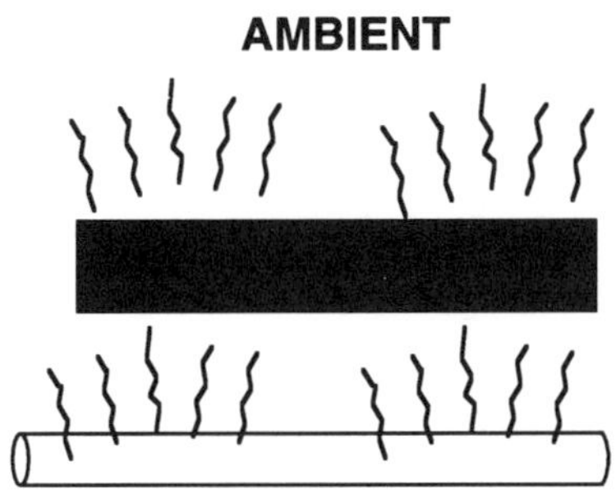

As current is passed through the conductor, the heat that is produced (I^2R) must be dissipated through the insulation to the ambient. Ambient is that which encompasses on all sides.

Ambient temperature is the temperature surrounding the conductor into which the heat of the conductor is dissipated. If the ambient temperature is higher it slows down the dissipation of the heat.

A conductor in free air, as shown below, can carry more amperage than one that is installed in a conduit, as the conduit will entrap the heat. The conductors installed in attics, around a furnace, or a boiler encounters higher ambient temperatures that requires lowering the current flow on the conductor to prevent insulation damage.

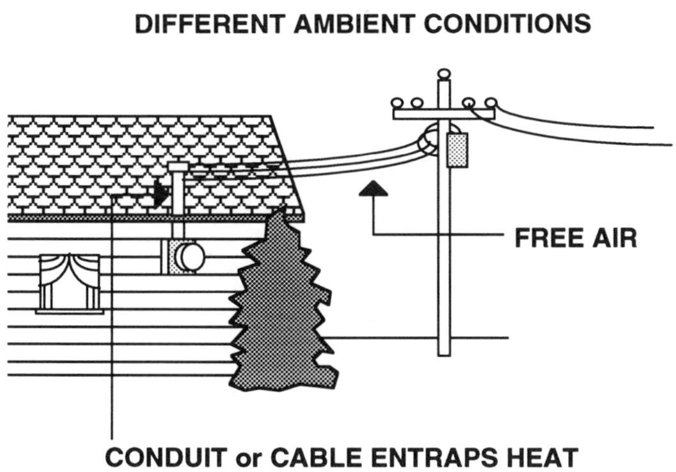

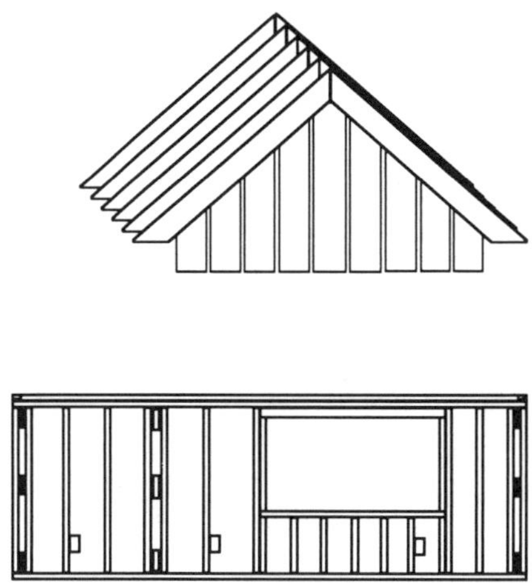

It is important to understand how a conductor dissipates heat.

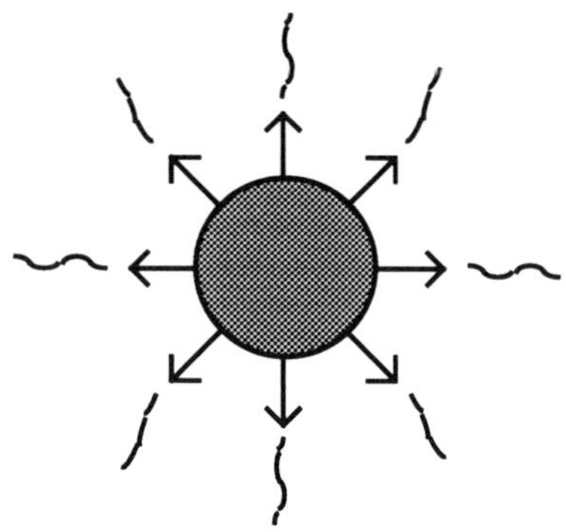

The heat produced from the current flowing through the conductor is dissipated through the skin in all directions.

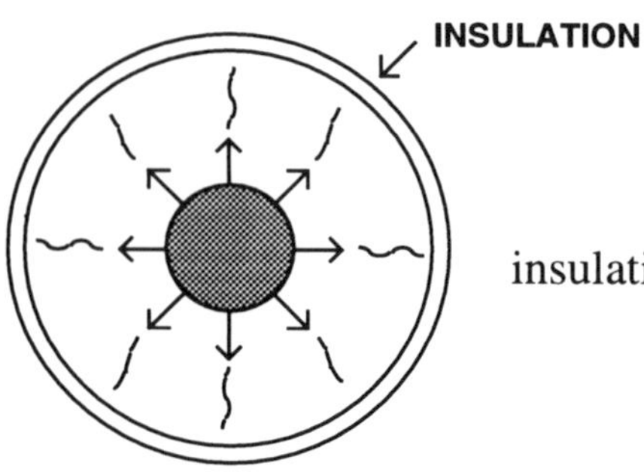

The heat dissipates through the insulation into the surrounding ambient.

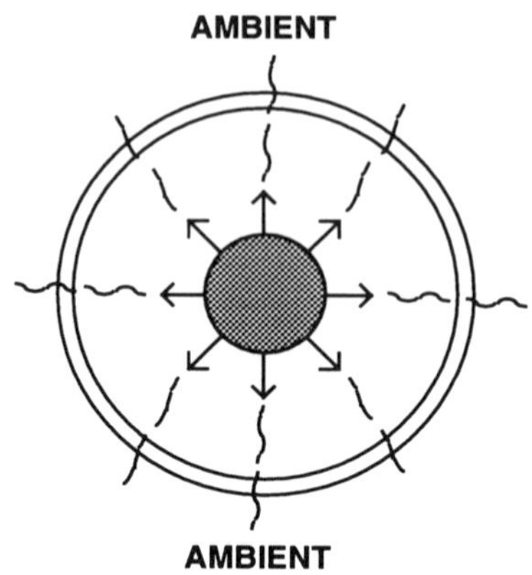

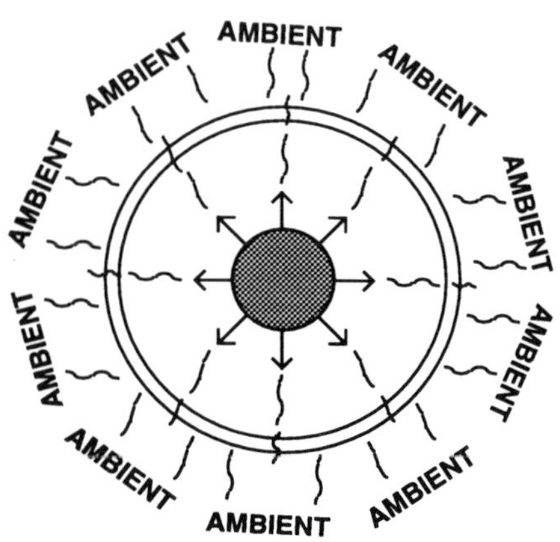

If the surrounding ambient temperature is higher it opposes the dissipation of the heat through the insulation. The ampacity of the conductor must be reduced for the higher ambient temperature. If the ampacity is not reduced the heat is contained longer in the conductor and the insulation, thus causing insulation damage. Even though the ambient is cooler when a conductor is buried, it is slower in dissipating heat the deeper it is buried.

Except for mechanical abuse, the greatest hazard that conductors must endure is *heat*. Conductor insulation can be damaged by excessive heat in various ways, depending on the type of insulation and the degree of overheating. Continued exposure to excessive heat causes insulation to become soft, perhaps to melt, and in extreme cases to burn.

This heat comes from two sources: From the ambient air surrounding the conductors and from the current the conductors must carry. There is a point where an increase in current causes excessive heat even though conducting materials such as copper or aluminum have a low resistivity.

For many years natural rubber was used to insulate conductors, but age along with heat caused such rubber insulation to dry out, crack, and become brittle. Today we have better quality rubber and thermoplastic materials that not only permit thinner insulation on conductors, but also withstand temperature, resulting in higher ampacities of conductors.

The type letter on the insulation indicates its insulation, maximum operating temperature, and application provisions.

#10 RHW

The #10 RHW, the "R" indicates rubber insulation. The "H" indicates 75°C - 167°F maximum operating temperature (insulation rating). The "W" indicates moisture resistant.

#10 THHN

The #10 THHN, the "T" indicates thermoplastic insulation. The "HH" indicates 90°C - 194°F maximum operating temperature (insulation rating). The "N" indicates nylon covering.

The #10 THW has a maximum operating temperature of 75°C which is 167°F.

The #10 THHN has a maximum operating temperature of 90°C which is 194°F. A "HH" rated insulation will allow more heat to be dissipated faster than an "H" rated insulation thus raising the ampacity (the current the conductor can carry safely without damage).

The maximum operating temperature is the insulation rating of the conductor and must not be exceeded. Proper designing is a very important factor.

You must first understand what words mean; such as ampacity, ambient temperature, insulation rating, etc.

#10 TW 30 ampacity
60°C - 140°F temperature rating

A #10 TW conductor has an ampacity of 30 amperes. The insulation rating is 60°C or 140°F.
This does *not* mean that a TW insulation can be installed where the ambient temperature reaches 140°F.

What this means is: If a #10 TW conductor is loaded to the allowable ampacity, 30 amperes in an ambient that has a temperature of 30°C or 86°F, the temperature of the *insulation* will reach 60°C or 140°F.

Table 310-16 the table of ampacity is aimed at designating a level of current that will permit the conductor to reach its thermal limit, but not exceed it.

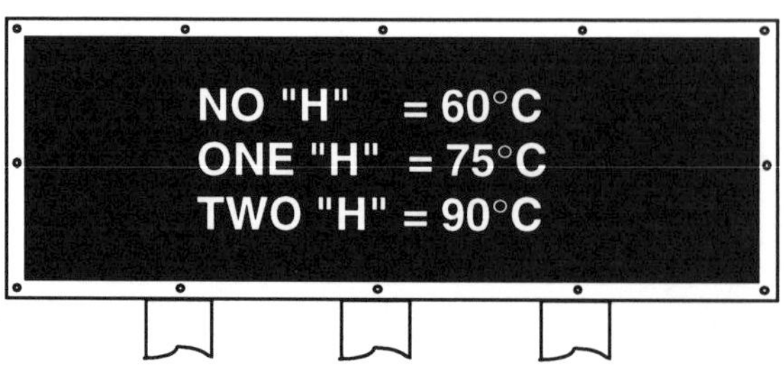

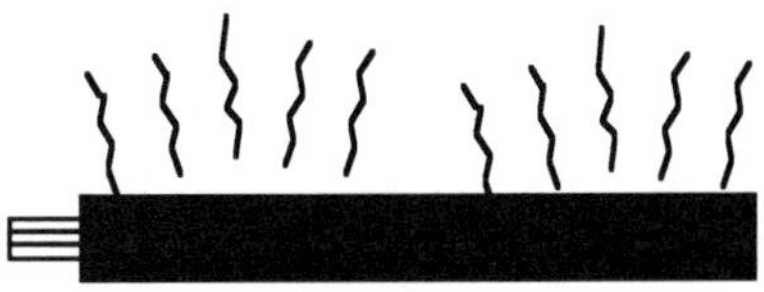

#10 TW 30 amps of current flowing

The 30 amps of current flowing produces heat in the conductor which must dissipate through the insulation to the ambient.

With the ambient temperature at 86°F and with 30 amperes of current flowing through the conductor, a thermometer placed on the *insulation* would read 140°F which is maximum operating temperature for this type insulation (TW).

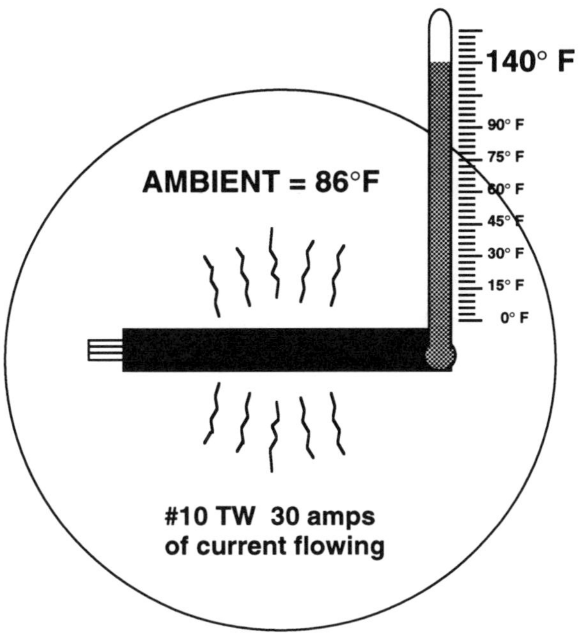

For a #10 TW conductor, any current above 30 amps or any ambient temperature above 86°F will cause insulation damage, as you will exceed the maximum operating temperature of the conductor; 140°F.

Maximum operating temperature = Full ampacity at 86°F.

140°F - 86°F = 54°F for the 30 amperes of current flow in the #10 TW conductor.

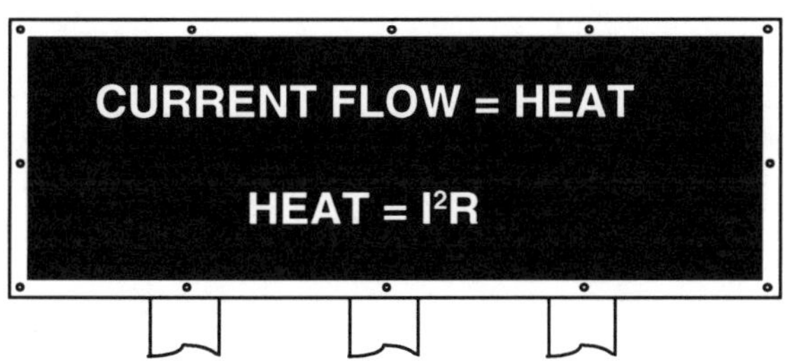

Shown below are the three different insulations; TW, THW and THHN. A #10 TW will allow 30 amperes of current flow through this insulation with an ambient temperature of 86°F without insulation damage. A #10 THW will allow 35 amperes of current flow at 86°F. A #10 THHN will allow 40 amperes at 86°F.

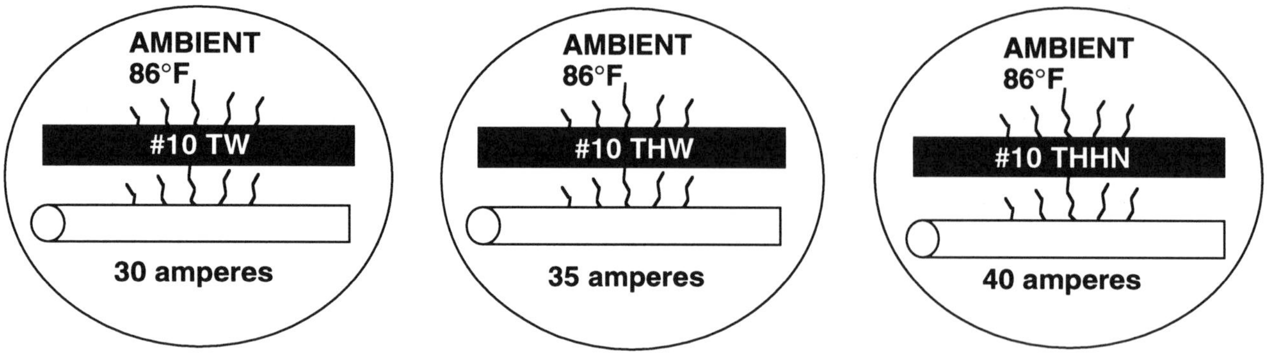

As you can see different insulation types have different ampacities. The better grade of insulation will allow the dissipation of the heat faster so the current carrying capacity (ampacity) is higher.

But, ampacities are based on 30°C which the equivalent is 86°F. If the ambient where the conductor is being placed is higher than 30°C or 86°F, the normal ampacity shall be *lowered.*

With a higher ambient it slows down the dissipation of the heat, so the heat produced by the current flow must be reduced to prevent damage.

AMPACITY IS BASED ON AN AMBIENT OF 86°F

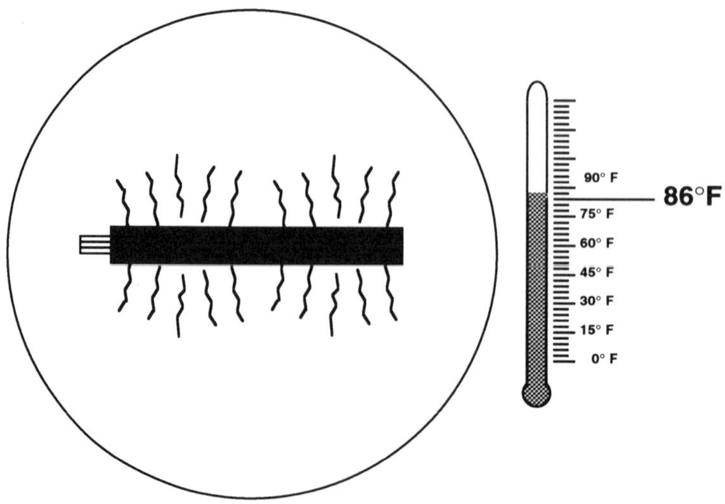

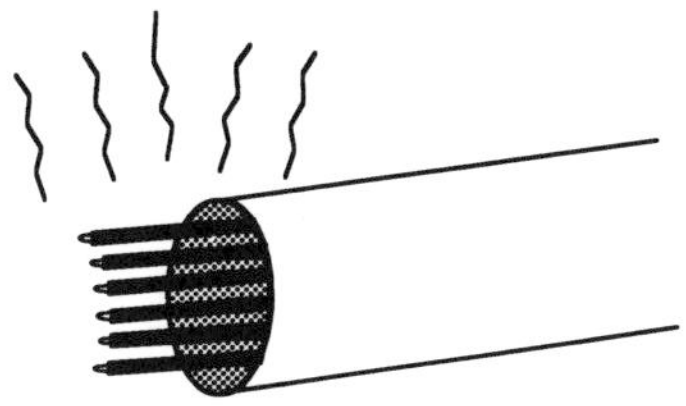

When there are more than three current-carrying conductors in a raceway or cable, the ampacity of each conductor must be reduced as indicated in Note 8 to compensate for heating effects and reduced heat dissipation due to reduced ventilation of individual conductors.

Note 8 also applies when paralleling conductors per section 310-4.

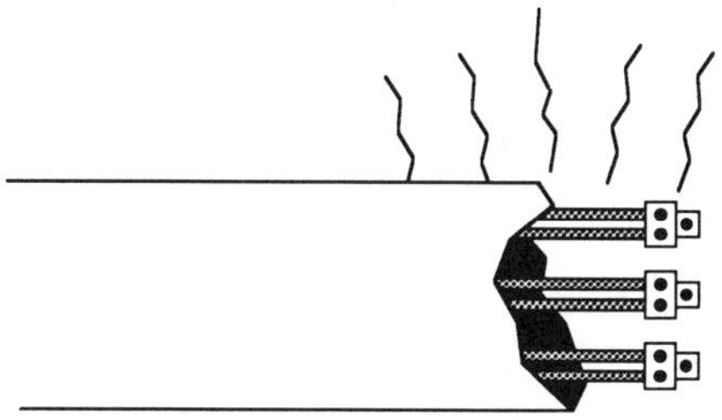

It is wrong to think since you connected two conductors in parallel on one lug that you now only have one conductor. Heat is measured by $W = I^2R$. In parallel you have *two* conductors carrying current producing heat.

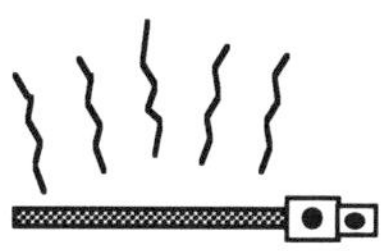

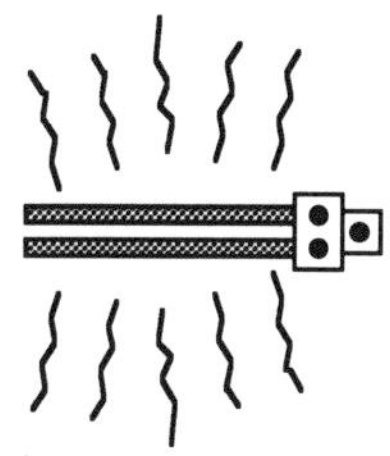

It is wrong to think that by using a larger size conduit than required would satisfy the Note 8 reduction of ampacity. The larger conduit would have more volume area, but it's like heating a rock; it may take a little longer but it will still reach the same temperature.

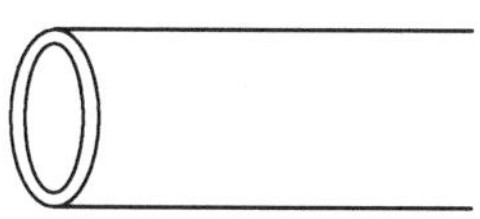

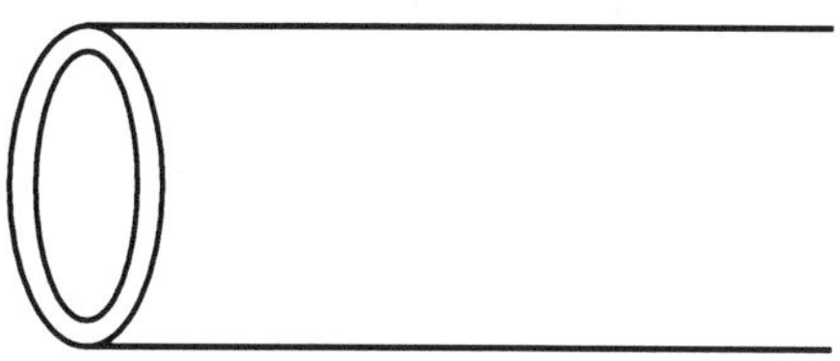

To avoid applying Note 8 ampacity adjustment deration you can install two separate conduits as shown below. Now you only have 3 current carrying conductors in each conduit.

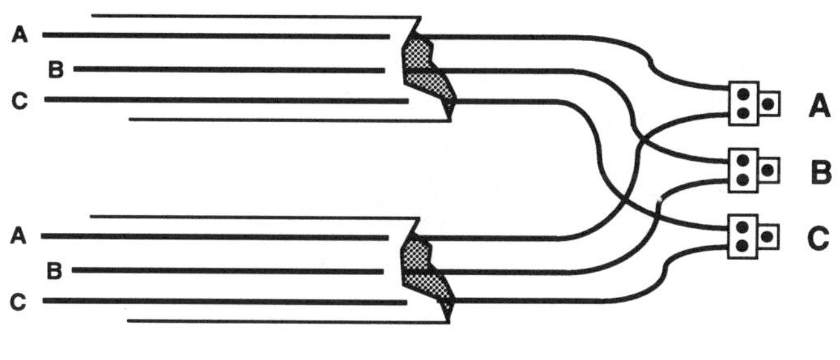

Proper design is to parallel smaller conductors rather than installing one large conductor.

The reason for paralleling conductors is to get the most ampacity per circular mil area.

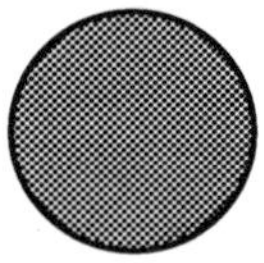

#12 CONDUCTOR
6530 cm

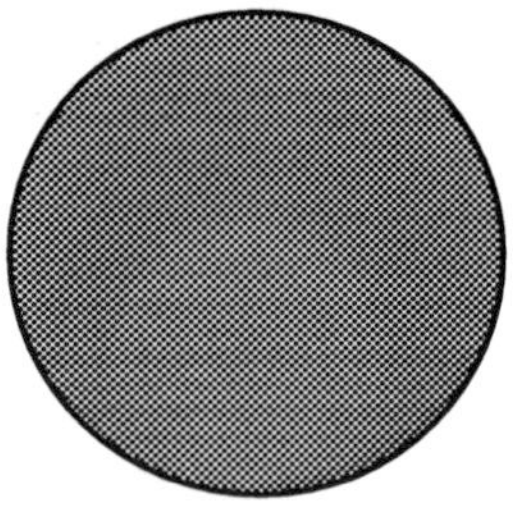

#6 CONDUCTOR
26240 cm

A #6 conductor has a cross sectional area approximately four times larger than a #12 (6530cm x 4 = 26120cm).

But, the ampacity of a #6 conductor is *not* four times larger than a #12 conductor. A #12 TW has an ampacity of 25 and a #6 TW has an ampacity of 55. About twice as large.

A #6 conductor has only twice the surface area (circumference) of a #12 and not four times. The heat produced in the wire can only be dissipated through the surface area.

Using smaller conductors to obtain the required current carrying ampacity reduces the total cross section of the conductor which produces a higher resistance and a greater voltage drop.

#12 CONDUCTOR
1.93Ω per k/ft

#6 CONDUCTOR
.491Ω per k/ft

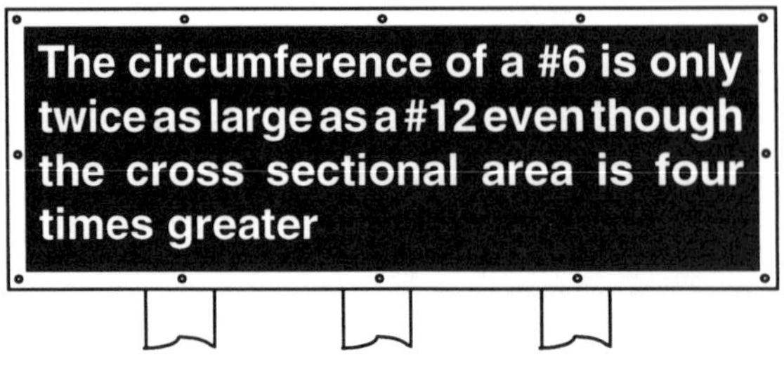

A #1/0 conductor at 105600 cm is ten times larger in cross sectional area than a #10 conductor at 10380 cm. The #1/0 does not have ten times the ampacity of a #10.

#10 TW
10380 cm
30 ampacity

#1/0 TW
105600 cm
125 ampacity

A #700 THW kcmil conductor has an ampacity of 460 amps. Two #4/0 THW conductors paralleled have an ampacity of 460 amps.

A #700 kcmil has 700,000 cm whereas two #4/0 have a combined circular mil area of 423,200 (2 x 211,600); 40% less conductor with the same ampacity. This is why the paralleling of conductors is to be considered in electrical designing.

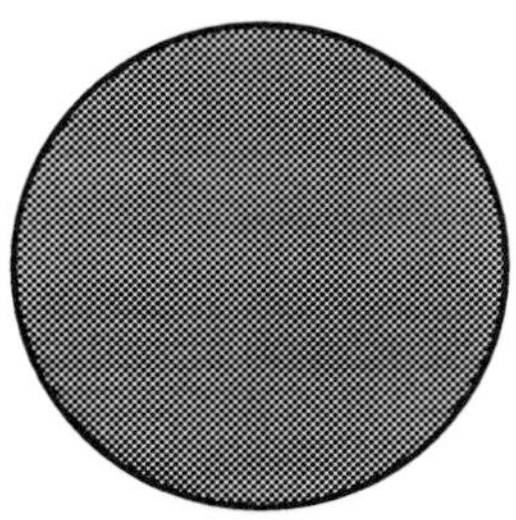

#700 THW kcmil
460 ampacity

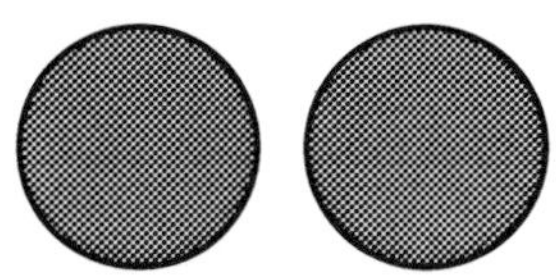

#4/0 THW
230 ampacity each

It is seldom economical to use conductors of larger sizes. It's more economical to parallel smaller conductors as you can see why.

THE TERMINATIONS

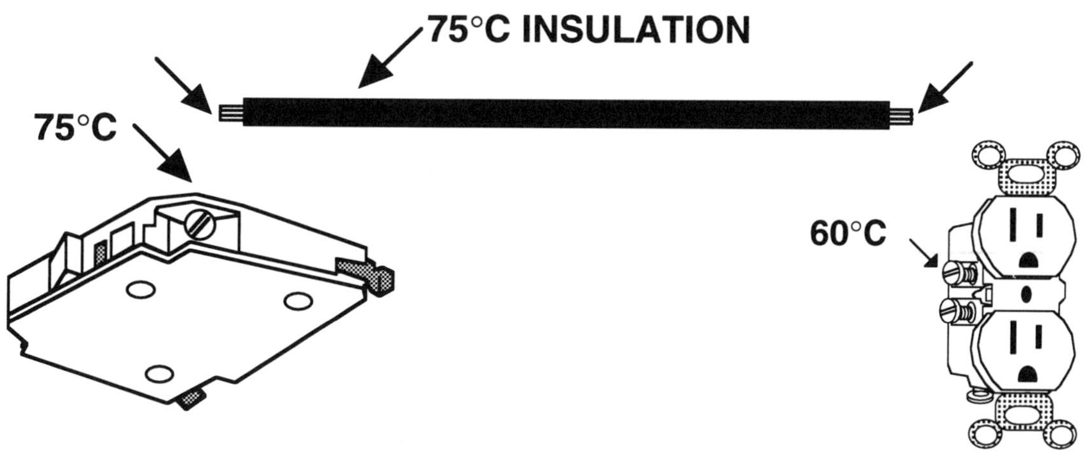

CONDUCTOR TERMINATIONS

The frequent misapplication of conductor temperature ratings occurs when the rating of the equipment is ignored when connecting the conductor.

Shown below are the ampacities and temperature ratings for a #10 conductor.

#10 TW 30 ampacity
60°C - 140°F temperature rating

#10 THW 35 ampacity
75°C - 167°F temperature rating

#10 THHN 40 ampacity
90°C - 194°F temperature rating

Conductors carry a specific temperature rating based on the *type* of insulation on the conductor.

Conductor sizing must be determined to where the conductors will terminate and the rating of the termination.

If a termination is rated for 60°C, this means that the temperature at that termination may rise up to 60°C when the equipment is loaded to its *ampacity*. Any additional heat at the connection above the 60°C conductor insulation rating could cause damage to the conductor insulation.

When a conductor is chosen to carry a specific load, the designer must know the termination ratings of the equipment in the entire circuit.

When connecting the 90°C THHN insulation to 60°C rated equipment, the THHN cannot be loaded over the 60°C ampacity.

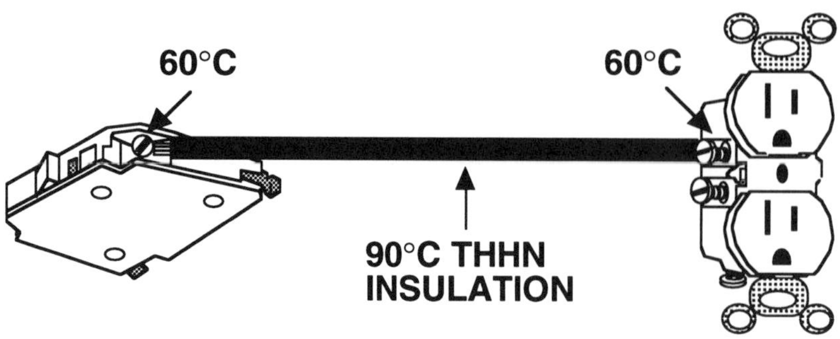

If the temperature rating of the equipment is not considered and the THHN is loaded to the 90°C ampacity it will lead to overheating at the termination and *premature opening* of the breaker.

The breaker has a heat sink which determines when it will open, the added extra heat is a determining factor in the premature opening.

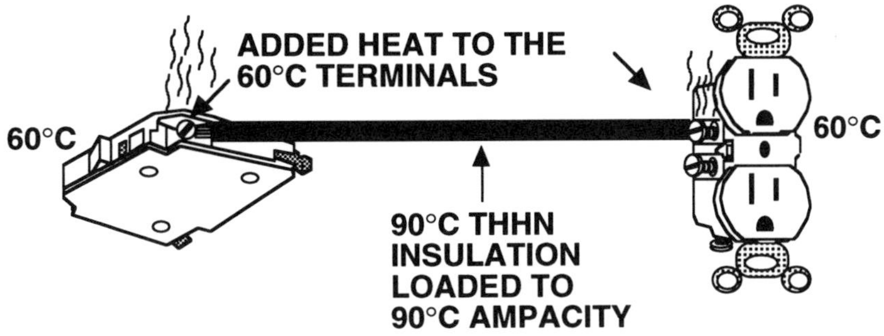

Electrical equipment such as circuit breakers, panelboards, receptacles, switches, terminal blocks, wire nuts, lugs, etc. have *temperature ratings* that must be observed.

The termination provisions are based on the use of 60°C ampacities for wire sizes #14 - #1 and 75°C ampacities for wire sizes #1/0 and larger. If using wire sizes #14-#1 in equipment marked "75°C only" or "60/75°C", it is intended that 75°C wire may be used at its full ampacity.

If the equipment is not marked, it is 60°C rated for 100 amps and less. Over 100 amps is 75°C rated. If the equipment is rated higher, it would have to be marked to indicate the higher temperature.

100 amps or less	Over 100 amps
Wire size #14 - #1	Wire size #1/0 and larger
60°C ampacity	75°C ampacity

The Code states in section 110-14c the temperature rating associated with the ampacity of the conductor shall be so selected not to exceed the lowest temperature rating of any conductor, device or connected termination.

This Code section is often violated, even ignored. A THHN 90°C conductor *cannot* be connected to a 60°C terminal and the conductor loaded to the 90°C ampacity.

#10 THHN 40 ampacity
90°C - 194°F temperature rating

The THHN conductor is the most popular insulation used in circuits. Because of its thinner insulation more conductors can be installed in a conduit compared to conductors with a thicker insulation.

Where the violation occurs is when the installer loads the THHN to the 90°C ampacity!

The only way a THHN could be loaded to a 90°C ampacity would be if every piece of electrical equipment (lugs, receptacle, switch, panelboard, overcurrent device, etc.) in the *entire* circuit is rated at 90°C.

The first thing the installer must be made aware of is that electrical equipment terminations are rated at 60°C, 75°C, or 60/75°C. If there is no marking on the equipment it is 60°C for 100 amps or less. Over 100 amps is 75°C rated.

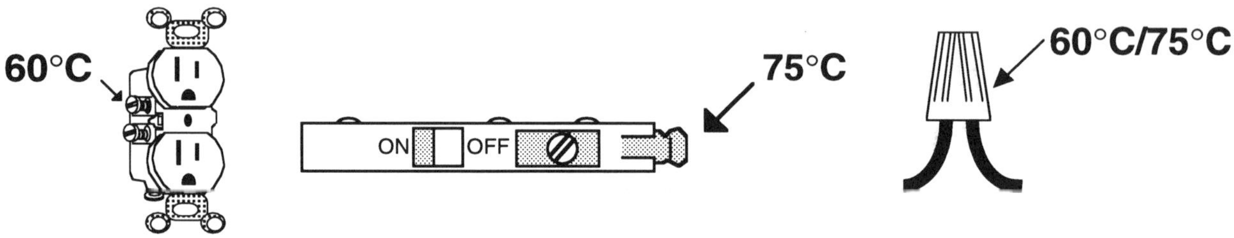

Circuit breakers are listed as suitable for 60°C only, 75°C only, or 60/75°C wire. Very little electrical equipment is tested for use with conductors operating over 75°C, although some large bolted pressure contact switches are suitable for conductors operating at 90°C.

When a conductor is chosen for the circuit, the designer or installer must know the termination ratings for the equipment in the *entire* circuit.

90°C RATED

The lug may be rated for 90°C but this does not give permission to the installer to use a 90°C conductor at 90°C ampacity. Every electrical component in the entire circuit must be 90°C rated.

When terminals are tested at 60°C or 75°C, the use of 90°C conductors loaded to their higher ampacity ratings can damage the terminals on circuit breakers, switches, etc. Electrical equipment has experienced termination failures even when the load current did not exceed the current rating of the circuit breaker or switch, etc.

When a 60°C rated circuit breaker terminal is fed by a conductor operating at 90°C, there will be substantial heat conducted from the 90°C metal to the 60°C terminal that can damage the terminal over a period of time, even though the load current does not exceed the 60°C terminal rating and does not exceed the 90°C ampacity.

The definition of ampacity is the current in amperes a conductor can carry *continuously* under the conditions without exceeding its (insulation) temperature rating. The greater the ampacity of a higher-temperature rated conductor establishes the ability of the conductor insulation to withstand the higher heat being produced by the higher current through the wire. But, it is *wrong* to assume that the equipment the conductor is being connected to is also capable of withstanding this higher heat.

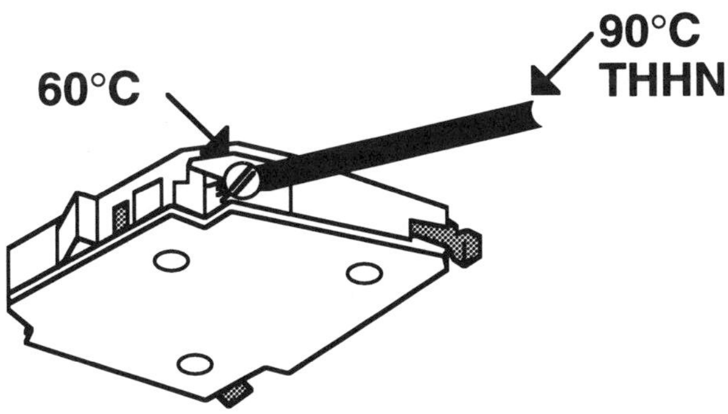

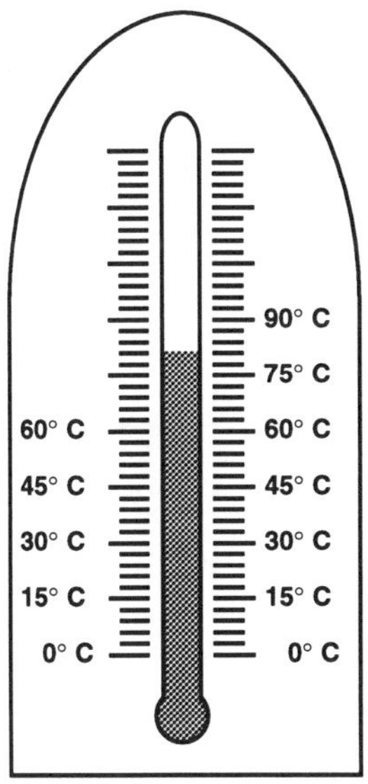

When two metallic parts are tightly connected having different operating temperatures (60°C - 90°C) the higher-temperature part (90°C) will give heat to the lower temperature part (60°C) and raise its temperature to above 60°C.

A 75°C temperature marking on a circuit breaker normally intended for wire sizes #14-#1 does not in itself allow the 75°C wire to be used at a 75°C ampacity unless the enclosure in which the circuit breaker is installed is also marked 75°C.

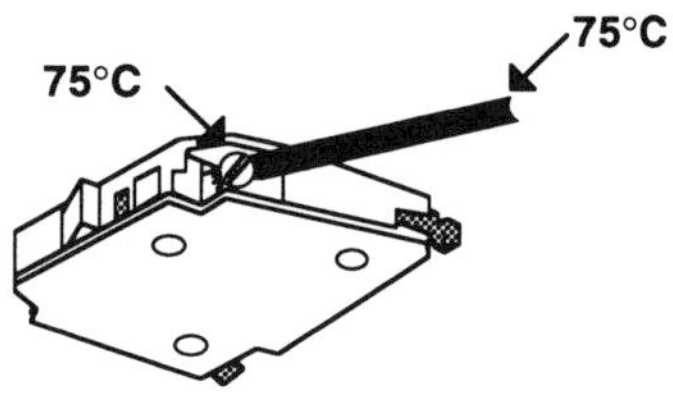

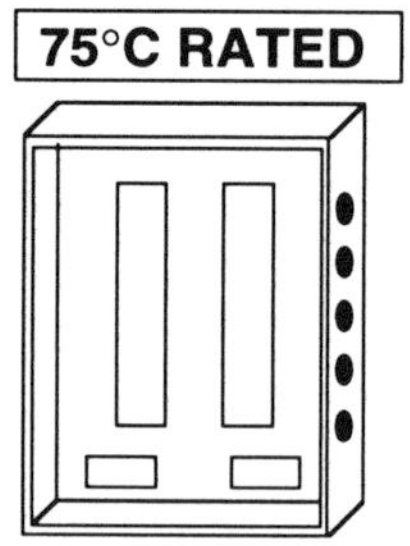

A conductor has *two* ends. Conductors must be sized by giving consideration to where each end will terminate and the temperature rating of the termination equipment.

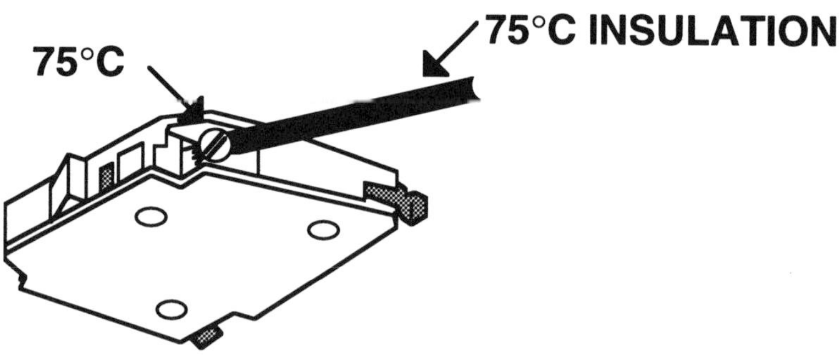

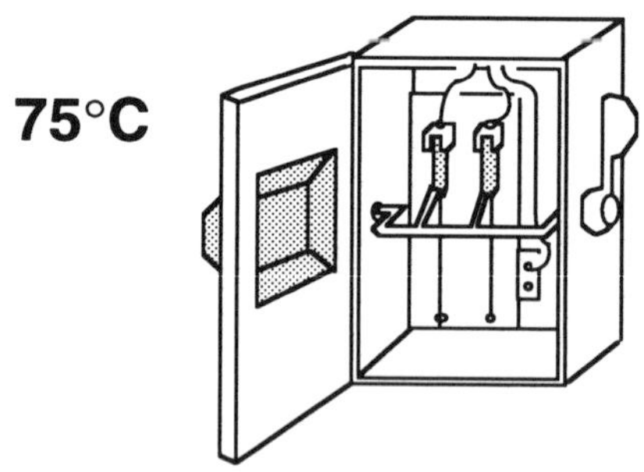

If a termination is rated for 75°C, this means that the temperature at that termination may rise up to 75°C when the equipment is loaded to its full ampacity. If 60°C conductors were used, the additional heat at the termination above 60°C conductor insulation rating could result in insulation damage.

Often, manufacturers are asked when electrical equipment will be available with terminations that will allow THHN 90°C conductors to be loaded to the 90°C ampacity. The answer is complex and requires not only significant equipment redesign to handle the extra heat, but also significant changes in the product testing/listing standards would also have to occur.

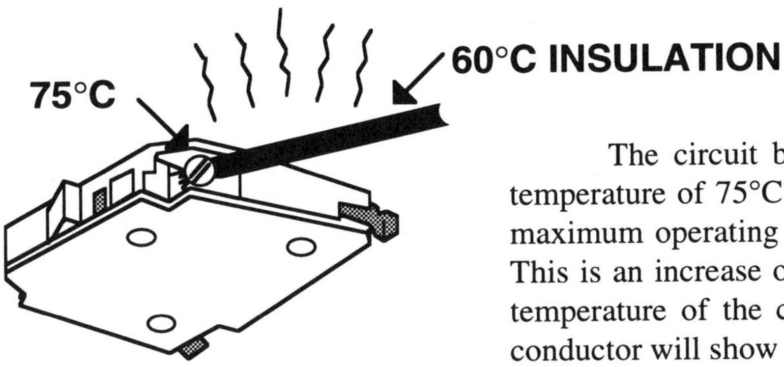

The circuit breaker lug fully loaded will reach a temperature of 75°C which is 167°F. The 60°C conductor maximum operating temperature is 60°C which is 140°F. This is an increase of 27°F above the *maximum* operating temperature of the conductor. Over a period of time the conductor will show signs of insulation damage such as the insulation becoming hard and brittle close to the termination.

The panelboard has a 150 amp circuit breaker with a temperature of 75°C. It requires a conductor with an ampacity of at least 150 amperes. Table 310-16 shown below lists the ampacities of different conductors. We must select a conductor with an ampacity of at least 150 from the *75°C column* which is a #1/0. A #1 THHN 90°C conductor also has an ampacity of 150. But, a #1 THHN *cannot* be loaded to the 150 ampacity because the equipment is only rated at 75°C. Using a #1 THHN conductor in this application would lead to possible overheating at the termination or premature opening of the circuit breaker. A 60°C insulated conductor is not acceptable regardless of the size since the temperature at the termination (75°C) could rise to a value greater than the insulation rating of 60°C.

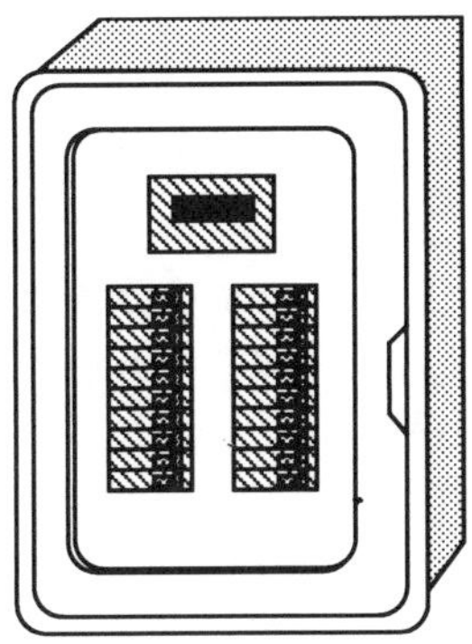

Table 310-16. Allowable Ampacities of Insulated Conductors

Size	Temperature Rating of Conductor. See Table 310-13.						Size
AWG kcmil	60°C (140°F)	75°C (167°F)	90°C (194°F)	60°C (140°F)	75°C (167°F)	90°C (194°F)	AWG kcmil
	TYPES TW†, UF†	TYPES FEPW†, RH†, RHW†, THHW†, THW†, THWN†, XHHW†, USE†, ZW†	TYPES TA, TBS,SA, SIS, FEP†, FEPB†, MI RHH†, RHW-2, THHN†, THHW†, THW-2, THWN-2 XHHW†, USE-2	TYPES TW†, UF†	TYPES RH†, RHW†, THHW†, THW†, THWN†, XHHW†, USE†	TYPES TA, TBS, SA, SIS, THHN†, THHW†, THW-2, THWN-2, RHH†, RHW-2, USE-2, XHH, XHHW, XHHW-2, ZW-2	
	COPPER			ALUMINUM OR COPPER-CLAD ALUMINUM			
18	....	....	14	....	....	....	....
16	....	....	18	....	....	....	....
14	20†	20†	25†	....	....	....	....
12	25†	25†	30†	20†	20†	25†	12
10	30	35†	40†	25	30†	35†	10
8	40	50	55	30	40	45	8
6	55	65	75	40	50	60	6
4	70	85	95	55	65	75	4
3	85	100	110	65	75	85	3
2	95	115	130	75	90	100	2
1	110	130	150	85	100	115	1
1/0	125	150	170	100	120	135	1/0
2/0	145	175	195	115	135	150	2/0
3/0	165	200	225	130	155	175	3/0
4/0	195	230	260	150	180	205	4/0

A circuit breaker is rated 75°C and the receptacle to which the branch circuit conductor is connected is rated 60°C. The conductor must have an insulation rating of at least 75°C (due to the rating of the breaker), and the conductor must be sized based on the ampacity of a 60°C conductor (due to the rating of the receptacle).

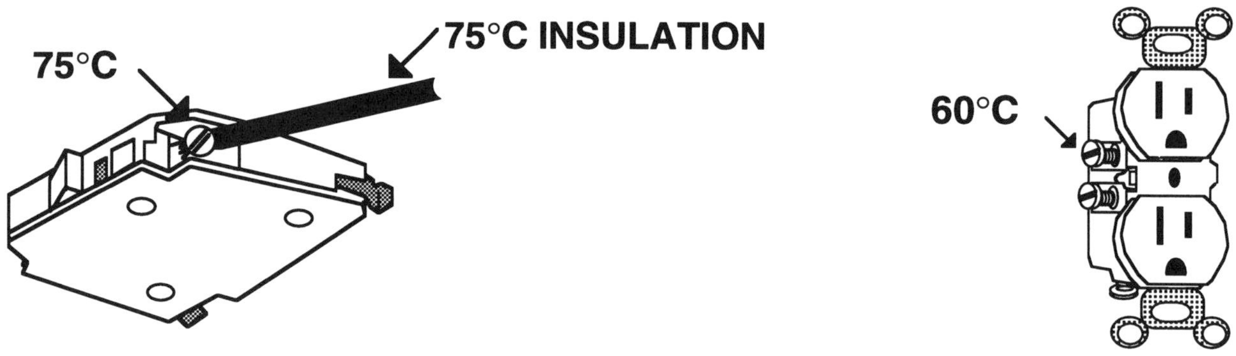

The general rule is 100 amps or less is wire sizes #14 - #1 with 60°C ampacity and for over 100 amps the wire sizes are #1/0 and larger with 75°C ampacity. But, there are two exceptions: (1) Conductors with higher temperature ratings such as 75°C can be used but at the ampacity of the equipment temperature rating as shown above. (2) Equipment termination provisions can be used with higher rated conductors at the ampacity of the higher rated conductors, provided the equipment is listed at the higher rating. This means a 20 amp receptacle can be rated 75°C even though it is 100 amps or less.

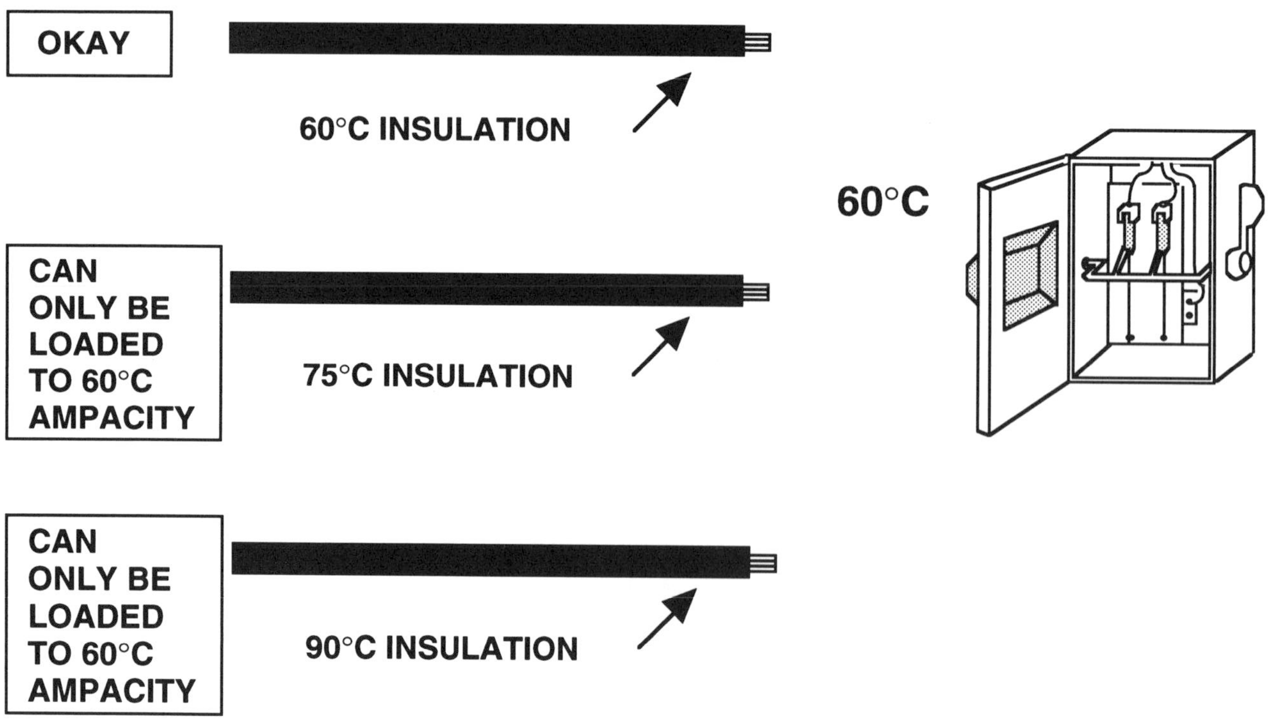

NO — 60°C INSULATION

OKAY — 75°C INSULATION

CAN ONLY BE LOADED TO 75°C AMPACITY — 90°C INSULATION

75°C

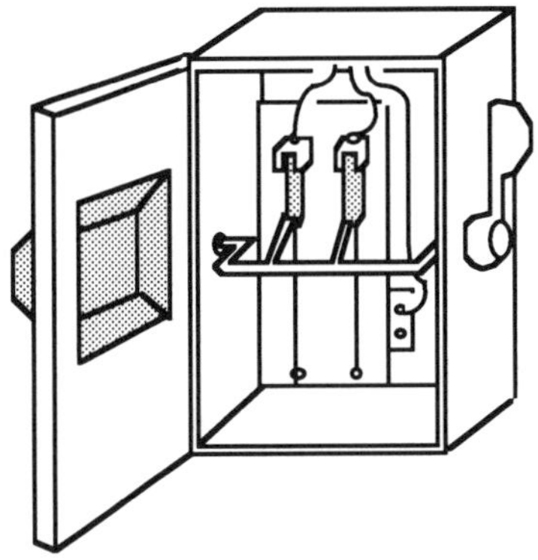

OKAY — 60°C INSULATION

CAN ONLY BE LOADED TO 60°C OR 75°C AMPACITY — 75°C INSULATION

CAN ONLY BE LOADED TO 60°C OR 75°C AMPACITY — 90°C INSULATION

60°/75°C

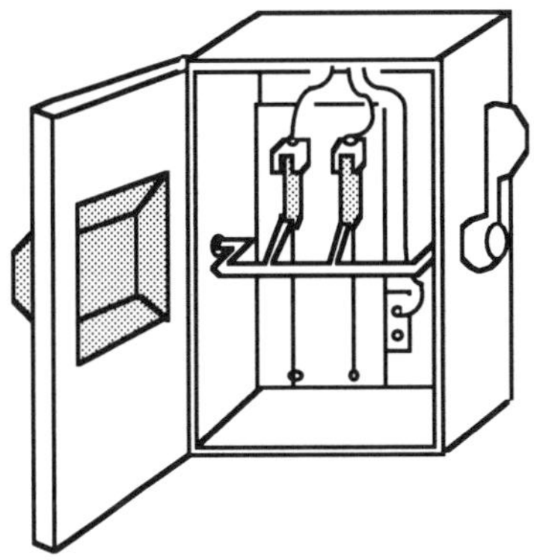

NO — 60°C INSULATION

NO — 75°C INSULATION

OKAY *The equipment must have a 90°C rating. — 90°C INSULATION

***90°C**

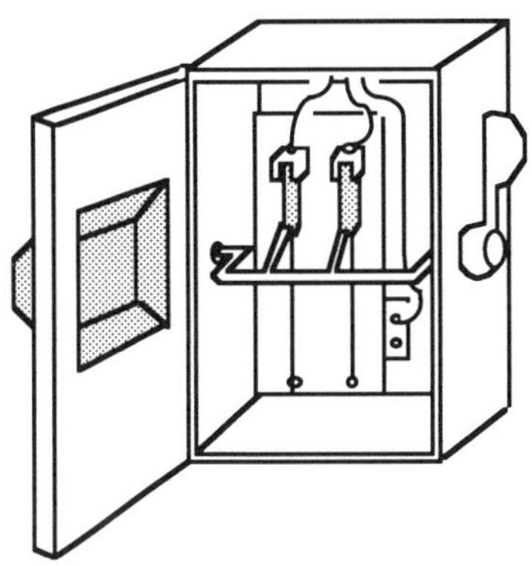

90°C CONDUCTORS

THE ADVANTAGES OF 90°C CONDUCTORS

As you have seen different types of insulations have different ampacities and temperature ratings. Insulation also changes the size of the conductor and this is important when considering conduit fill.

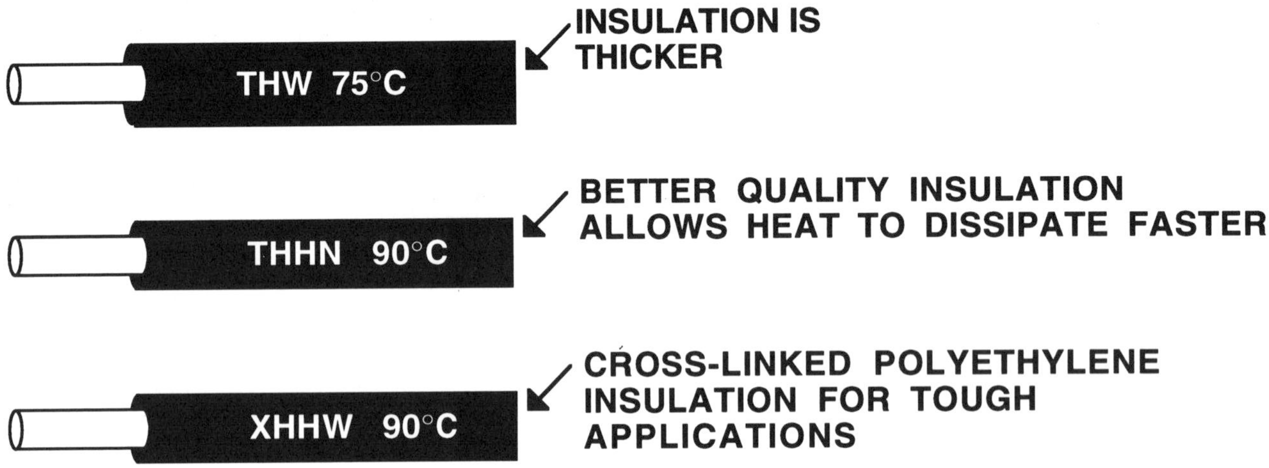

A #12 THW has a thicker insulation .026 sq.in. compared to the thinner insulation of a #12 THHN at .0133 sq.in.

You could install 8 - #12 THW's in a 3/4" conduit or 16 - #12 THHN's.

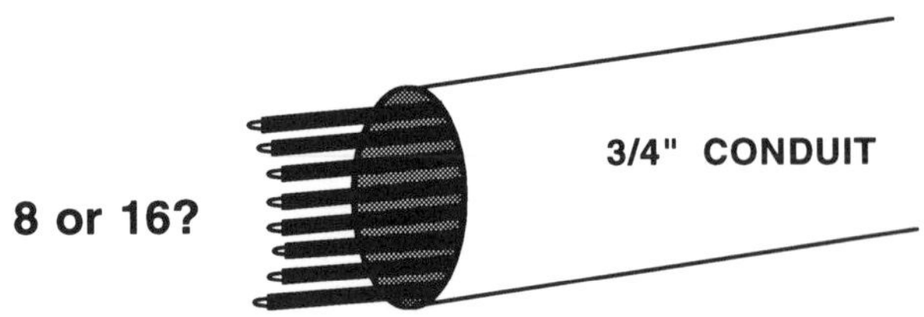

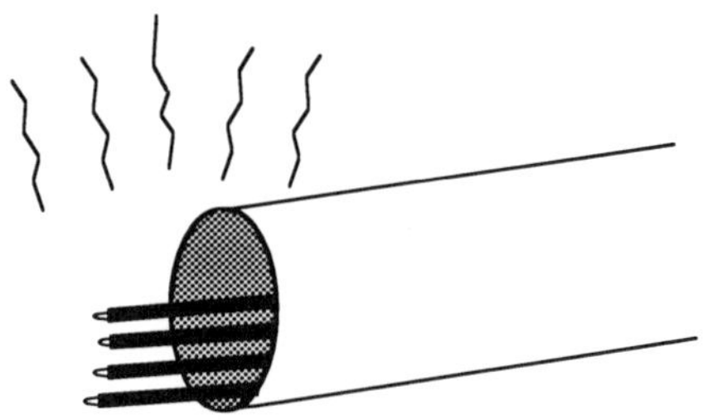

Code Table 310-16 Note 8 states that where the number of current-carrying conductors in a raceway or cable exceeds *three,* the allowable ampacity must be reduced to compensate for the increased heating effect due to reduced ventilation of an enclosed group of closely spaced conductors.

8. Ampacity Adjustment Factors.

(a) More than Three Conductors in a Raceway or Cable. Where the number of conductors in a raceway or cable exceeds three, the ampacities shall be reduced as shown in the following table:

Number of Conductors	Percent of Values in Tables as Adjusted for Ambient Temperature if Necessary
4 through 6	80
7 through 9	70
10 through 20	50
21 through 30	45
31 through 40	40
41 through 60	35

Selecting THHN 90°C insulation is an advantage in designing when applying the derating factors (correction factor and/or Note 8) to the normal ampacity of the conductor.

Example: The circuit is installed in a conduit that has a total of seven current carrying conductors. The equipment in the circuit has a rating of 60°C.

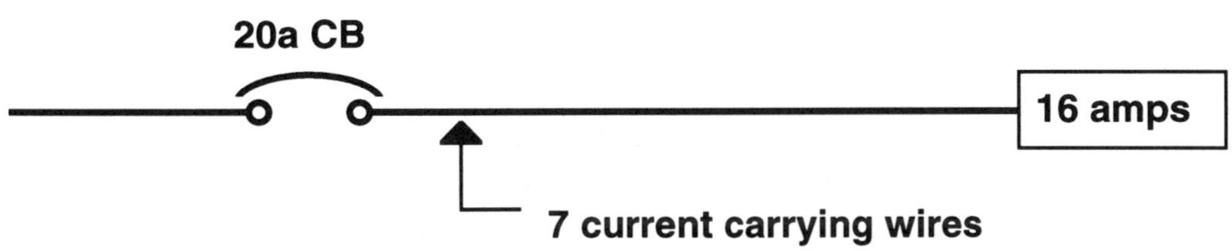

The Code requires the normal ampacity to be reduced for the seven current carrying conductors.

A #12 TW ampacity is 25 x 70% =17.5 amps the maximum current permitted.

A #12 THW ampacity is 25 x 70% = 17.5 amps is the maximum current permitted.

A #12 THHN has a normal ampacity of 30 x 70% = 21 amps is the maximum current permitted on the conductor. A #12 TW or #12 THW with a derated ampacity of 17.5 is considered protected by the next higher size breaker which is a 20. The advantage of a #12 THHN is it has a derated ampacity of 21 which is above the breaker rating of 20 amps. The THHN insulation eliminates any chance of overload damage to the insulation.

DERATED AMPACITY = 17.5

#12 TW 25 ampacity
60°C - 140°F temperature rating

DERATED AMPACITY = 17.5

#12 THW 25 ampacity
75°C - 167°F temperature rating

DERATED AMPACITY = 21

#12 THHN 30 ampacity
90°C - 194°F temperature rating

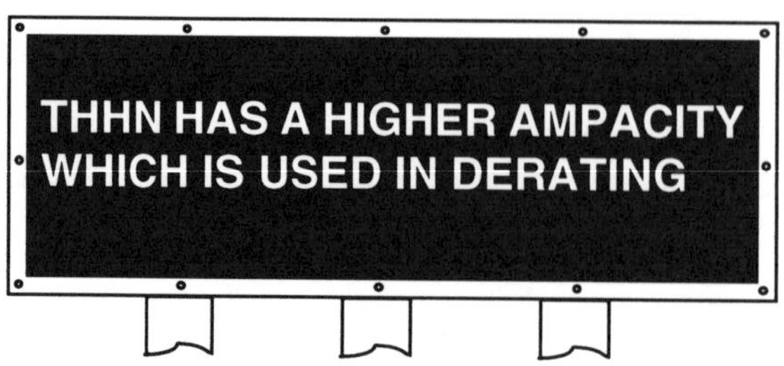

Example: A branch circuit has a noncontinuous load of 29 amps. The circuit is installed in a conduit that has a total of four current carrying conductors. The equipment in the circuit has a rating of 60°C.

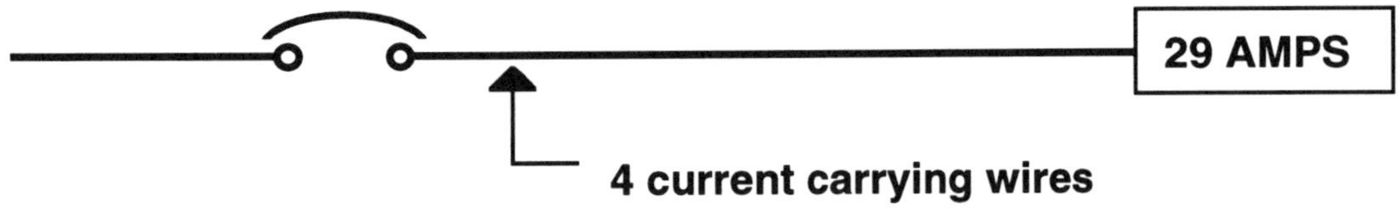

A #10 TW ampacity is 30 x 80% =24 amps the maximum current permitted. A #10 TW would violate the Code in this example as the load is 29 amps.

A #10 THW ampacity is 35 x 80% = 28 amps is the maximum current permitted. A #10 THW would also be a violation as the load is 29 amps.

A #10 THHN has a normal ampacity of 40 x 80% = 32 amps is the maximum current permitted on the conductor. Next go to the 60°C column and select a conductor that will carry the 29 amp load. A #10 THHN can be used in this condition. The THHN conductor can safely carry 32 amps without damage to the *conductor insulation*. But, the 60°C column shows a #10 conductor cannot be loaded to more than 30 amps without possible damage to the *termination*.

Otherwise you would have to use a #8 TW (40a x 80% = 32a) or a #8 THW (50a x 80% = 40a). By selecting THHN insulation you can use a smaller #10 conductor.

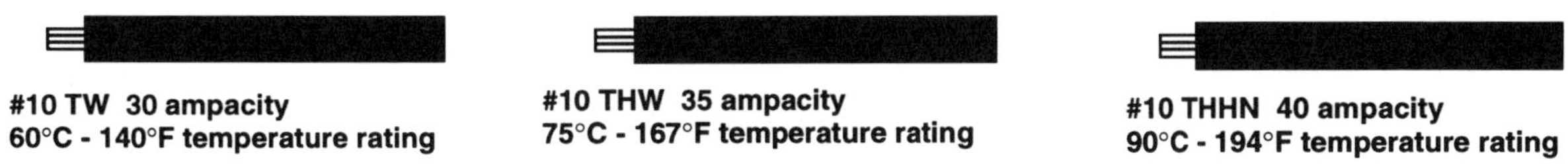

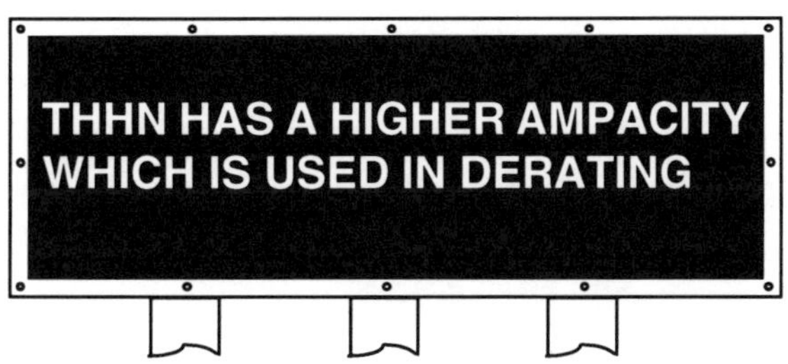

90°C THHN VOLTAGE DROP

The 90°C THHN and XHHW conductors have higher ampacities which benefits the designer when applying the derating factors from the heat conditions. The 90°C conductors have a smaller square inch area which permits more conductors in a conduit.

The designer must also calculate the voltage drop in the 90°C conductor compared to a 60°C or 75°C.

In reality the installer would select a #8 THHN as the conductor for the branch circuit shown below. Table 310-16 lists the ampacity of a #8 THHN at 55 amps and the load is only 50. Let's apply correct designing and calculate the minimum size conductor the Code would permit. All the equipment in the circuit is 60°C rated.

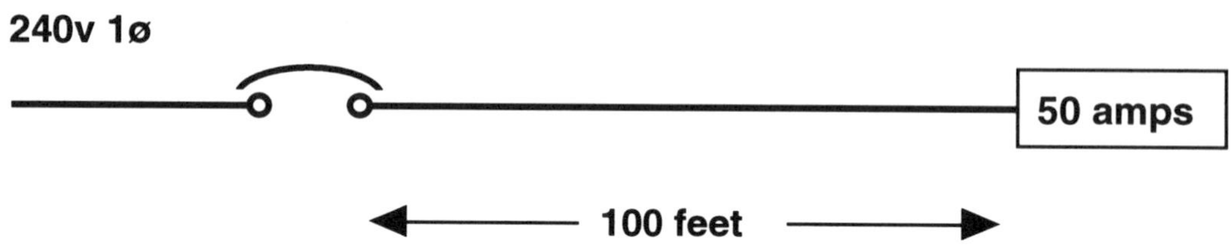

From Table 310-16 the conductor choices would appear to be a #6 TW @ 55 ampacity, a #8 THW @ 50 ampacity or a #8 THHN @ 55 ampacity. In reality the most common wire sold over the counter is the THHN insulation.

Now let's apply what we have learned so far. Since the equipment is 60°C rated which ever conductor insulation is selected would have to be used at a 60°C *ampacity*. A #8 THHN could *not* be used as a #8 @ 60°C as it can only be loaded to a maximum of 40 amps on 60°C rated equipment and the load is 50 amps.

Let's calculate the voltage drop for the branch circuit that is 100 feet in distance.

$$\text{Wire size} = \text{CM} = \frac{2 \times 12.9 \times 100' \times 50\text{a}}{7.2 \text{ vd}} = 17{,}917 \text{ cm required}$$

Table 8 would require a *minimum* #6 conductor with a circular mil area of 26240. A #8 conductor has a circular mil area of 16510, not large enough for this branch circuit.

Always remember, by selecting a #8 THHN rather than a #6 TW because of the higher ampacity a #8 has a smaller circular mil area which has a higher *resistance* and a *greater voltage drop*.

#8 CONDUCTOR
16510 cm
.778Ω per k/ft.

#6 CONDUCTOR
26240 cm
.491Ω per k/ft.

A #6 THHN will satisfy the voltage drop requirement and the 60°C ampacity which is 55.

Also the advantage of using a #6 THHN would be a big plus for derating of heat conditions.

The same circuit installed in an ambient of 140°F requires a correction factor to be applied.

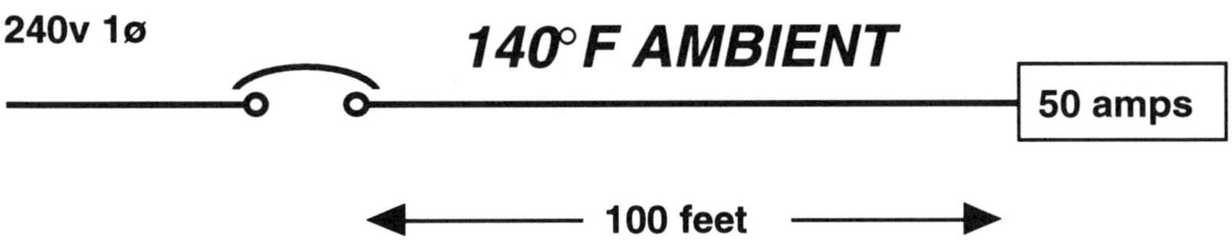

A #6 THHN has an ampacity of 75 x .71 correction factor for 140°F = 53.25 is the maximum current that can be applied to a #6 THHN in a 140°F ambient.

Now check Table 310-16 for the ampacity of a 60°C conductor. It will carry 55 amps.

A #6 TW could *not* be used for this circuit even though it satisfies the voltage drop requirements. A #6 TW has an ampacity of 55, but there is no correction factor for 140°F which means you could not use a 60°C insulation. A #6 THW has an ampacity of 65 x .58 correction factor = 37.7 is the maximum current that can be applied to a #6 THW in a 140°F ambient. Since the load is 50 amps, a #6 THW conductor is a violation. The Code would require a minimum #6 THHN for this branch circuit.

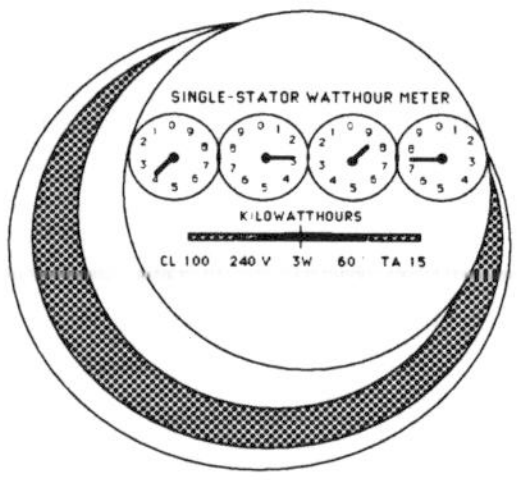

Often neglected in electrical design is the power loss in the conductors.

POWER LOSS = VD X I

Voltage drop times amperage equals power (watts) loss. This is the energy wasted by heating the wire while passing a current through a resistance.

I (amperes) represents the load in amperes which cannot be changed in most cases. VD represents the voltage drop which cannot be eliminated as all wire has some resistance. Voltage drop can be reduced by installing larger conductors which have less resistance.

With heavy commercial and industrial loads that operate continuously, increasing the size of the wire is an area of design the designer should take a close look at, as it can provide a power savings in the monthly electric bill.

The formula for \$ COST \$ = $\frac{\text{WATTS x HOURS USED x RATE PER HOUR}}{1000}$

A #6 THHN was the *minimum* size conductor permitted by the Code for the 50 amp load in an ambient of 140°F.

The voltage drop would be: $VD = \frac{2 \text{ x } 12.88384 \text{ x } 100' \text{ x } 50a}{26240cm}$ = 4.91 volts dropped

Power Loss = VD x I   4.91v x 50a = 245.5 watts

If the 50 amp load operated 8 hours per day, 30 days a month, at 8¢ a kwh the cost of power loss would be: $\frac{245.5w \text{ x } 8 \text{ hours x } .08}{1000}$ = .157¢ per day x 30 days = \$4.71 per month.

If a *#2 THHN* was selected instead of the Code minimum #6, the voltage drop for the #2 would be: $VD = \frac{2 \text{ x } 12.87384 \text{ x } 100' \text{ x } 50a}{66360cm}$ = 1.94 volts dropped

The power loss for the #2 THHN would be: 1.94v x 50a = 97 watts

The cost of power loss for the #2 THHN would be: $\frac{97w \text{ x } 8 \text{ hours x } .08}{1000}$ = .062 per day

.062 x 30 days = \$1.86 per month cost of power loss in conductors.

The savings in power loss between a #6 @ \$4.71 and a #2 @ \$1.86 is \$2.85 per month or \$34.20 per year for one 50 amp load.

SUMMARY FOR DERATING OF 90°C CONDUCTORS

THHN WIRE SIZE	60°C Termination Maximum Current	75°C Termination Maximum Current	**THHN** AMPACITY USED FOR DERATING
#14	**20†**	**20†**	**25†**
#12	**25†**	**25†**	**30†**
#10	**30**	**35†**	**40†**
#8	**40**	**50**	**55**
#6	**55**	**65**	**75**
#4	**70**	**85**	**95**
#3	**85**	**100**	**110**
#2	**95**	**115**	**130**
#1		**130**	**150**
#1/0		**150**	**170**
#2/0		**175**	**195**
#3/0		**200**	**225**
#4/0		**230**	**260**
#250		**255**	**290**
#300		**285**	**320**
#350		**310**	**350**
#400		**335**	**380**
#500		**380**	**430**

•Electrical equipment *100 amps and less* is 60°C rated. To have any rating above 60°C the equipment would have to be marked 75°C, 60°/75°C or 90°C. If there is no marking it's 60°C.

•Electrical equipment *above* 100 amps is 75°C rated. To have any rating above 75°C the equipment would have to be marked 90°C. If there is no marking it's 75°C.

Example: A #8 THHN with 8 current carrying conductors in a conduit. 55a x 70% Note 8 = 38.5 ampacity. The 60°C column shows the maximum current allowed on the *60°C termination* is 40 amps. The #8 *THHN conductor* current cannot exceed the derated ampacity of 38.5 amps.

Example: A #8 THHN with 4 current carrying conductors in a conduit. 55a x 80% Note 8 = 44 ampacity. The 60°C column shows the *maximum* current allowed on the 60°C *termination* is 40 amps. Even though the #8 THHN has a derated ampacity of 44 amps, which means the *conductor* could be loaded to 44 amps in this condition without damage to the conductor insulation, the maximum load cannot exceed 40 amps or it could cause damage to the 60°C termination.

Example: A #14 THHN with 4 current carrying conductors in a conduit. 25a x 80% Note 8 = 20 ampacity. The 60°C column shows the maximum current allowed is 20, but the maximum overcurrent protection (fuse or breaker) is 15† so 15 amps is the maximum current on this conductor unless otherwise specified in the Code. An example would be a motor circuit which is permitted to use a larger overcurrent device than 15 amps for the #14 conductor per Code section 240-3f.

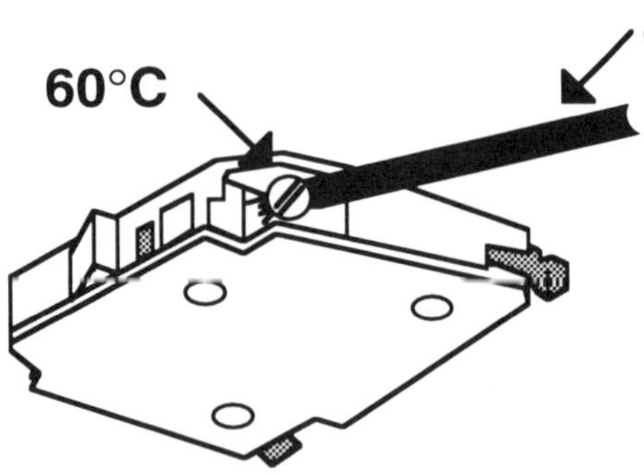

The THHN insulation can never be loaded above the AMPACITY. Any current above the ampacity could cause damage to the INSULATION of the conductor.

The termination can NEVER be loaded HIGHER than shown in the 60°C column for that size of conductor. Any current above the 60°C column amount could cause TERMINATION damage.

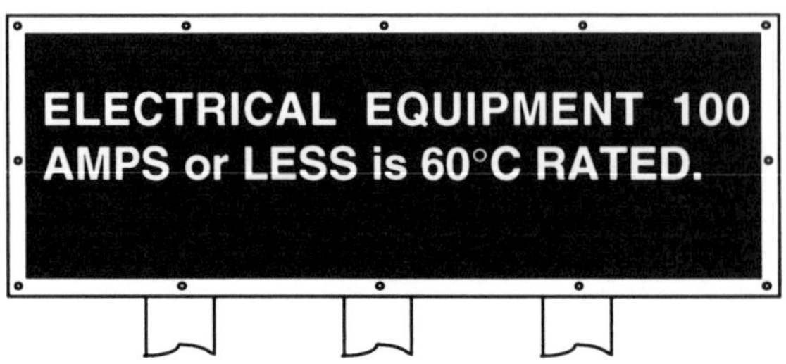

Example: A #1/0 THHN conductor installed in an ambient of 105°F. 170a x .87 correction factor = 147.9 ampacity. The 75°C column shows the maximum current allowed is 150 amps on the termination, but the #1/0 THHN conductor cannot be loaded above the derated ampacity of 147.9 amps. At 105°F a #1/0 THHN could *not* be used for a 150 amp load.

Example: A #3/0 THHN conductor installed in an ambient of 104°F. 225a x .91 correction factor = 204.75 ampacity. The 75°C column shows the maximum current allowed is 200 amps on the termination. A #3/0 THHN could be used for a 200 amp load at 104°F.

Example: A #3/0 THHN conductor installed in an ambient of 105°F. 225a x .87 correction factor = 195.75 ampacity. The 75°C column shows the maximum current allowed is 200 amps on the termination, but the #3/0 THHN conductor cannot be loaded above the derated ampacity of 195.75 amps. At 105°F a #3/0 THHN could *not* be used for a 200 amp load.

104°F and 105°F are *not* high ambient temperatures. This is an area where the installer is very neglectful. Electrical designing of conductors requires sizing the conductors to the *highest* ambient temperature the conductors could *ever* be subjected to in the circuit.

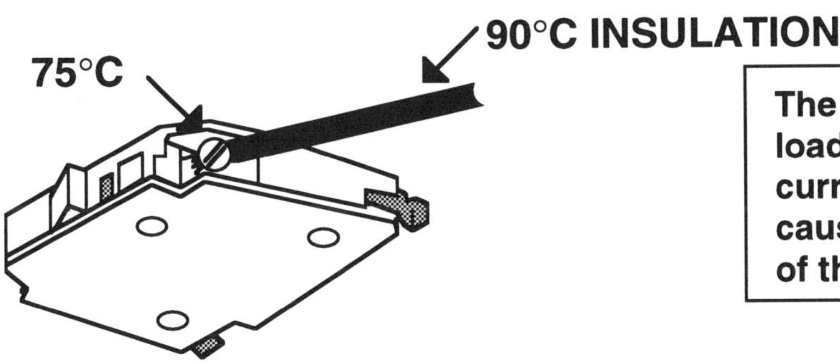

The THHN insulation can never be loaded above the AMPACITY. Any current above the ampacity could cause damage to the INSULATION of the conductor.

The termination can NEVER be loaded HIGHER than shown in the 75°C column for that size of conductor. Any current above the 75°C column amount could cause TERMINATION damage.

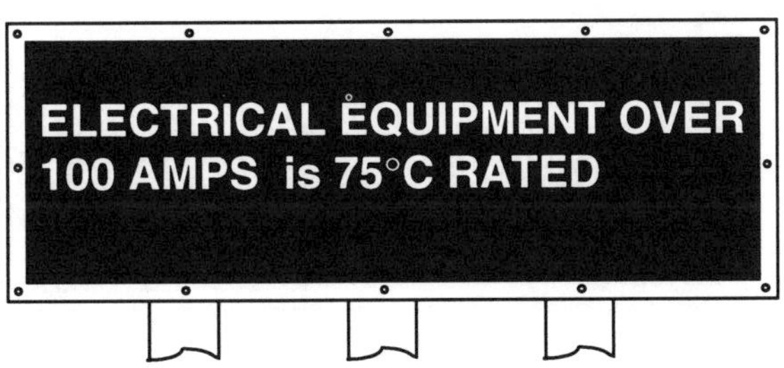

Example: Size the feeder to this 100 amp panel. The equipment is rated at 60°C.

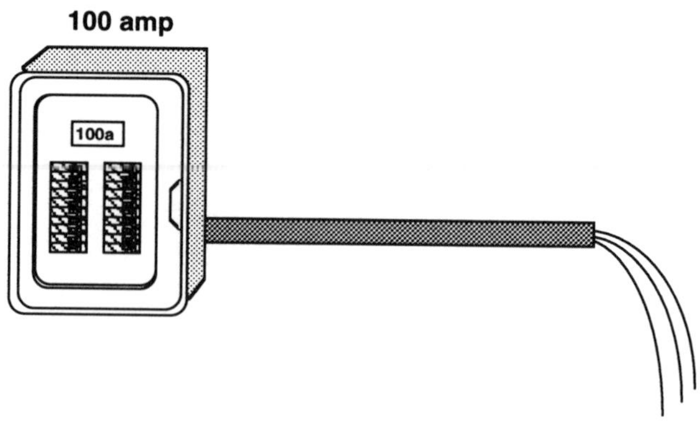

Using 60°C insulation a #1 TW has a normal ampacity of 110 amps. The load is 100 amps.

Using 90°C insulation, a #3 THHN has a normal ampacity of 110 amps. It can only be *loaded* to the 60°C ampacity. A #3 @ 60°C has a normal ampacity of 85. The load is 100 amps.

A #2 THHN has a normal ampacity of 130 amps. It can only be loaded to the 60°C ampacity. A #2 @ 60°C has a normal ampacity of 95. The load is 100 amps.

A #1 THHN has a normal ampacity of 150 amps. It can only be loaded to the 60°C ampacity. A #1 @ 60°C has a normal ampacity of 110 amps. The load is 100 amps.

A #1 TW or THHN is the minimum size conductor permitted.

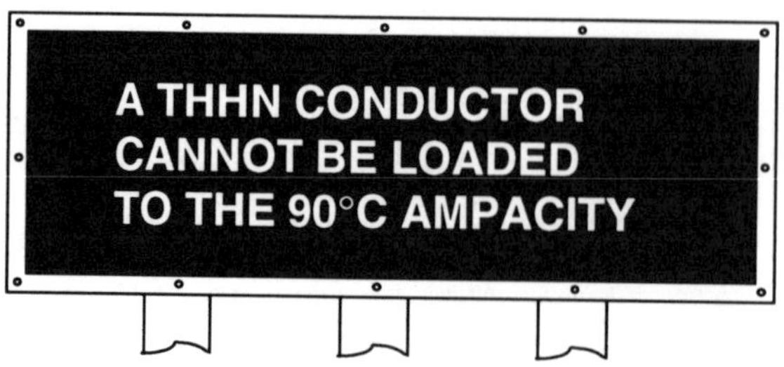

Example: Size the feeder to this 100 amp panel. The equipment is rated at *75°C*.

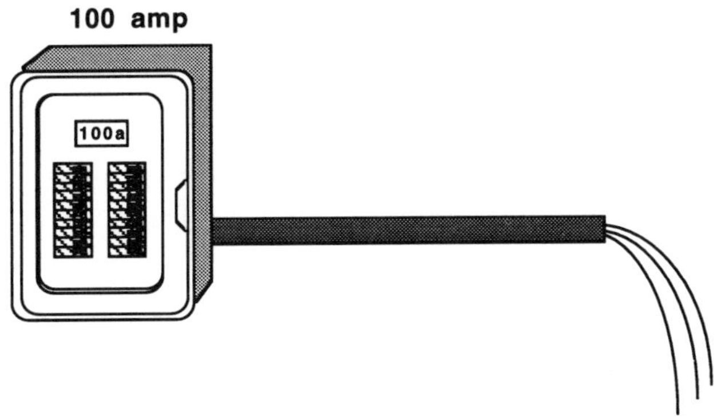

Using 75°C insulation a #3 THW has a normal ampacity of 100 amps. The load is 100 amps.

Using 90°C insulation, a #3 THHN has a normal ampacity of 110 amps. It can only be *loaded* to the 75°C ampacity. A #3 @ 75°C has a normal ampacity of 100. The load is 100 amps.

A #3 THW or THHN is the minimum size conductor permitted.

The advantage of using 75°C rated equipment rather than 60°C is a #3 conductor can be used for the 100 amp load rather than a #1 conductor.

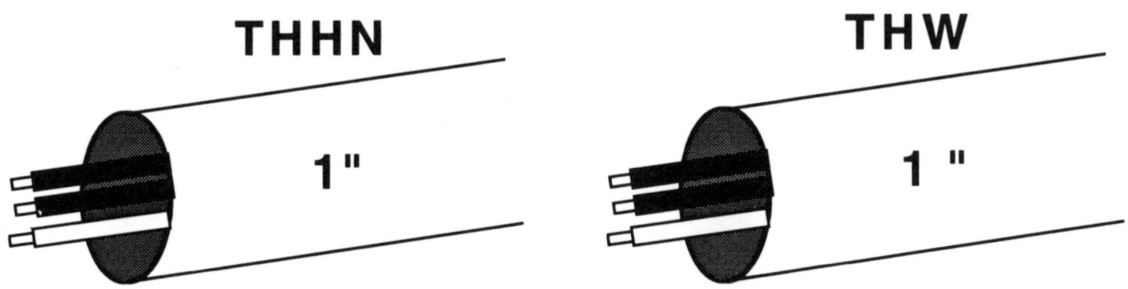

Example: Size the feeder to this 100 amp panel. The equipment is rated at 60°C.

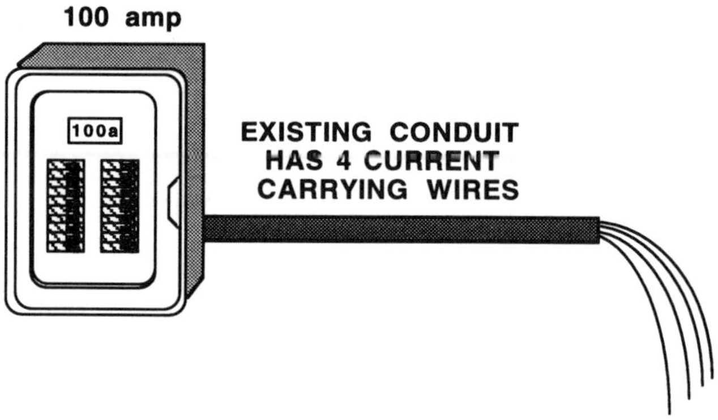

Using 60°C insulation a #1/0 has a normal ampacity of 125 amps x 80% (note 8) = 100a.

Using 90°C insulation, a #2 THHN has a normal ampacity of 130 amps x 80% (note 8) = 104a. The #2 THHN would be protected by the 100a breaker, but a #2 THHN can only be loaded to the 60°C ampacity. A #2 @ 60°C has a normal ampacity of 95. The load is 100 amps.

A #1 THHN has a normal ampacity of 150 x 80% (note 8) = 120 amps but, it can only be loaded to the 60°C ampacity. A #1 @ 60°C has an ampacity of 110. The load is 100 amps.

The advantage of using a #1 THHN would be you can install four in a 1 1/4" EMT conduit. Four #1/0 TW conductors require a 2" EMT conduit.

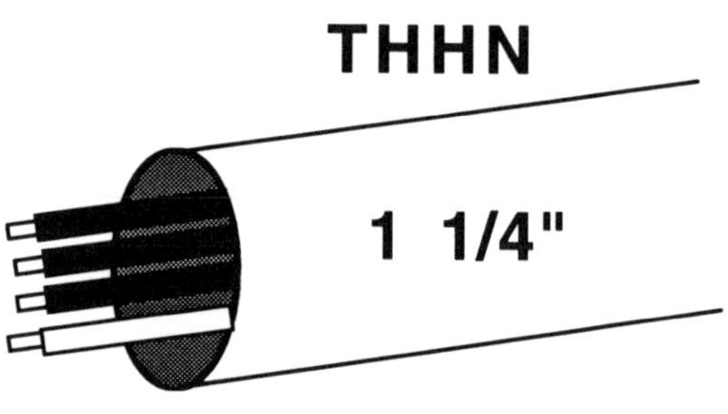

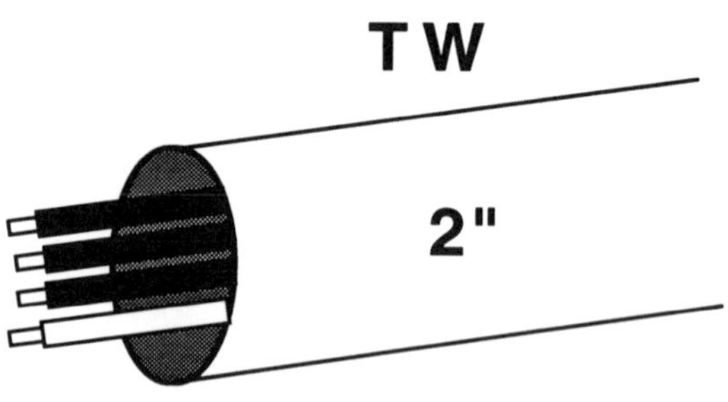

Table 310-16 for ampacity is based on an ambient temperature of 86°F or 30°C. If the conductors are installed in an ambient that is above or below these temperatures, the ampacity must be adjusted by the *correction factor* that applies from the table.

When the terminals are recognized by UL for use with 60°C conductors, if 75°C or 90°C conductors are connected to the 60°C rated terminals, they *cannot* be loaded over the 60°C ampacity.

Example: Size the feeder to this 100 amp panel. The equipment is rated at 60°C. The ambient temperature reaches 96°F.

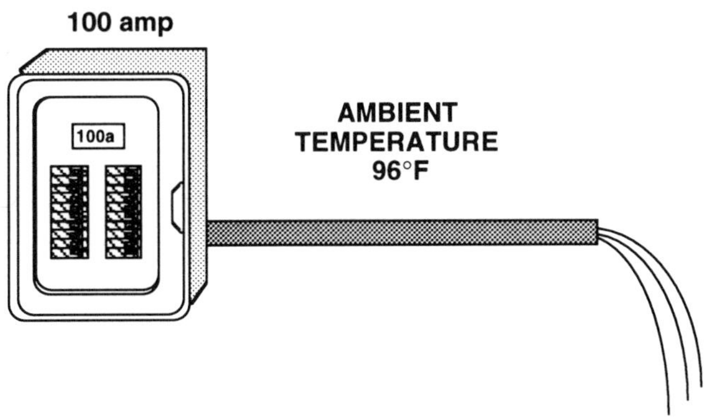

A #1/0 TW has a normal ampacity of 125 x .82 (correction factor) = 102.5 ampacity.

A #3 THHN has a normal ampacity of 110. The 90°C insulation allows a higher correction factor for 96°F of .91. Normal ampacity 110 x .91 (correction factor) = 100.1 ampacity. The #3 THHN would be protected by the 100a breaker, but the #3 THHN can only be loaded to the 60°C ampacity. A #3 @ 60°C has a normal ampacity of 85. The load is 100 amps.

A #2 THHN has a normal ampacity of 130 x .91 = 118.3 ampacity, but the #2 THHN can only be loaded to the 60°C ampacity. A #2 @ 60°C has a normal ampacity of 95. The load is 100 amps.

A #1 THHN has a normal ampacity of 150 x .91 = 136.5 ampacity, but the #1 THHN can only be loaded to the 60°C ampacity. A #1 @ 60°C has a normal ampacity of 110. The load is 100 amps.

A #1/0 TW or a #1 THHN is the *minimum* size conductor permitted by the Code. 96°F is not that unusual of a temperature. It would be very easy to reach this ambient in most circuits.

Table 310-16. Allowable Ampacities of Insulated Conductors
Rated 0-2000 Volts, 60°to 90°C (140°to 194°F)
Not More Than Three Conductors in Raceway or Cable or Earth
(Directly Buried), Based on Ambient Temperature of 30°C (86°F).

Size	Temperature Rating of Conductor. See Table 310-13.						Size
	60°C (140°F)	75°C (167°F)	90°C (194°F)	60°C (140°F)	75°C (167°F)	90°C (194°F)	
AWG kcmil	TYPES TW†, UF†	TYPES FEPW†, RH†, RHW†, THHW†, THW†, THWN†, XHHW†, USE†, ZW†	TYPES TA, TBS,SA, SIS, FEP†, FEPB†, MI RHH†, RHW-2, THHN†, THHW†, THW-2, THWN-2 XHHW†, USE-2	TYPES TW†, UF†	TYPES RH†, RHW†, THHW†, THW†, THWN†, XHHW†, USE†	TYPES TA, TBS, SA, SIS, THHN†, THHW†, THW-2, THWN-2, RHH†, RHW-2, USE-2, XHH, XHHW, XHHW-2, ZW-2	AWG kcmil
	COPPER			ALUMINUM OR COPPER-CLAD ALUMINUM			
18	....	....	14	....	....	....	....
16	....	....	18	....	....	....	....
14	20†	20†	25†	....	....	....	....
12	25†	25†	30†	20†	20†	25†	12
10	30	35†	40†	25	30†	35†	10
8	40	50	55	30	40	45	8
6	55	65	75	40	50	60	6
4	70	85	95	55	65	75	4
3	85	100	110	65	75	85	3
2	95	115	130	75	90	100	2
1	110	130	150	85	100	115	1
1/0	125	150	170	100	120	135	1/0
2/0	145	175	195	115	135	150	2/0
3/0	165	200	225	130	155	175	3/0
4/0	195	230	260	150	180	205	4/0
250	215	255	290	170	205	230	250
300	240	285	320	190	230	255	300
350	260	310	350	210	250	280	350
400	280	335	380	225	270	305	400
500	320	380	430	260	310	350	500
600	355	420	475	285	340	385	600
700	385	460	520	310	375	420	700
750	400	475	535	320	385	435	750
800	410	490	555	330	395	450	800
900	435	520	585	355	425	480	900
1000	455	545	615	375	445	500	1000
1250	495	590	665	405	485	545	1250
1500	520	625	705	435	520	585	1500
1750	545	650	735	455	545	615	1750
2000	560	665	750	470	560	630	2000

AMPACITY CORRECTION FACTORS

Ambient Temp. °C	For ambient temperatures other than 30°C (86°F), multiply the ampacities shown above by the appropriate factor shown below.						Ambient Temp. °F
21-25	1.08	1.05	1.04	1.08	1.05	1.04	70-77
26-30	1.00	1.00	1.00	1.00	1.00	1.00	78-86
31-35	.91	.94	.96	.91	.94	.96	87-95
36-40	.82	.88	.91	.82	.88	.91	96-104
41-45	.71	.82	.87	.71	.82	.87	105-113
46-50	.58	.75	.82	.58	.75	.82	114-122
51-55	.41	.67	.76	.41	.67	.76	123-131
56-60	....	.58	.71	....	.58	.71	132-140
61-70	....	.33	.58	....	.33	.58	141-158
71-80	....	....	.41	....	....	.41	159-176

† Unless otherwise specifically permitted elsewhere in this Code, the **overcurrent protection** for conductor types marked with an **obelisk** (†) shall not exceed 15 amperes for #14 AWG, 20 amperes for #12 AWG, and 30 amperes for #10 AWG **copper**; or 15 amperes for #12 AWG and 25 amperes for #10 AWG **aluminum and copper-clad aluminum** after any **correction factors for ambient temperature and number of conductors have been applied**.

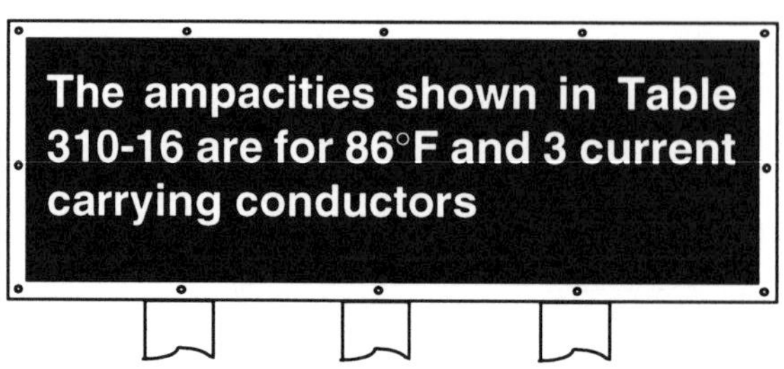

By using 75°C rated equipment rather than 60°C will benefit in smaller conductors.

Example: Size the feeder to this 100 amp panel. The equipment is rated at *75°C.*

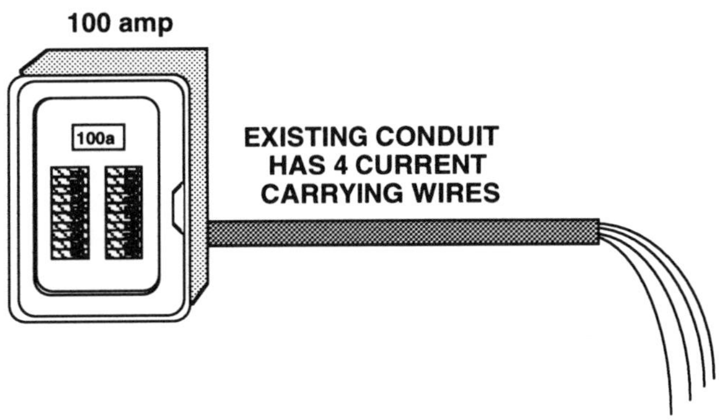

Using 75°C insulation, a #1 has a normal ampacity of 130 amps x 80% (note 8) = 104a.

Using 90°C insulation, a #2 THHN has a normal ampacity of 130 amps x 80% (note 8) = 104a. The #2 THHN would be protected by the 100a breaker, but a #2 THHN can only be loaded to the 75°C ampacity. A #2 @ 75°C has an ampacity of 115. The load is 100 amps.

A #1 THW or a #2 THHN would be the minimum size conductor permitted.

By using 75°C rated equipment for the 100 amp load rather than 60°C equipment, a #1 conductor can be used rather than a #1/0.

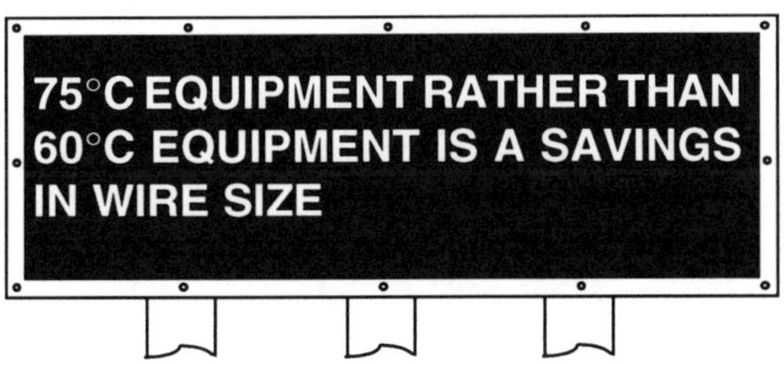

Example: Size the feeder to this 100 amp panel. The equipment is rated at 60°C. Both ampacity adjustments are required; note 8 and ambient correction.

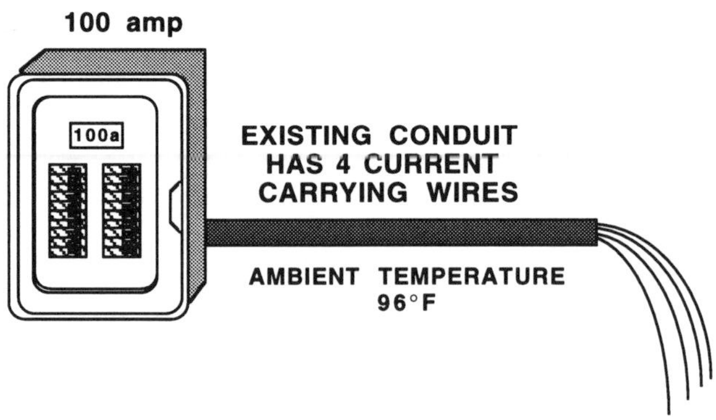

A #3/0 TW 60°C insulation has a normal ampacity of 165 x 80% (note 8) = 132a x .82 (correction factor) = 108.24 ampacity. The load is 100 amps.

A #1/0 THHN 90°C insulation has a normal ampacity of 170 x 80% (note 8) = 136a x .91 (correction factor) = 123.76 ampacity, but a #1/0 THHN can only be loaded to the 60°C ampacity. A #1/0 @ 60°C has an ampacity of 125. The load is 100 amps.

A #3/0 TW or a #1/0 THHN is the minimum size conductor permitted.

The four #3/0 TW conductors require a 2" EMT conduit. The four #1/0 THHN conductors require a 1 1/2" EMT conduit.

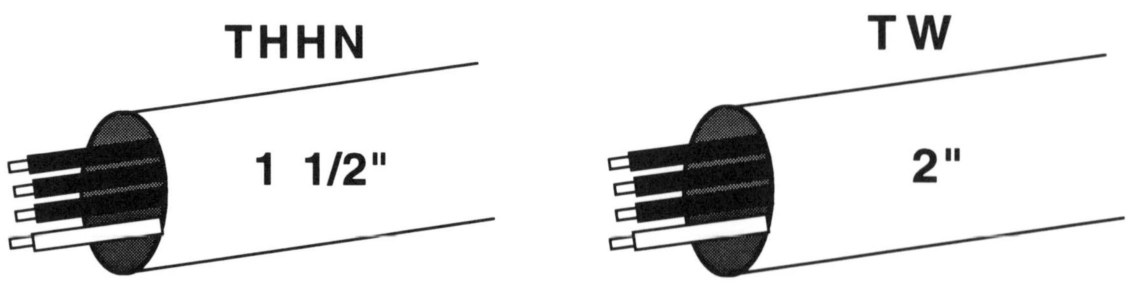

Example: Size the feeder to this 100 amp panel. The equipment is rated at *75 °C*. Both ampacity adjustments are required; note 8 and ambient correction.

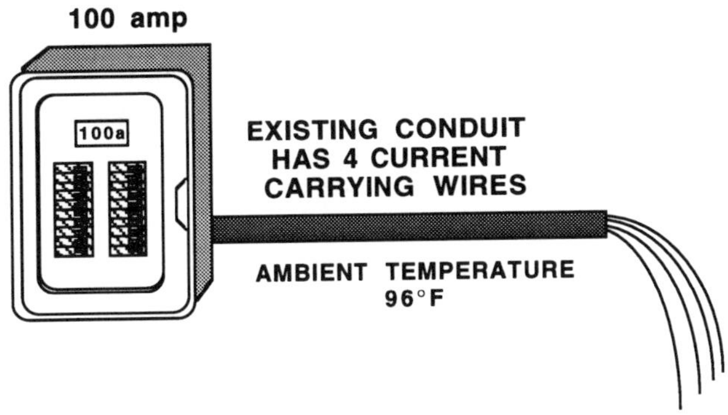

A #1/0 THW 75°C insulation has a normal ampacity of 150 x 80% (note 8) = 120a x .88 (correction factor) = 105.6 ampacity. The load is 100 amps.

A #1 THHN 90°C insulation has a normal ampacity of 150 x 80% (note 8) = 120a x .91 (correction factor) = 109.2 ampacity, but a #1 THHN can only be loaded to the 75°C ampacity. A #1 @ 75°C has an ampacity of 130. The load is 100 amps.

By using 75°C equipment rather than 60°C, a #1/0 THW conductor can be used rather than a #3/0 TW or a #1 THHN rather than a #1/0 THHN.

The advantage is four #1 THHN conductors can be installed in 1 1/4" EMT conduit. Four #1/0 THW conductors require a 2" EMT conduit.

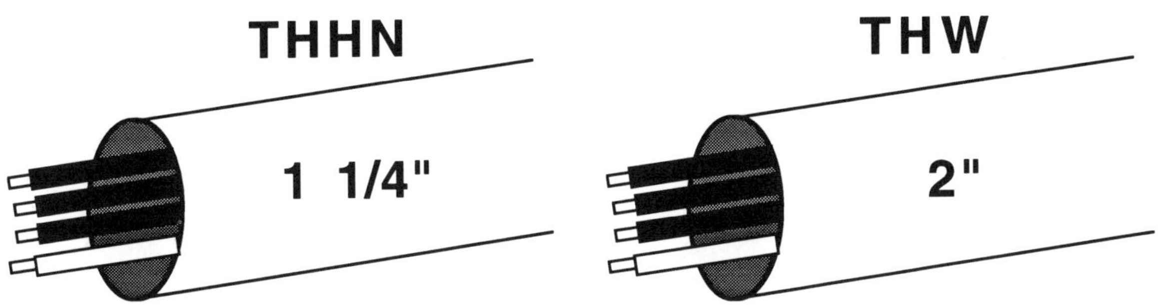

As you can see, the equipment rating, ambient temperature and the number of current carrying conductors are all very important factors in selecting the correct conductor size. For a 100 amp load the conductors vary in size from a #3 to a #3/0 using very moderate factors of 96°F and only four conductors carrying current.

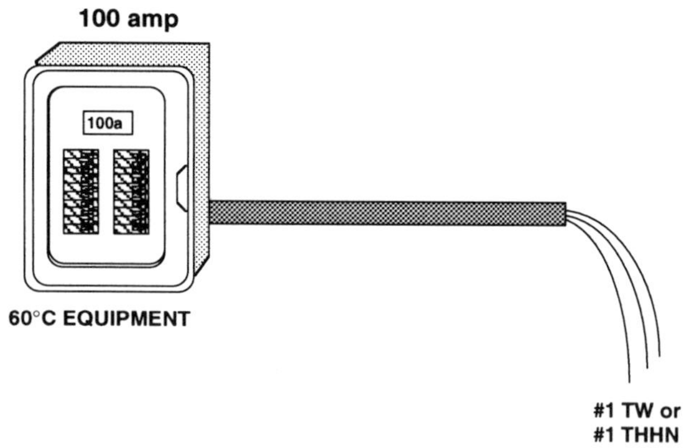

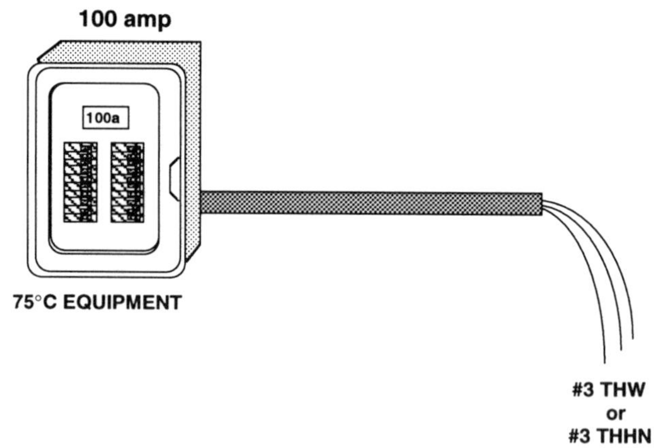

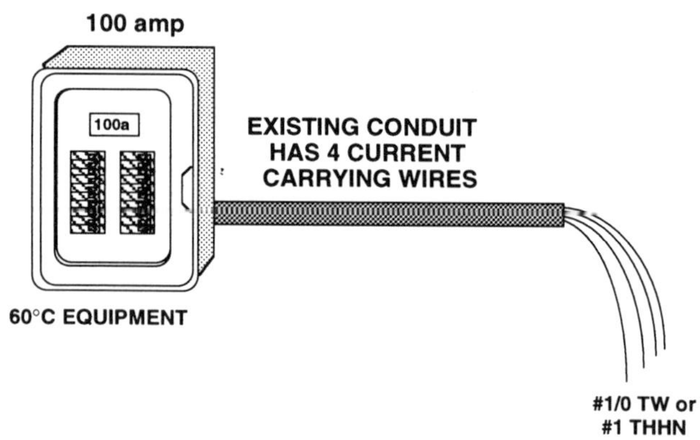

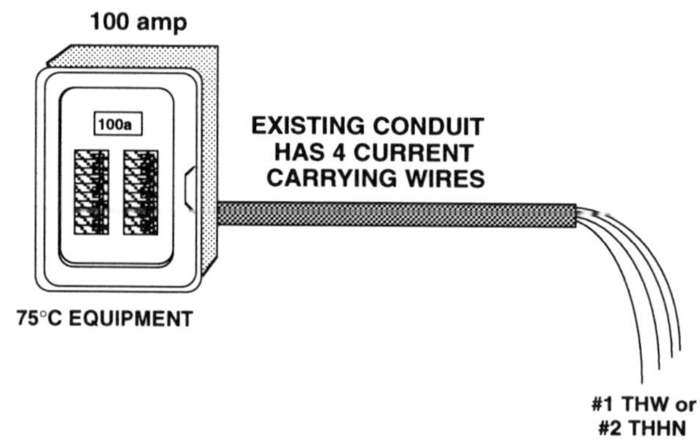

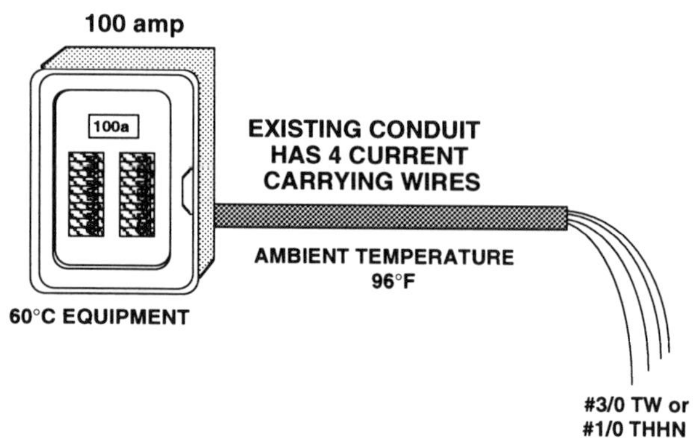

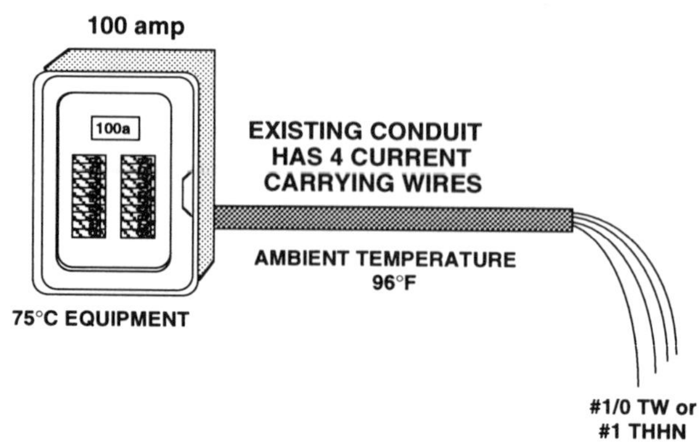

CONTINUOUS LOAD

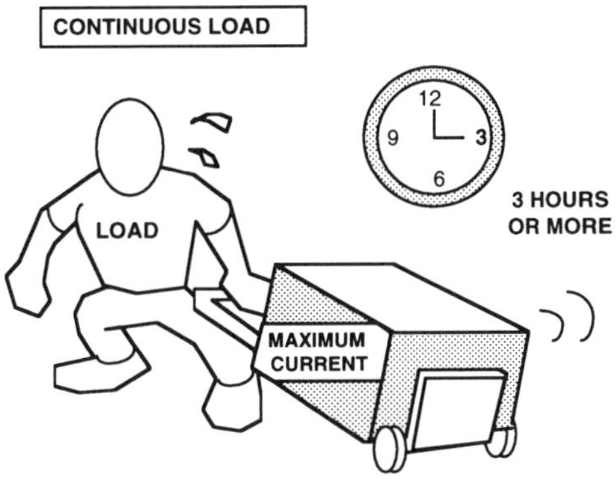

CONTINUOUS LOAD

Let's first start with a noncontinuous loaded circuit. In general, a receptacle circuit is considered a noncontinuous circuit.

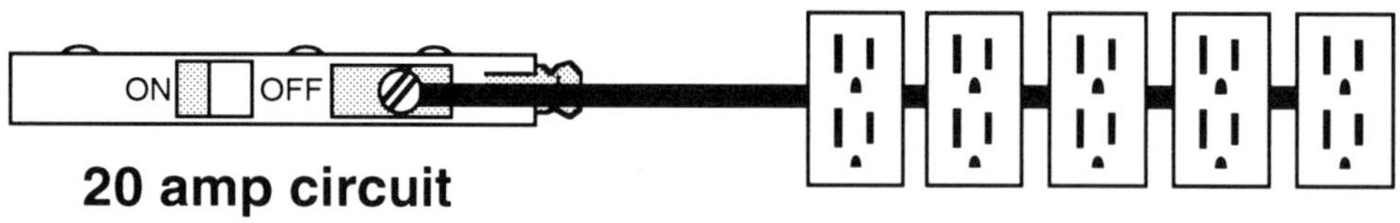

This 20 amp rated circuit can be loaded to 20 amps per Table 210-24 maximum loading. A #12 conductor with an ampacity of 20 can safely carry 20 amperes *continuously* per the definition of ampacity. A 20 amp rated circuit breaker means the breaker will carry a *continuous* current of 20 amps indefinitely at 40°C (104°F) in *free air.*

•The key is a wire can carry the rated ampacity continuously and the overcurrent device can carry the current continuously in *free air.*

The overcurrent device must be installed in an *enclosure* and this is what has the affect on the continuous loading element.

As noted, a breaker is rated at continuous current. When using the breaker for a continuous load, not only the breaker but also the *assembly* that houses the breaker must be listed for continuous loading, which is three hours or more. The assembly will be *ventilated* to allow the heat to escape.

A continuous load is a load where the maximum current is expected to continue for three hours or more.

The Code states that the total load on any overcurrent device located in a panelboard shall not exceed 80% of its rating where the load will continue for three hours or more under normal conditions.

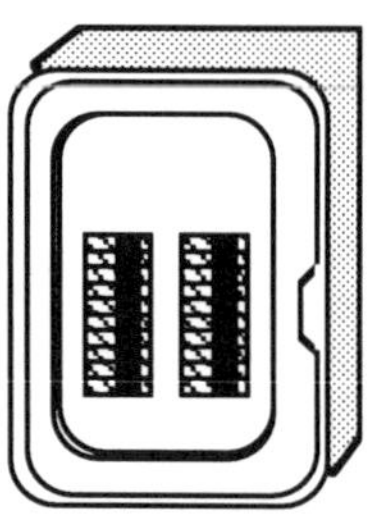

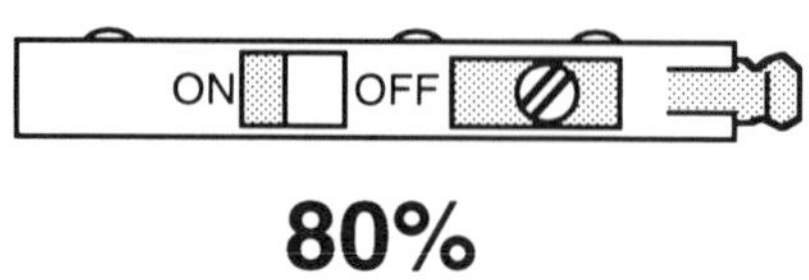

Ambient temperature is the temperature at which the continuous current rating is based. Ambient temperature is the temperature of the air immediately around the overcurrent device which can affect the thermal (overload) opening of the device.

A 20 amp rated circuit breaker means that the breaker will carry a continuous current of 20 amps indefinitely at 40°C (104°F) in free air.

Current flowing through the breaker creates most of the thermal heat used for the tripping action. The ambient temperature surrounding the breaker either adds to or subtracts from this available heat.

Most breakers are listed at 40°C (104°F) which means it will carry its rated current continuously when the air surrounding it is the same temperature as the rated temperature. At lower than rated ambient temperature, a breaker may carry more than its rated current which can cause damage to the conductors and equipment if not properly addressed. When applied at higher than rated temperatures, a breaker will carry less than its rated current. This should be kept in mind when selecting overcurrent devices to be used at extreme high and low temperatures.

As noted, a breaker is rated at continuous current. When using the breaker for a continuous load, not only the breaker but also the *assembly* that houses the breaker must be listed for continuous loading which is three hours or more. The assembly will be ventilated to allow the heat to escape.

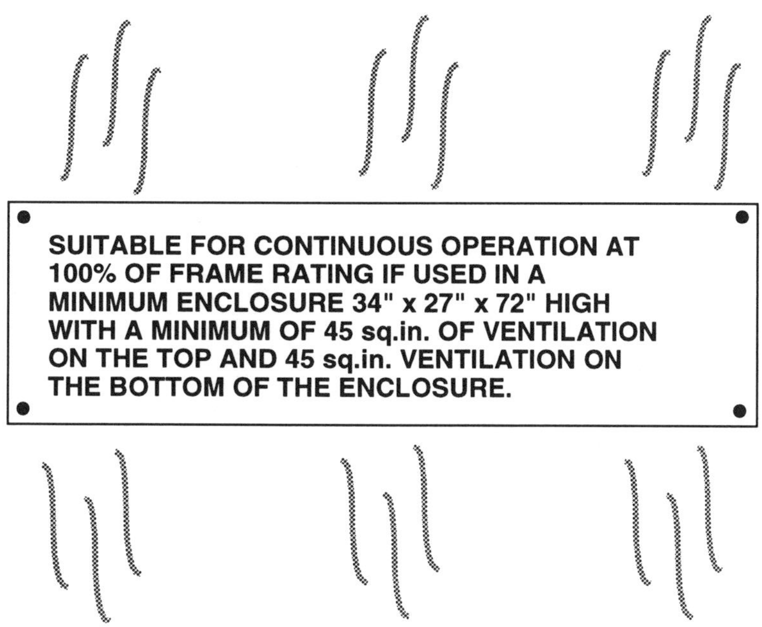

The reason for the 80% current limitation on the circuit breaker is when several circuits containing continuous loads, such as lighting, are contained in the same panelboard and with the cover enclosed, *heat* is entrapped which has a definite affect on the tripping of the circuit breaker.

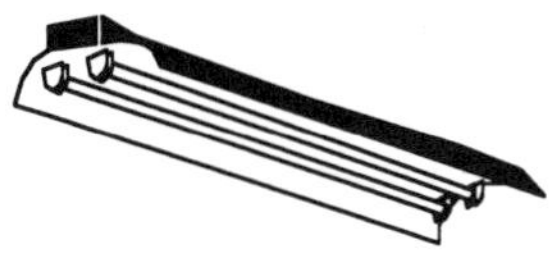

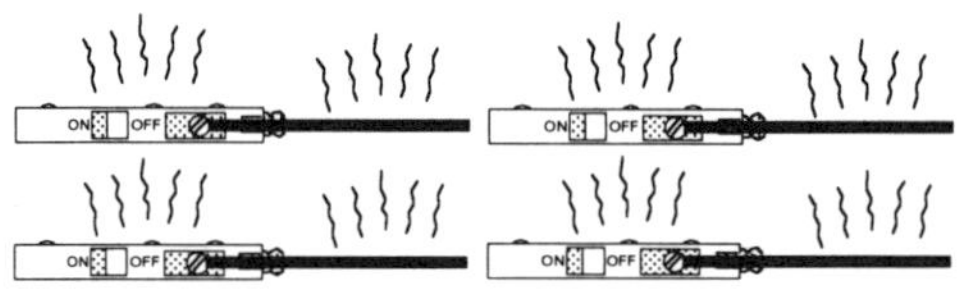

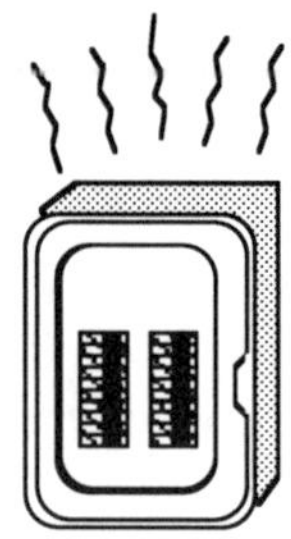

An overcurrent device such as a circuit breaker does not trip on *current* alone. Example; an amp clamp clamped on a conductor could read 17 amperes and a 20 amp rated circuit breaker could trip.

20 amp rated

Amp clamp reads 17 amps

The current carrying conductor produces heat but so does the *ambient temperature*. This is the temperature surrounding the conductor or equipment (circuit breaker).

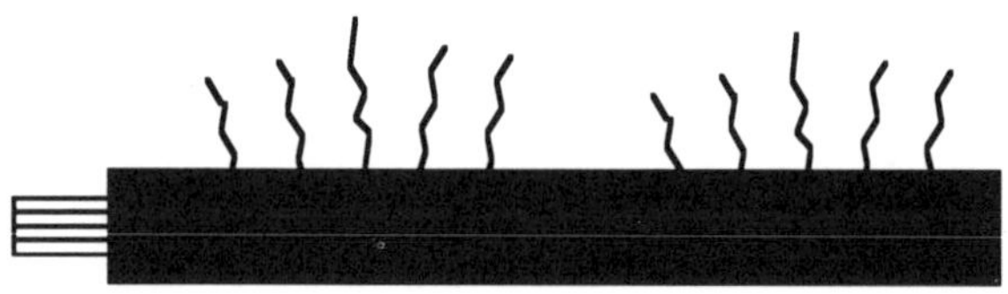

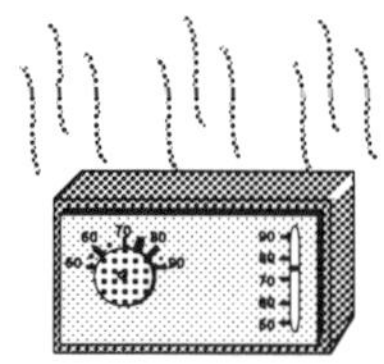

Ambient temperature is the temperature surrounding the conductor or equipment

The circuit breaker has a bimetallic element that trips on the amount of *heat.* The conductor and the ambient temperature, combined with heat, determines when it trips.

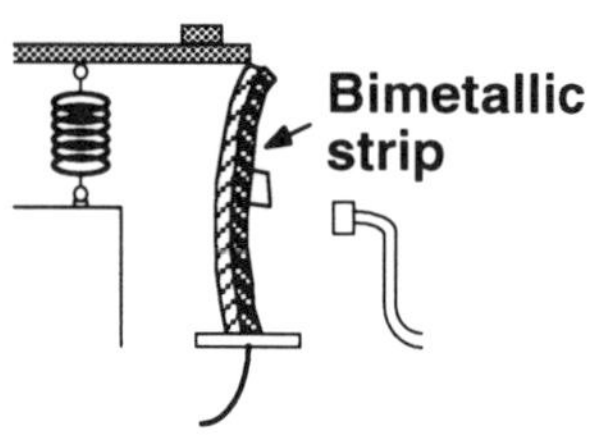

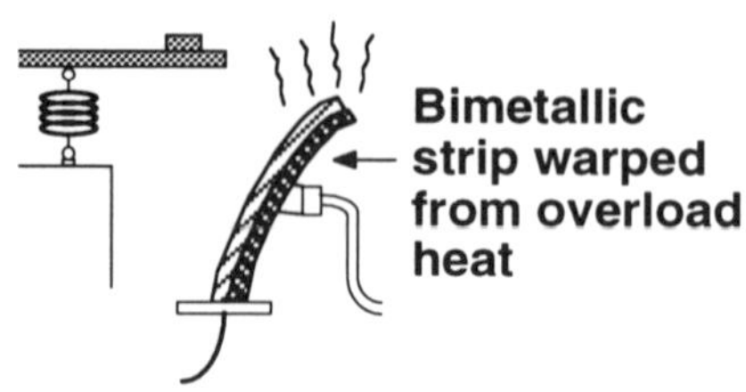

Even though only 17 amps of current is flowing through the circuit, this combined with the added ambient heat from the other circuits with continuous loads raises the heat thus tripping the circuit breaker.

The Code states that for branch circuit and feeder conductors the rating of the *overcurrent device* shall be 125% of the continuous load.

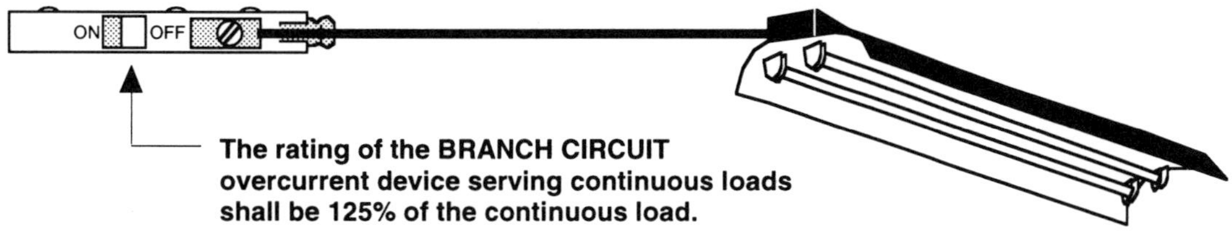

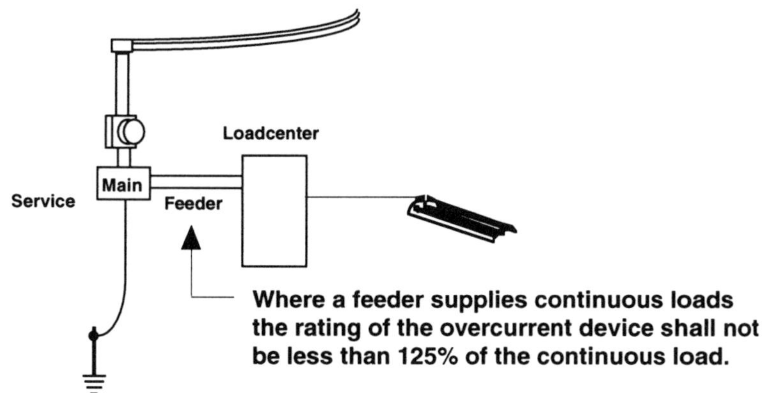

The reason for the 125% increase for continuous loading is *not* because the wire, fuse, or circuit breaker can't carry the load at 100%.

The definition of ampacity makes this clear: The current, in amperes, a conductor can carry *continuously* under the conditions of use without exceeding its temperature rating.

A fuse or circuit breaker can also carry 100% of its current rating continuously without damage to itself, but the *heat* conducted from the heat sink of the fuse or circuit breaker is conducted into the conductors at the terminals causing excessive heat which can damage the conductor insulation.

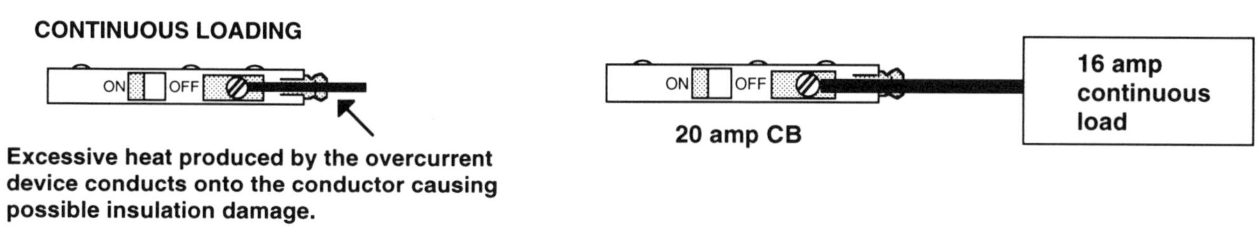

384-16c also says a 20 amp CB can only be loaded 80%. 20a x 80% = 16 amp continuous load. 220-10b requires the overcurrent device to be increased 125% for continuous load. 16 amp continuous load x 125% = 20 amp CB. 80% and 125% are reciprocals.

Example: A 50 amp NONcontinuous load on a branch circuit is shown sized below.

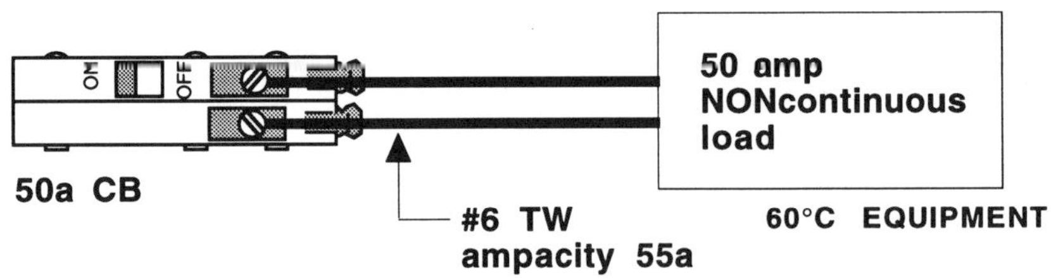

If the 50 amp load was a *continuous* load, the 50a x 125% = 62.5. Section 240-6 lists standard sizes for overcurrent devices; 62.5 amp is not a standard size breaker. The next higher standard size is a 70 amp. 384-16c states the breaker can only be loaded 80% 70a x 80% = 56 amps. The continuous load is 62.5 amps so the correct overcurrent device is a 70 amp.

Since the branch circuit is a continuous load and a #4 TW will carry 70 amps continuously, is a #4 TW the correct size conductor?

240-3 states conductors shall be protected against overcurrent in accordance with their ampacities. A #4 TW is required to be protected at its ampacity which is 70. Section 240-6 lists a 70 amp overcurrent device as a standard size.

The continuous load requires a 70 amp circuit breaker which will properly protect a #4 TW conductor with a 70 ampacity.

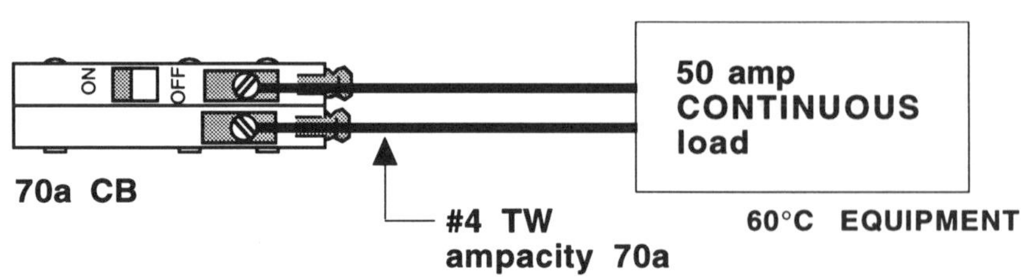

The advantage and disadvantage of a THHN is shown below. The #4 THHN with an ampacity of 95 is protected by a 70a breaker. The #4 THHN can be loaded to the 60°C ampacity which is 70 amps. The *load* is 50 x 125% = 62.5 amps.

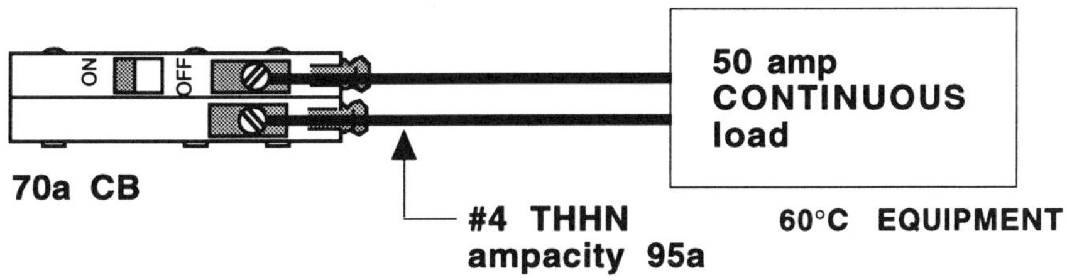

The everyday violation is the lack of knowledge when installing circuits using THHN insulation. THHN has two advantages: (1) Thinner insulation permits more conductors in a conduit. (2) It has a higher ampacity. The disadvantage is the THHN cannot be *loaded* to the 90°C ampacity. This is where the violation occurs. The installer takes the load from the nameplate and then turns to Table 310-16 and selects a THHN conductor that has ampacity equal to or higher than the load.

Always remember by selecting a #4 THHN rather than a #4 TW because of the higher ampacity that the Code states in section 110-14c the temperature rating associated with the ampacity of the conductor shall be so selected not to exceed the lowest temperature rating of any conductor, device or connected termination.

100 amps or less	**Over 100 amps**
Wire size #14 - #1	**Wire size #1/0 and larger**
60°C ampacity	**75°C ampacity**

Shown below is an example of how everyday installation of THHN conductors is violated.

For an 80 amp continuous load the installer turns to Table 310-16 and selects a THHN conductor that will carry 80 amps, a #4 THHN has an ampacity of 95. The installer feels real good about a #4 THHN, it even has 15 amps above the 80 amp load.

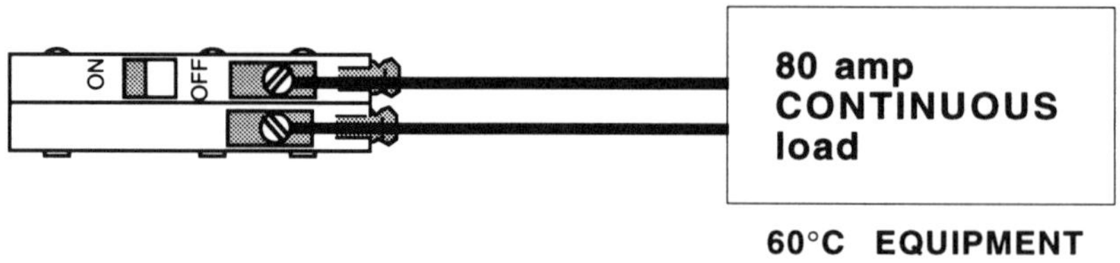

The first step is to size the breaker. Take the 80 amp load x 125% (220-10b) = 100 amp breaker size. 384-16c states a breaker can be loaded 80% for continuous loading (100a cb x 80% = 80 amp load).

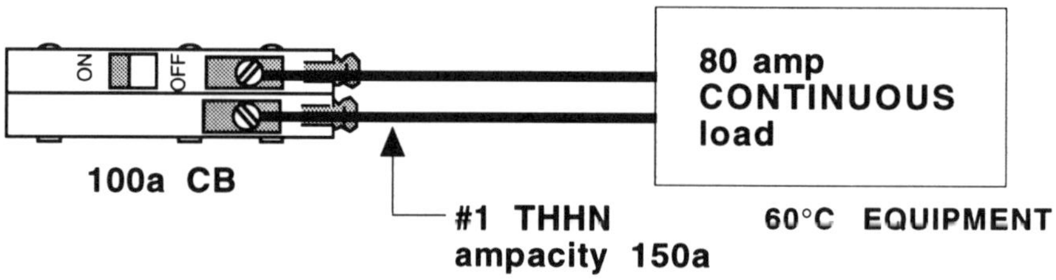

The next step is to select a THHN conductor that will be protected by the 100 amp breaker. Table 310-16 lists a #3 THHN with an ampacity of 110 which can be protected by a 100a breaker. A #3 THHN can only be loaded to the 60°C ampacity of 85 amps and the load is 100 amps. A #2 THHN has an ampacity of 130 which would be protected by a 100a breaker.

A #2 THHN can only be loaded to the 60°C ampacity of 95. The load is 100 amps.

The *minimum* size THHN conductor the Code allows for an 80 amp continuous load is a #1.A #1 THHN has an ampacity of 150a which can be loaded to the 60°C ampacity of 110. The load is 80 x 125% = 100 amps.

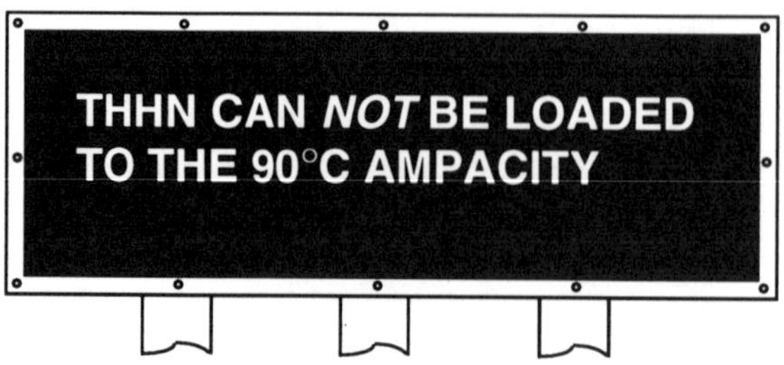

A set of service entrance conductors usually follow the same step-by-step calculation for sizing feeder conductors.

There is no justification to require service conductor ampacity to be increased because of a continuous load. The overcurrent device rating is what has to be increased for a continuous load and then the conductors may have to be oversized to be protected in accordance with section 240-3.

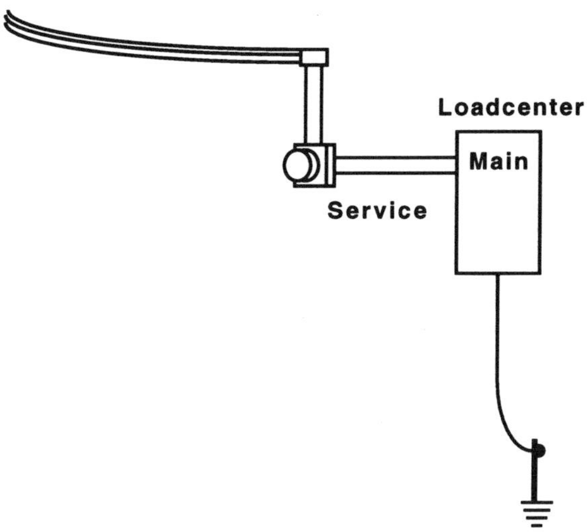

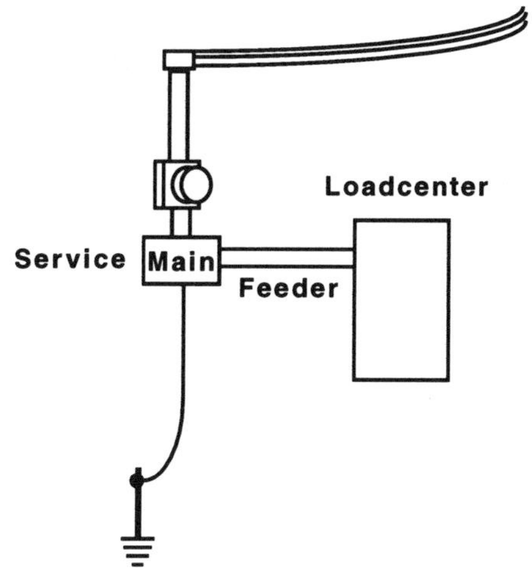

Example: The service load is calculated to be 130 amperes of which 84 amps is noncontinuous and 46 amps is continuous. 75°C equipment.

Noncontinuous load	=	84a
Continuous load 46a x 125%	=	57.5a
		141.5 amps

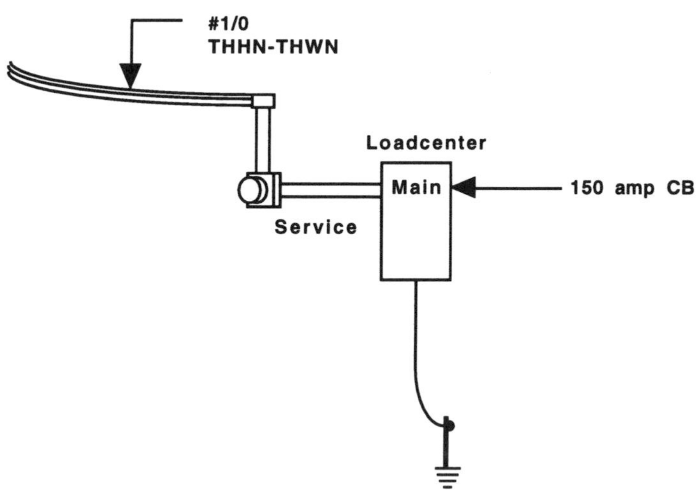

Section 240-6 lists a 150 amp overcurrent device as the next higher standard size. A #1/0 THHN has an ampacity of 170 and can be loaded to the 75°C ampacity of 150. The load is 141.5 amps. The 150a breaker can be loaded 80% for *continuous load*, 150a x 80% = 120 amps. The continuous load is 46 x 125% = 57.5 amps.

Example: The service load is calculated to be 140 amperes of which all is NONcontinuous loading. 75°C equipment.

NONcontinuous load 140a = 140 amps

Section 240-6 lists a standard size overcurrent device at 150 amps. A #1/0 THHN has an ampacity of 170 and can be loaded to the 75°C ampacity of 150 amps.

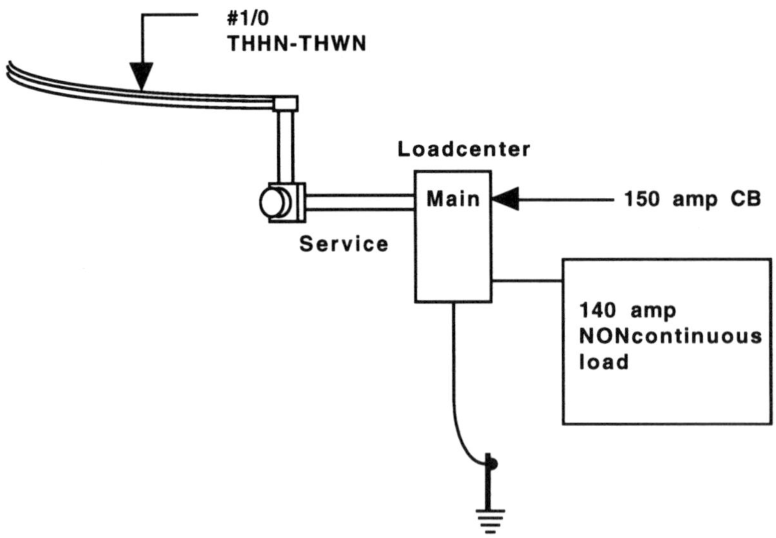

Example: The service load is calculated to be 140 amperes of which all is CONTINUOUS loading. 75°C equipment.

Continuous load 140a x 125% = 175 amps

Section 240-6 lists a standard size overcurrent device at 175 amps. A #2/0 THHN has an ampacity of 195 and can be protected by a 175a breaker. The #2/0 THHN can be loaded to the 75°C ampacity of 175. The load is 175 amps.

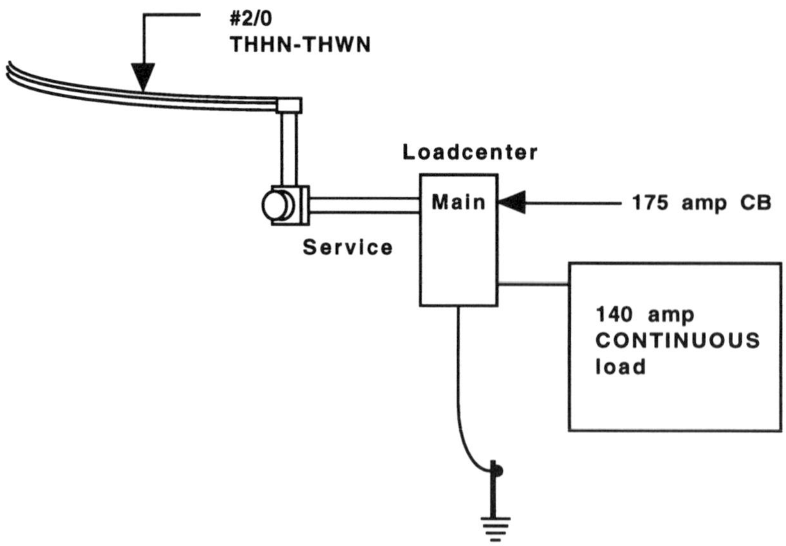

Example: The service load is calculated to be 96 amperes of which all is NONcontinuous loading. 60°C equipment.

NONcontinuous load 96a = 96 amps

Section 240-6 lists a standard size overcurrent device at 100 amps.

Here is a good example of how THHN conductors are misused. The load is 96 amps noncontinuous and the next higher size breaker is 100 amps.

The designer goes to Table 310-16 and selects a #3 THHN which has an ampacity of 110 and the designer feels satisfied he has selected the correct size conductor, as the 110 ampacity would be easily protected by the 100 amp breaker and the load is only 96 amps. *BUT*, a #3 THHN can only be loaded to the 60°C ampacity which is 85 and the load is 96 amps.

A #2 THHN has an ampacity of 130, but it can only be loaded to the 60°C ampacity of 95 and the load is 96 amps.

The *smallest* THHN conductor that would be allowed to be connected to 60°C equipment and loaded to 96 amps would be a #1 THHN.

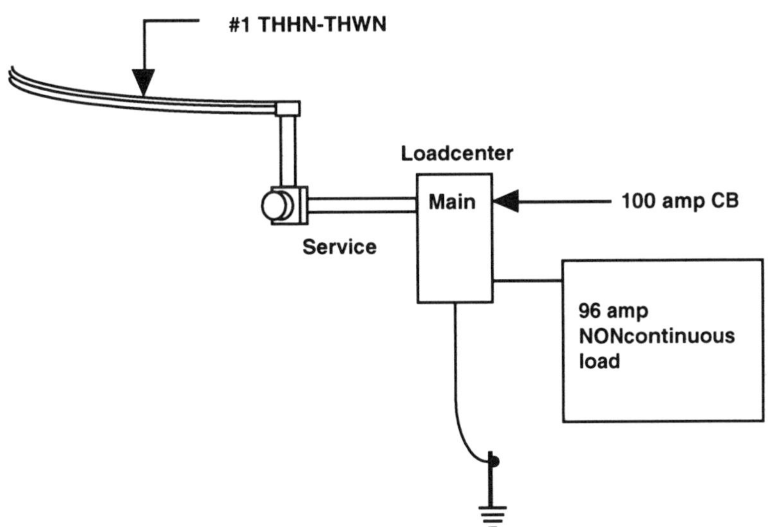

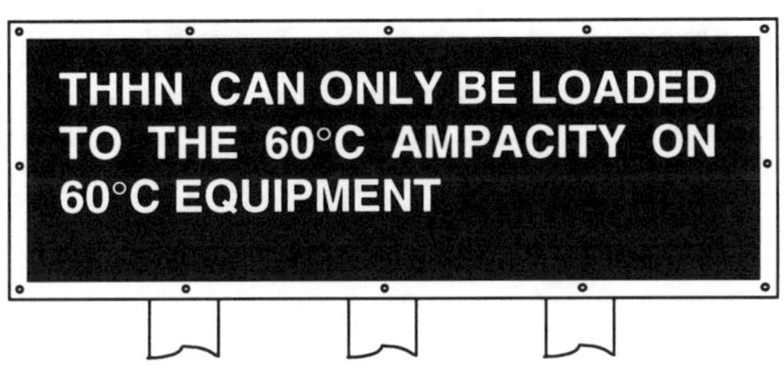

Example: The service load is calculated to be 98 amperes of which 56 amps is continuous and 42 amps is NONcontinuous. 60°C equipment.

Continuous load 56a x 125%	=	70 amps
NONcontinuous 42a	=	42a
		112a

Section 240-6 lists a standard size overcurrent device at 125 amps. A #1 THHN has an ampacity of 150, but it can only be loaded to the 60°C ampacity of 110. The load is 112 amps.

A #1/0 THHN has an ampacity of 170 and can be loaded to the 60°C ampacity of 125. The 125a breaker can be loaded 100 amps for *continuous* load (125a x 80%=100a). The continuous load is 56 x 125% = 70 amps.

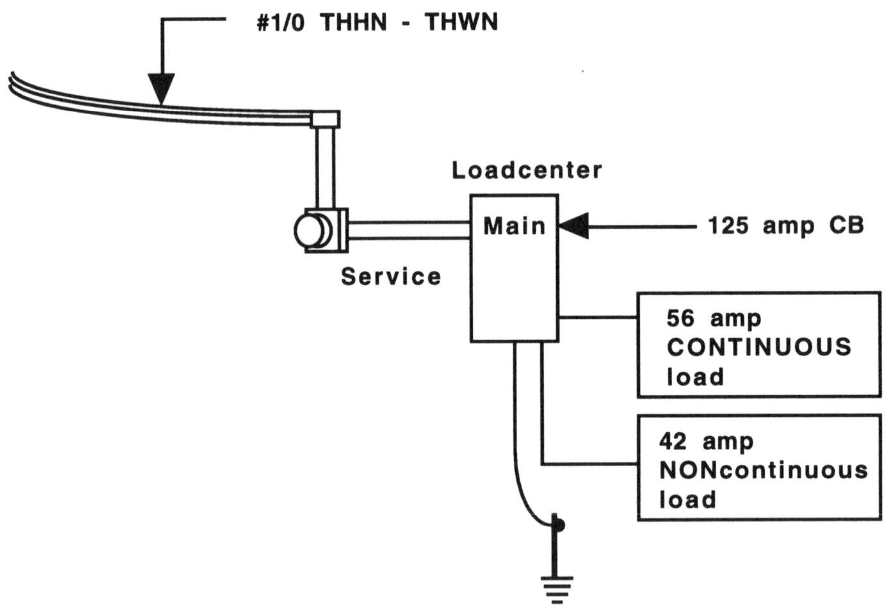

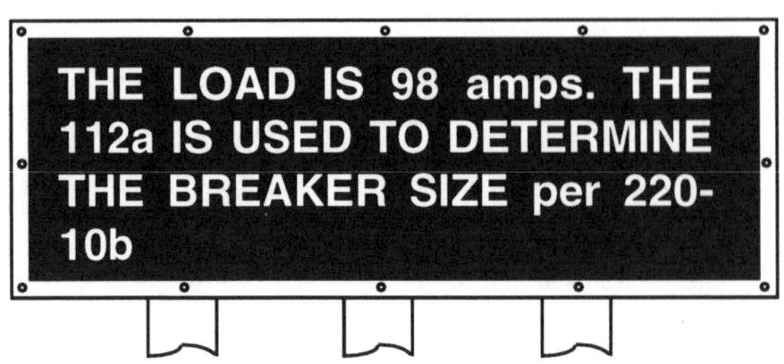

Example: The service load is calculated to be 895 amperes of which all is continuous loading. 75°C equipment.

Continuous load 895a x 125% = 1118.75 or 1119 amps

Section 240-6 lists the next higher standard size overcurrent device at 1200 amps. Section 240-3c requires where the overcurrent device is rated *over* 800 amps, the ampacity of the conductors shall be equal to or greater than the rating of the overcurrent device. This would require paralleling of conductors to meet the 1200 ampacity requirements. Three #500 THHN with an ampacity of 430a x 3 = 1290a would meet the requirement of 240-3c. They can be loaded to the 75°C ampacity which is 380a x 3 = 1140. The load is 895 x 125% = 1119 amps.

Example: The service load is calculated to be 895 amperes of which all is NONcontinuous loading.

Noncontinuous load 895a = 895 amps

Section 240-6 lists the next higher standard size at 1000 amps. Section 230-3c requires conductors with an ampacity of at least 1000 amps. This would require paralleling three #350 THHN with an ampacity of 350a x 3 = 1050 amps. They can be loaded to the 75°C ampacity which is 310a x 3 = 930. The load is 895.

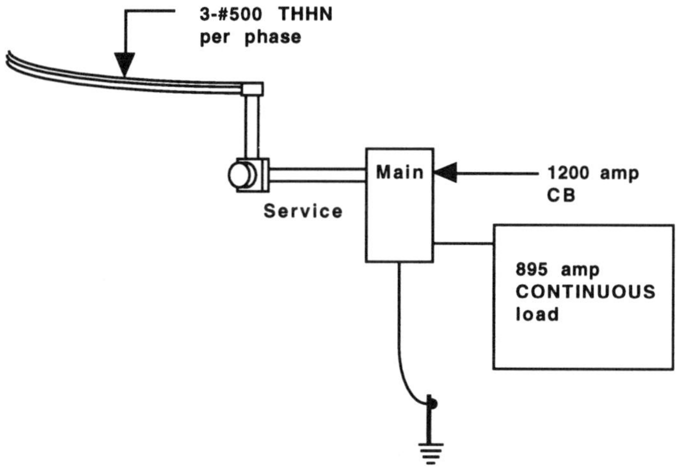

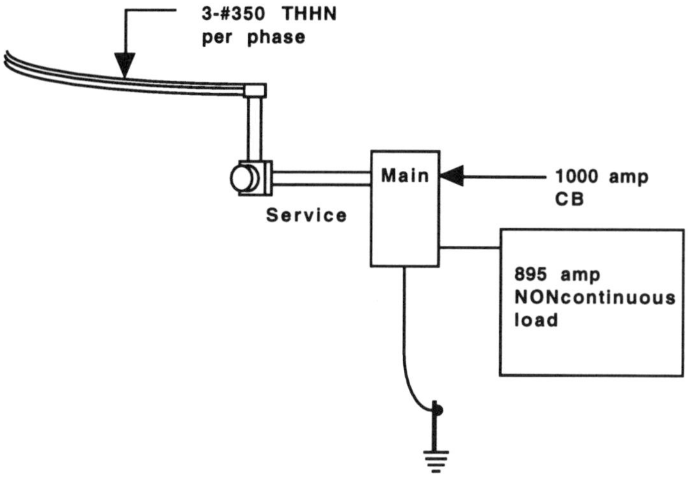

Example: Size the feeder to this 200 amp panel. *NONContinuous load*. The equipment is rated at 75°C.

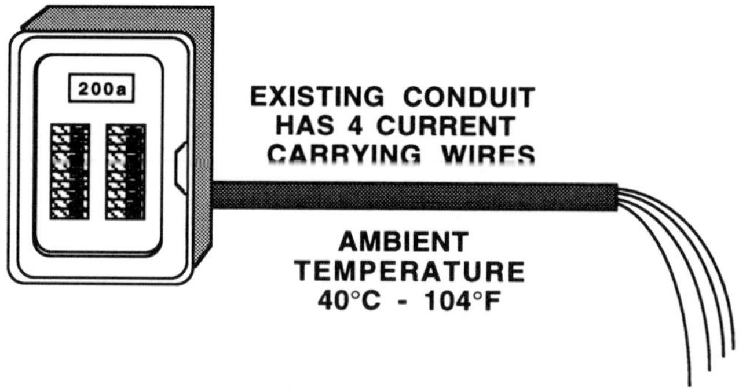

The first step is to select the overcurrent device 200 amp breaker.

A 3/0 THW 75°C insulation has an ampacity of 200, the conductor would be protected by a 200 amp breaker, but the derating factors would also be applied to the 200 ampacity.

A #300 kcmil THW has a normal ampacity of 285 x 80% (note 8) = 228a x .88 (correction factor) = 200.64 or 201 amps, the #300 kcmil THW with a derated ampacity of 201 would be protected by a 200 amp breaker.

A #250 kcmil THHN has a normal ampacity of 290 x 80% (note 8) = 232a x *.91* (correction factor) = 211 amps. The 90°C THHN can only be *loaded* to the 75°C ampacity. A #250 kcmil TW @ 75°C has an ampacity of 215 amps.

The minimum size conductors permitted for the conditions are #300 kcmil THW or #250 kcmil THHN.

Shown below is a 76 amp load which would require a #3 THHN conductor which has an ampacity of 110 amps, but can only be loaded to a 60°C ampacity of 85 amps.

The circuit breaker would be the next higher standard size of 80 amps.

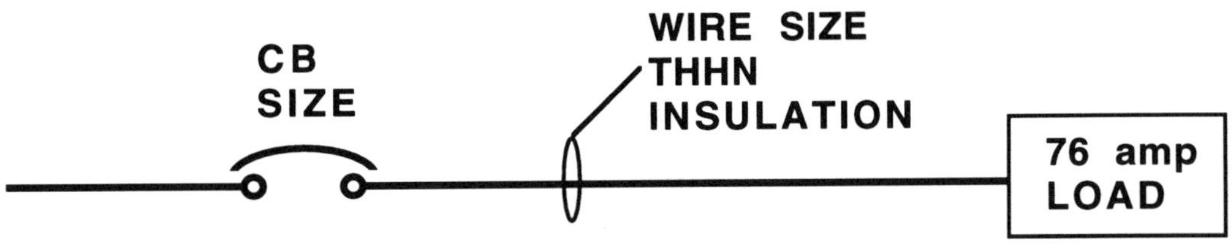

If the load was *continuous*, 76a x 125% = 95 amps. The next higher standard size breaker is a 100 amp. The breaker can only be loaded 80% x 100a = 80 amps. A #2 THHN has an ampacity of 130 amps, but can only be loaded to a 60°C ampacity of 95 amps.

If the noncontinuous load had *four current carrying conductors*, A #3 THHN has a normal ampacity of 110 amps x 80% (Note 8) = 88 ampacity. The 90°C THHN can only be *loaded* to the 60°C ampacity. A #3 TW @ 60°C has an ampacity of 85 amps. A #3 THHN is the smallest THHN conductor permitted for a noncontinuous load and four current carrying conductors.

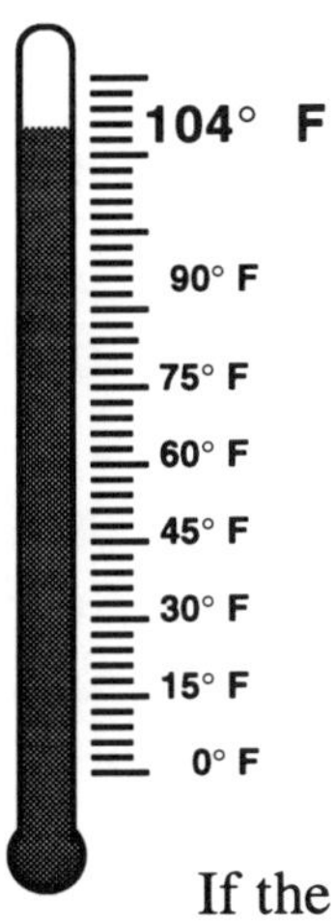

Ampacities from Table 310-16 are based @ 86°F. A more realistic ambient would be 96-104°F. At this temperature the #3 THHN would be derated twice, 110a x 80% =88a x .91 (correction factor) = 80.08 ampacity.

A #3 THHN can only be loaded to the 60°C ampacity. A #3 TW has an ampacity of 85 amps. The actual load is 76 amps.

A #3 THHN is the *smallest* conductor the Code would allow on a 76 amp noncontinuous load with four current carrying conductors and an ambient temperature of 96°F.

If the ambient temperature of this same circuit was raised to 132°F, a #3 THHN would be derated 110a x 80% (note 8) = 88a x .71 correction factor = 62.48 amps. The load is 76 amps.

Even though the 60°C #3 has an ampacity of 85, a #3 THHN under these heat conditions can only be loaded to 62.48 amps without incurring insulation damage.

A #2 THHN with a normal ampacity of 130 x 80% (note 8) = 104a x .71 = 73.84 derated ampacity. A #1 THHN with an ampacity of 150 x 80% = 120a x .71 = 85.2 derated ampacity.

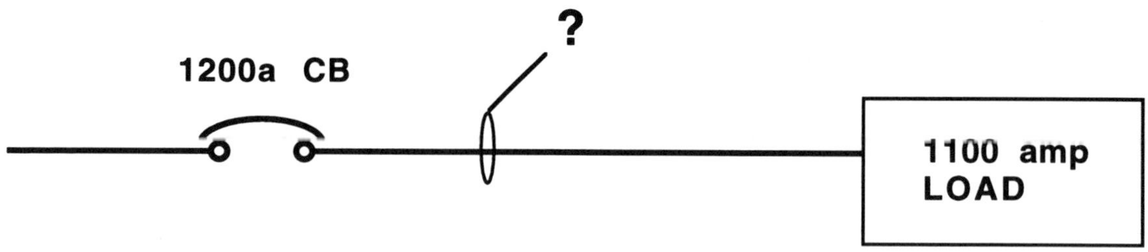

Example: If three #500 kcmil THW conductors were paralleled, the ampacity would be 3 x 380a = 1140 amps. Since you are *over* 800 amps you cannot go to the next higher size unless the conductor ampacity is equal to or higher than the breaker size of 1200 amps per section 240-3c.

If #500 kcmil THHN conductors were used instead of THW, the ampacity in parallel would be 430a x 3 = 1290 ampacity. A 1200 amp overcurrent device can be used and the conductors would be protected at their ampacity. Of course the 90°C THHN conductors could only be *loaded* to the 75°C ampacity of 1140 amps.

Example: The same load only now it's *continuous.*

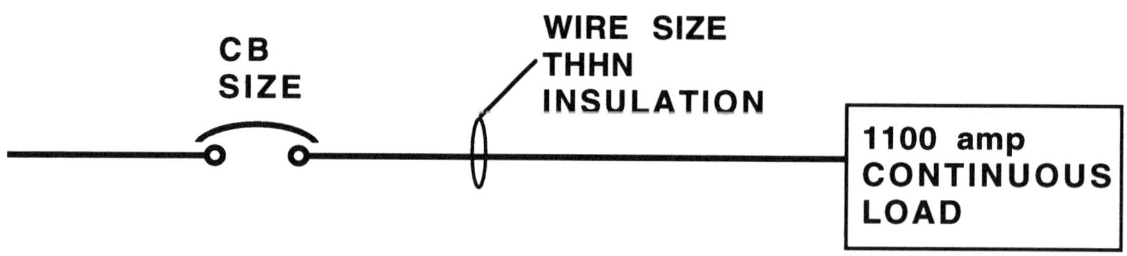

The first step is to determine the breaker size. For example, for a 1100 amp *continuous* load x 125% = 1375 amps. The Code states for continuous loading the overcurrent device shall not be loaded over 80%. In this example we could not select a lower standard size of 1200 amps because of the 80% load limitation required. 1200 amp overcurrent device x 80% = 960 amps maximum continuous load permitted, and the actual load is 1100 amps.

Section 240-6 lists the next higher standard size at 1600 amps. A 1600 amp overcurrent device x 80% = 1280 amps maximum continuous load permitted.

The overcurrent device would be a 1600 amp. The THHN minimum size would be 3 - #750 kcmil THHN @ 535a x 3 = 1605 ampacity. The conductors would be protected at 1600 amps. The 90°C THHN conductors could only be loaded to the 75°C ampacity of 475a x 3 = 1425 ampacity. The continuous load is 1100 x 125% = 1375 amps.

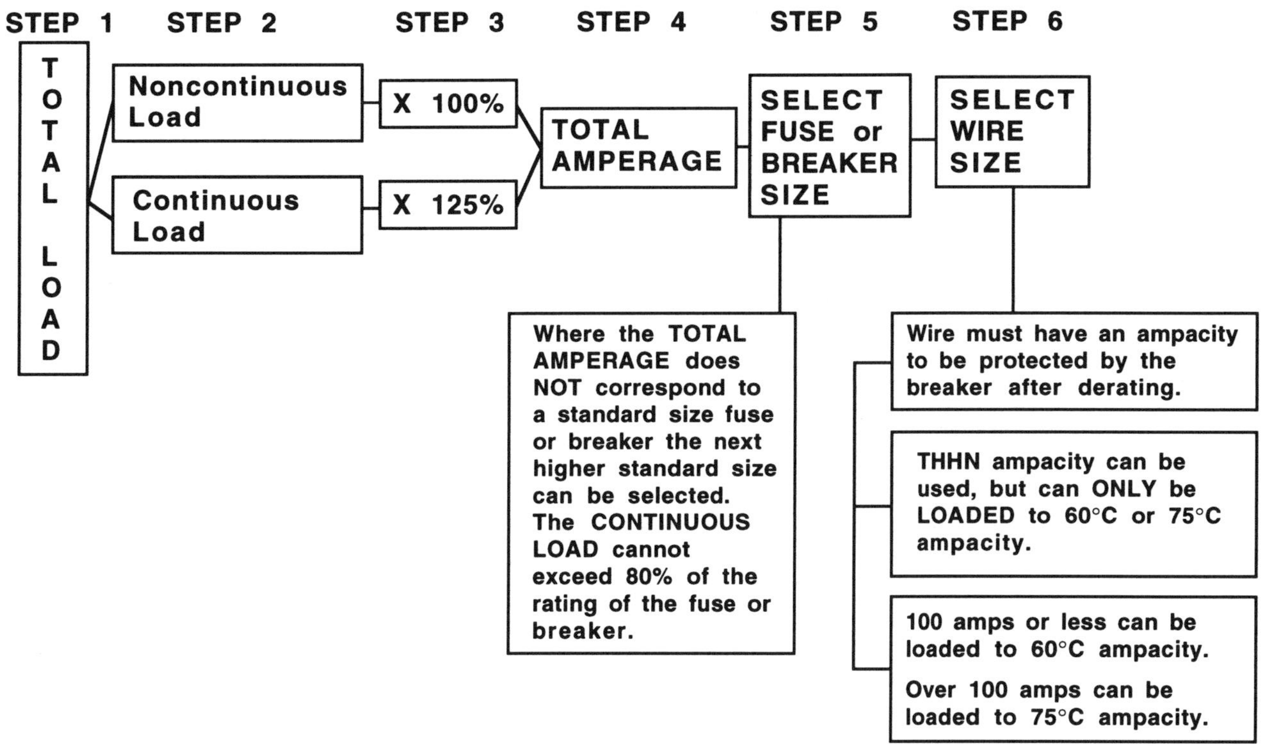

Example: What size breaker and THHN conductors are required for a 480 amp continuous load? 75°C equipment.

480 amp x 125% = 600 amp breaker (240-6). A #1500 kcmil THHN has an ampacity of 705 amps and can be protected by the 600 amp breaker. The #1500 kcmil THHN can only be loaded to the 75°C ampacity of 625. The load is 600 amps.

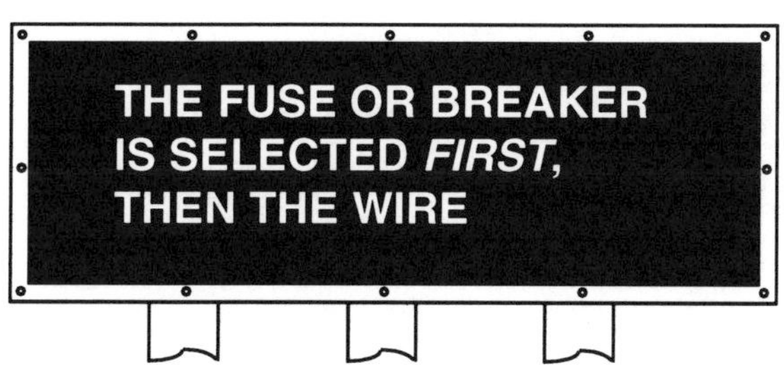

THE VARIABLES

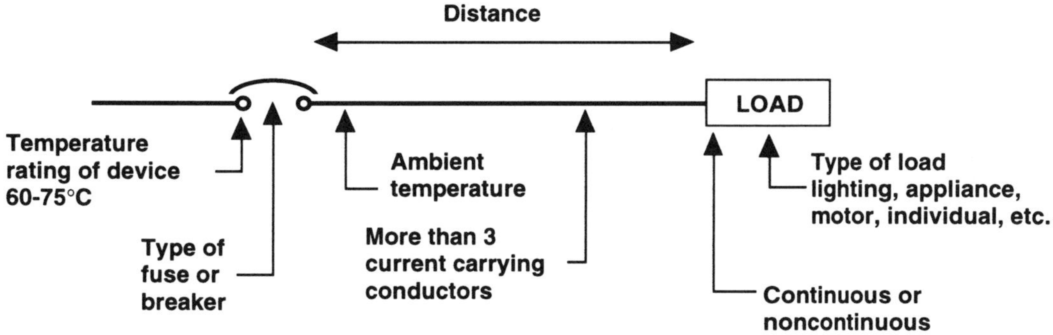

• VARIABLE 1 - CONTINUOUS LOAD

Continuous load applies *only* to the sizing of the fuse or breaker. The wire can carry the current continuously.

If the lights operate more than 3 hours this would be a *continuous load*. Code section 384-16c states the total load on the overcurrent device shall not exceed 80% of its rating. A 20 amp CB x 80% = 16 amps. A 20 amp CB would be okay as the total connected load is 16 amps.

The reason for the 80% current limitation on the circuit breaker is when several circuits containing continuous loads, such as lighting, are contained in the same panelboard and with the cover enclosed, *heat* is entrapped, which has a definite affect on the tripping of the circuit breaker.

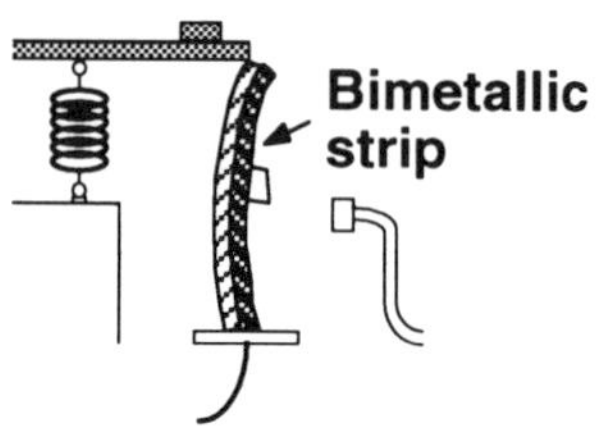

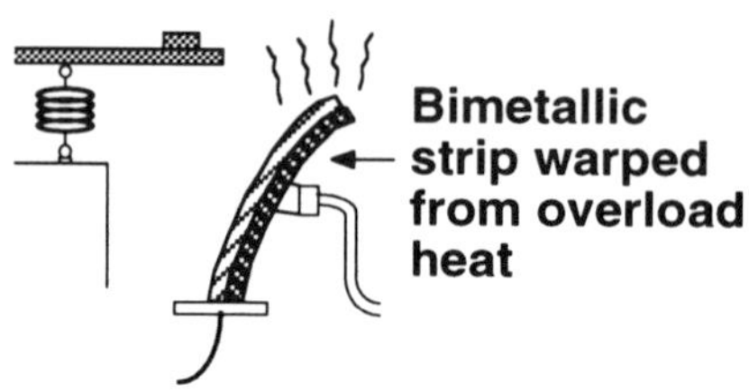

The circuit breaker has a bimetallic element that trips on the amount of *heat***.** The conductor and the ambient temperature combined with heat determines when it trips.

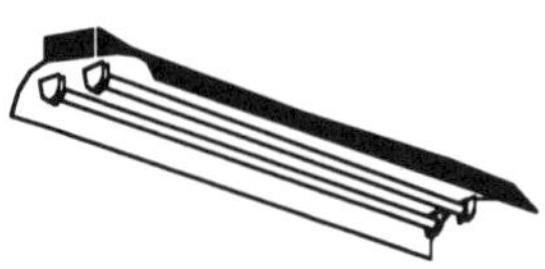

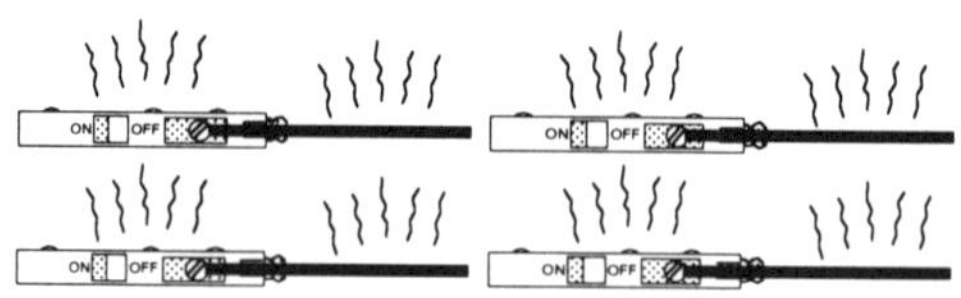

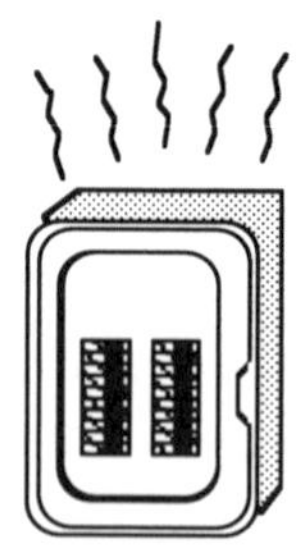

If the branch circuits are loaded to more than 80% it causes extra heat and combined with the surrounding heat (ambient) and the other circuits with continuous loads raises the heat thus tripping the 20 amp circuit breaker even though less than 20 amps is flowing through the circuit.

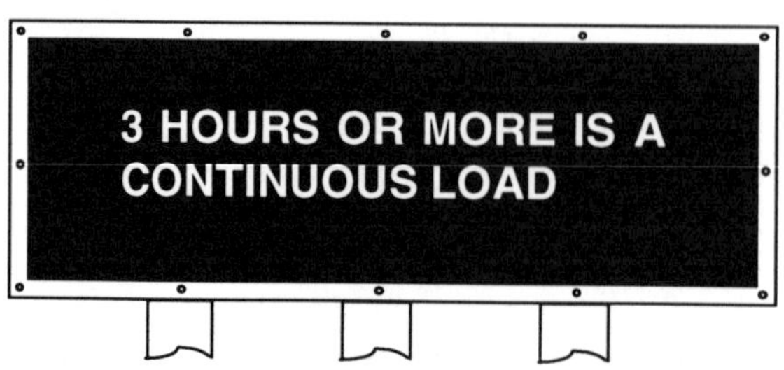

240-3 states conductors shall be protected against overcurrent in accordance with their ampacity.

A #12 TW conductor has an *ampacity* of 25 amps per Table 310-16.

The definition of *ampacity*: The current in amperes that a conductor can carry *continuously* under the conditions of use without exceeding its temperature rating.

The † from the footnotes of Table 310-16 state a #12 TW conductor is permitted to have a *maximum* overcurrent device (fuse or breaker) of 20 amps.

So for this *continuously* loaded lighting circuit a #12 TW wire protected by a 20 amp CB would meet the Code.

• VARIABLE 2 - VOLTAGE DROP

Once the load is determined, calculate how far the circuit can be installed on the *assumed* #12 wire.

The Code recommendation is to limit voltage drop on a branch circuit to no more than 3%. For a 120 volt branch circuit the voltage drop allowed is 120v x 3% = 3.6 volts.

How far does the Code permit a 16 amp load on a branch circuit to be run?

$$\text{Distance} = \frac{\text{CM} \times \text{VD}}{2 \times \text{K} \times \text{I}} = \frac{6530\text{cm} \times 3.6\text{v}}{2 \times 12.9294 \times 16\text{a}} = \textit{56.8 feet}$$

If the branch circuit was installed further than 56.8' from the overcurrent device it would be a violation of the recommendation of the Code using #12 wire. So the *length* of a branch circuit is a very important variable in electrical designing.

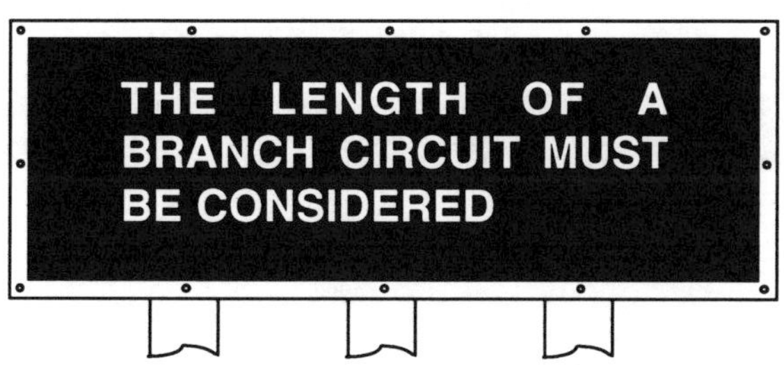

In some cases the branch circuit may be installed farther than 56.8 feet. This would require using a larger size wire.

Shown below is the calculated distance for a #10 wire.

$$\text{Distance} = \frac{\text{CM x VD}}{\text{2 x K x I}} = \frac{\text{10380cm x 3.6v}}{\text{2 x 12.8712 x 16a}} = \textit{90.7 feet}$$

Shown below is the calculated distance for a #8 wire.

$$\text{Distance} = \frac{\text{CM x VD}}{\text{2 x K x I}} = \frac{\text{16510cm x 3.6v}}{\text{2 x 12.84478 x 16a}} = \textit{144.6 feet}$$

#12 = 56.8 feet DISTANCE

#10 = 90.7 feet DISTANCE

#8 = 144.6 feet DISTANCE

• VARIABLE 3 - AMBIENT TEMPERATURE

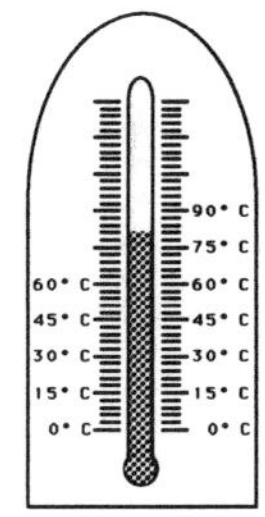

If the branch circuit conductors are going to be installed in an ambient temperature that exceeds 86°F, the derating of conductor ampacity is required. The Table of Ampacity 310-16 lists correction factors when ambient temperatures are above or below 86°F or 30°C.

A fuse or breaker is UL listed at an ambient of 40°C which is 104°F.

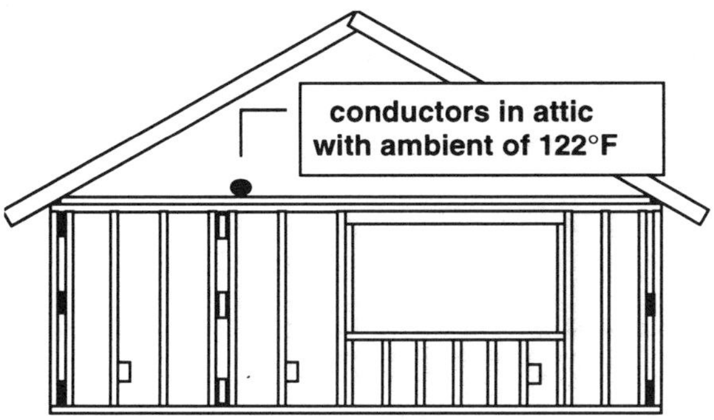

If the conductors were installed in an attic above the insulation where the ambient temperature reaches 122°F, a correction factor from Table 310-16 of .58 must be applied for TW insulation.

#12 TW normal ampacity is 25 x .58 = 14.5 amps is the maximum current the Code allows on a #12 TW wire where the ambient temperature reaches 122°F. Any current applied to the wire over 14.5 amps would result in insulation damage to the wire.

The connected load of these 8 lights is 16 amps. We need a TW conductor that will carry 16 amps safely without incurring insulation damage.

The formula for wire size is:

$$\frac{\text{LOAD}}{\text{CORRECTION FACTOR}} = \frac{16\text{ amps}}{.58} = 27.58\text{ required ampacity}$$

For ambient temperature we need conductors that have a minimum ampacity of 27.58 amps. Table 310-16 would require a #10 TW with an ampacity of 30.

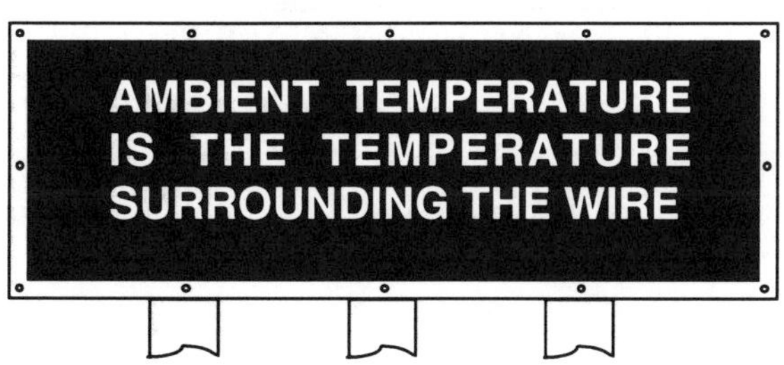

• VARIABLE 4 - NOTE 8

Table 310-16 Note 8 states that where the number of current-carrying conductors in a raceway or cable exceeds *three,* the allowable ampacity must be reduced to compensate for the increased heating effect due to reduced ventilation of an enclosed group of closely spaced conductors.

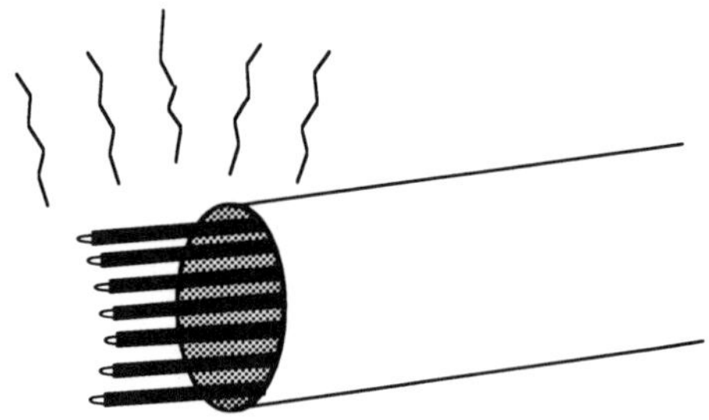

If the #12 TW conductors to the 8 lights were installed in a conduit with 5 other wires, making a total of 7 current carrying conductors in the conduit, Note 8 would require a 70% deration of ampacity.

#12 TW has an ampacity of 25 x 70% = 17.5 maximum load that can be applied to a conductor without incurring insulation damage from the excessive heating. The load is 16 amps.

$$\text{Wire size} = \frac{\text{LOAD}}{\text{Note 8}} = \frac{16\text{a}}{70\%} = 22.8 \text{ amps required}$$

A #12 TW conductor with an ampacity of 25 would meet the Code.

If the conductors were placed in an attic where the temperature could reach 122°F ambient and in a conduit that has a total of 7 current-carrying conductors, the ampacity is required to be reduced twice.

#12 TW = 25 ampacity x .58 correction factor = 14.5 amps x 70% Note 8 = 10.15 amps.

With both heat conditions the maximum current permitted on the #12 TW is 10.15 amps. The 8 lights have a connected load of 16 amps.

$$\text{Wire size} = \frac{\text{LOAD}}{\text{CORRECTION FACTOR X NOTE 8}} = \frac{16 \text{ amps}}{.58 \times .70} = 39.4 \text{ amps}$$

39.4 is the required ampacity. Table 310-16 would require a minimum #8 TW wire.

• VARIABLE 5 - COST

The first four variables are based on the minimum requirements of the Code; cost and future growth are not involved in the calculations.

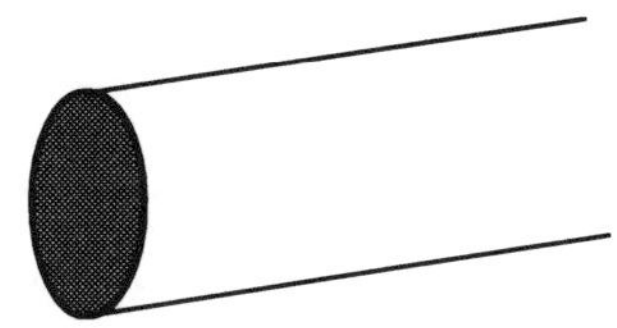

If the branch circuit was installed in metal conduit:

VARIABLES 1 and 2 requires for two #12 TW wires from Table C1 = 1/2" conduit

VARIABLE 3 requires for two #10 TW wires = 1/2" conduit

VARIABLE 4 requires for seven #8 TW wires = 1" conduit

The Code minimums are generally used by the designer when bidding work as this will make the cost of the job less. But, for example, an industrial location where the maintenance electrician will be maintaining and servicing this building for many years to come. Do we really want the Code minimum? This is the question that must be answered.

Installing a larger wire than the Code requires can result in a power savings in heat loss caused by the voltage drop. Voltage drop is a very neglected area in our electrical designing.

Some equipment will not operate efficiently at a lower voltage, thus requiring voltage drop to be calculated at 1% instead of the 3% permitted by the Code.

The designer should always consider using a larger conduit for the ease of pulling the wires and for future expansion. In some cases even pulling spare conductors for future growth.

Installing a green (grounding) conductor in metal conduit even though the Code does not require one (only in PVC). In case a loose connection in the metallic raceway should occur from compression fittings, set-screw fittings, etc. the green wire would maintain an effective grounding path for a fault current.

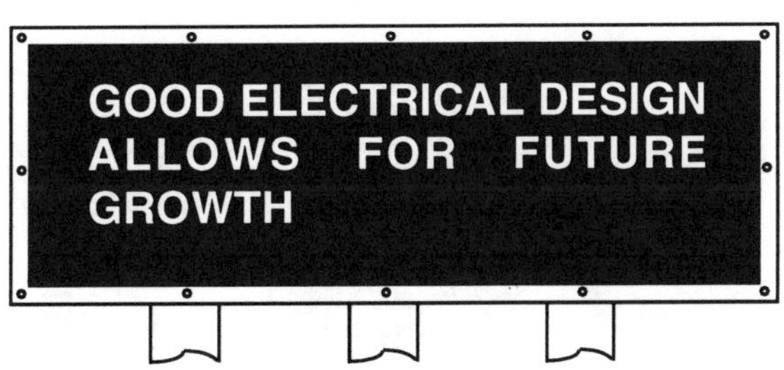

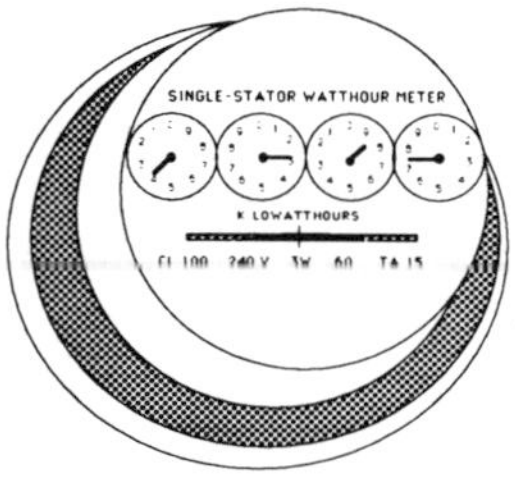

Often neglected in electrical design is the power loss in the conductors.

POWER LOSS = VD X I

Voltage drop times amperage equals power (watts) loss. This is the energy wasted by heating the wire while passing a current through a resistance.

I (amperes) represents the load in amperes which cannot be changed in most cases. VD represents the voltage drop which cannot be eliminated as all wire has some resistance. But, voltage drop can be reduced by installing larger conductors which have less resistance.

With heavy commercial and industrial loads that operate continuously, increasing the size of the wire is an area of design the designer should take a close look at, as it can provide a power savings in the monthly electric bill.

$$\text{The formula for \$ COST \$} = \frac{\text{WATTS x HOURS USED x RATE PER HOUR}}{1000}$$

If the 8 lights were installed in 1/2" conduit and #12 TW wire:

$$VD = \frac{2 \text{ x } K \text{ x } D \text{ x } I}{CM} = \frac{2 \text{ x } 12.9294 \text{ x } 56' \text{ x } 16a}{6530 \text{ cm}} = 3.55 \text{ volts dropped}$$

Watts Loss = 3.55v x 16a = 56.8 watts

$$\text{Cost} = \frac{\text{56.8w x 16 hours used per day x .08 cents per kwh}}{1000} = \text{.07 cents per day}$$

Based on a 30 day month = 30 x .07 cents = *$2.10 per month* is the cost of the heat loss.

The cost of the #12 TW wire, 1/2" conduit, fittings, less labor, etc. would be approximately $20.

It may seem like the savings is only pennies, but these pennies add up to big dollars when you design the entire system properly.

If the 8 lights were installed with #8 TW wire and 1" conduit, the cost could also be calculated.

$$VD = \frac{2 \times 12.84478 \times 56' \times 16a}{16510 \text{ cm}} = 1.39 \text{ volts dropped}$$

Watts Loss = 1.39v x 16a = 22.2 watts

$$\text{Cost} = \frac{22.2w \times 16 \text{ hours used per day} \times .08 \text{ cents per kwh}}{1000} = .028 \text{ cents per day}$$

Based on a 30 day month = 30 x .028 cents = *.84 cents* per month is the cost of the heat loss.

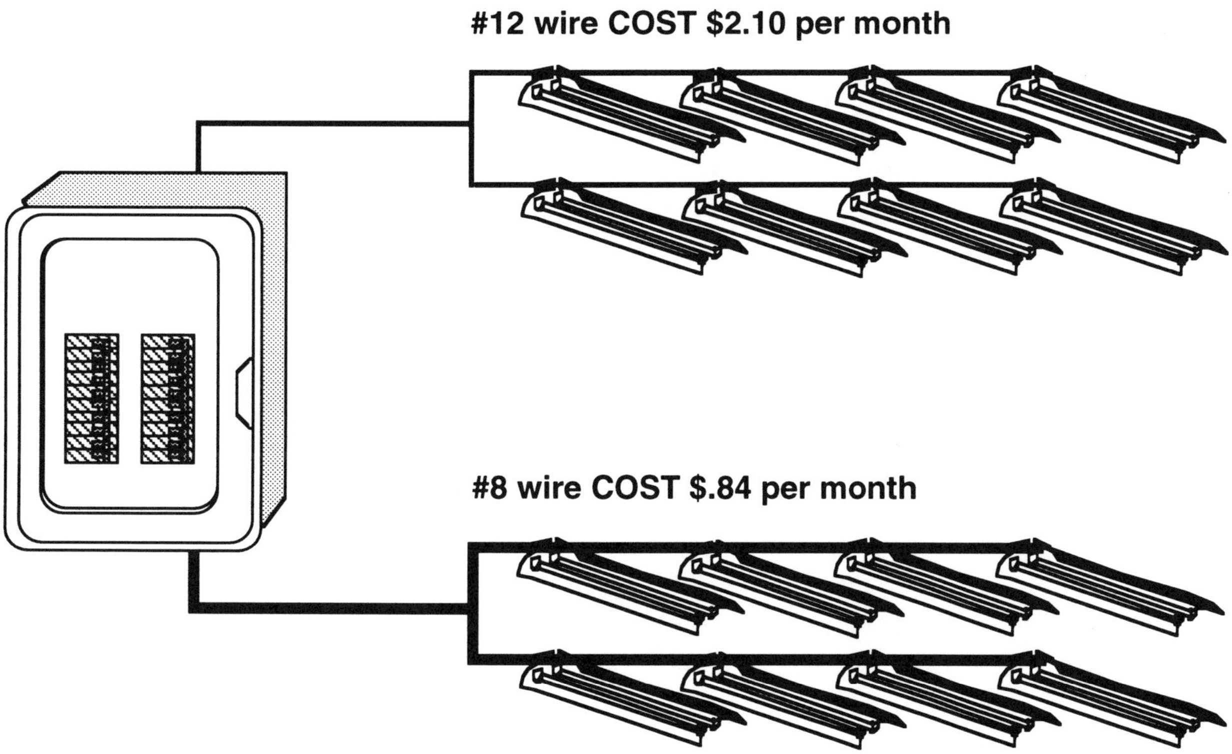

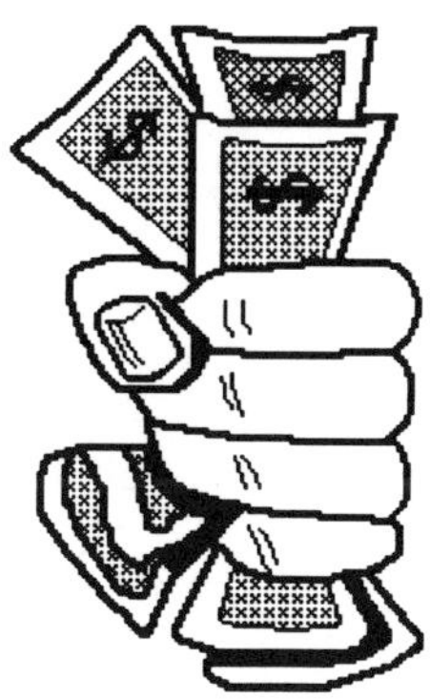

The cost of the #8 wire and the 1" conduit would be approximately $40 which is double the cost of the #12 wire in a 1/2" conduit.

$2.10 - .84 = $1.26 savings per month. 16 months of savings in the electric bill would pay for the extra cost of installing a larger branch circuit.

Count the branch circuits in a panelboard and then count the number of panelboards in your industrial plant and you can see how you can save some real energy by installing larger conductors than required by the Code minimum!

THE BRANCH CIRCUIT

THE BRANCH CIRCUIT

Definition: The circuit conductors between the final overcurrent device protecting the circuit.

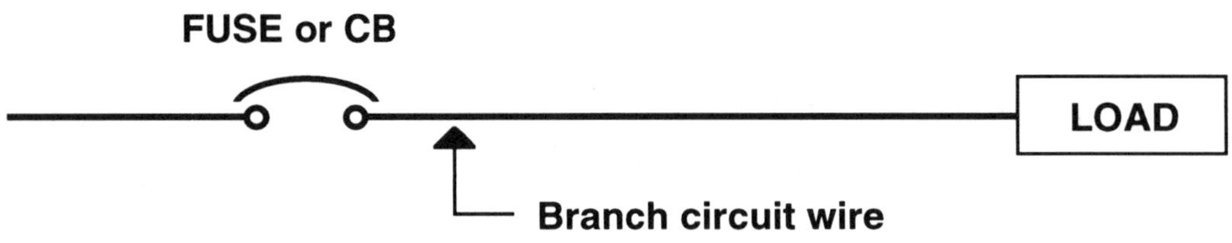

The sizing of the conductors, overcurrent protection, panelboard, etc. is often confusing to the electrician. Rightfully so, as there are many variables that enter into the proper design, such as type of circuit, type of load and type of occupancy.

The connected load (nameplate rating) is the first thing that must be determined, then apply the variables and select the wire size.

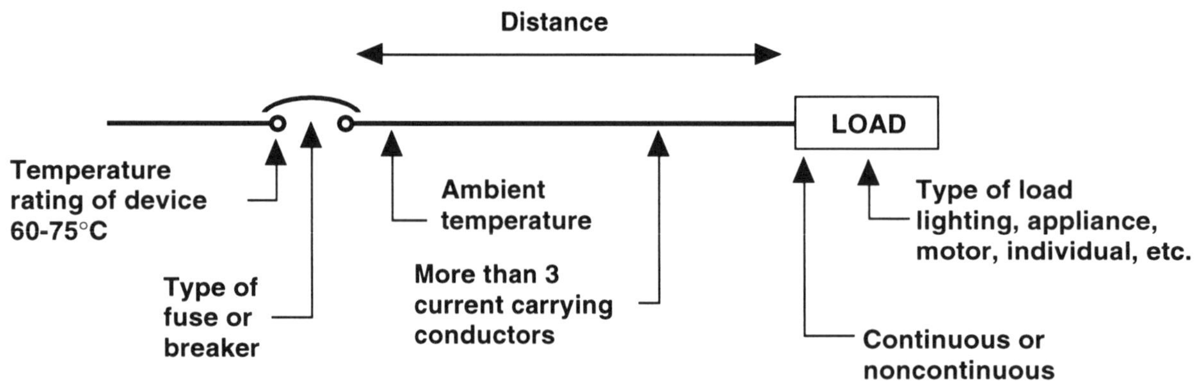

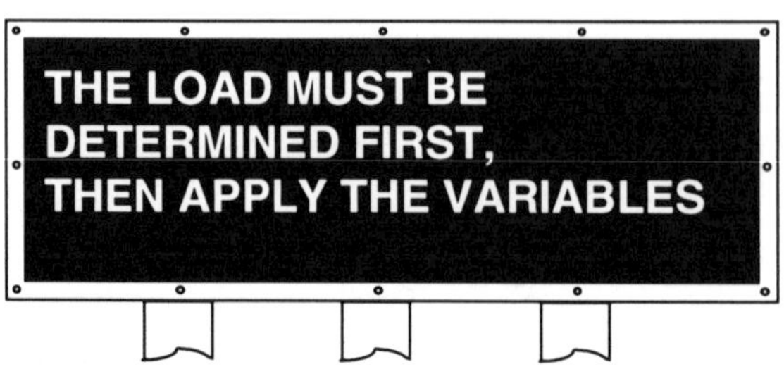

DIFFERENT TYPES OF BRANCH CIRCUITS

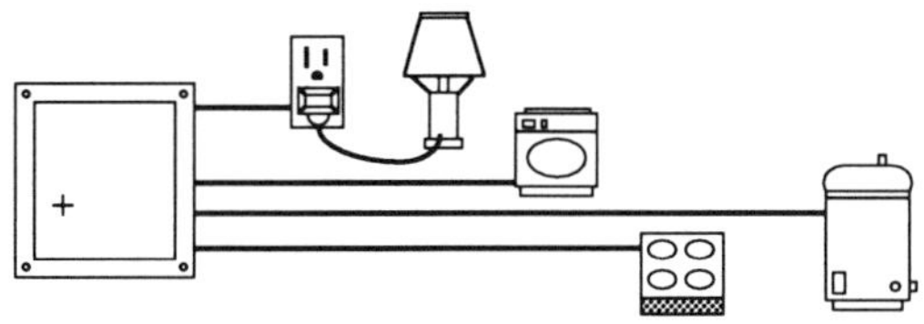

APPLIANCE BRANCH CIRCUIT

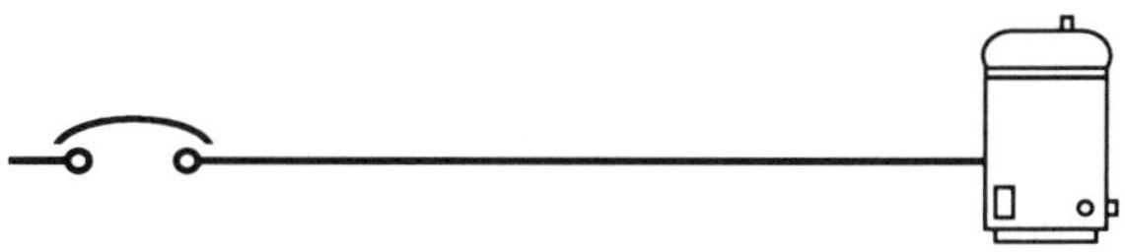

GENERAL PURPOSE

INDIVIDUAL BRANCH CIRCUIT

MULTIWIRE

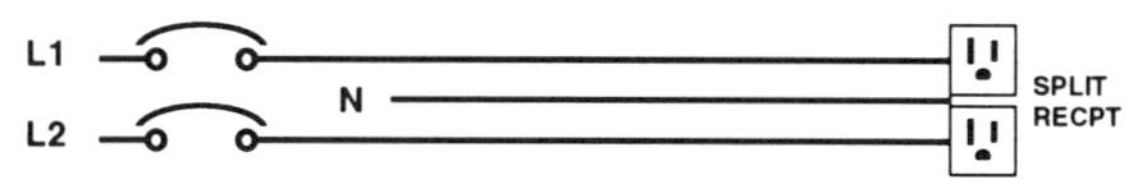

MOTOR BRANCH CIRCUIT

WELDER

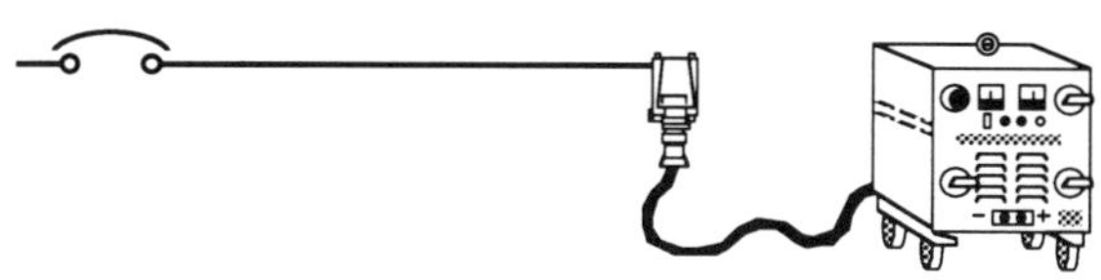

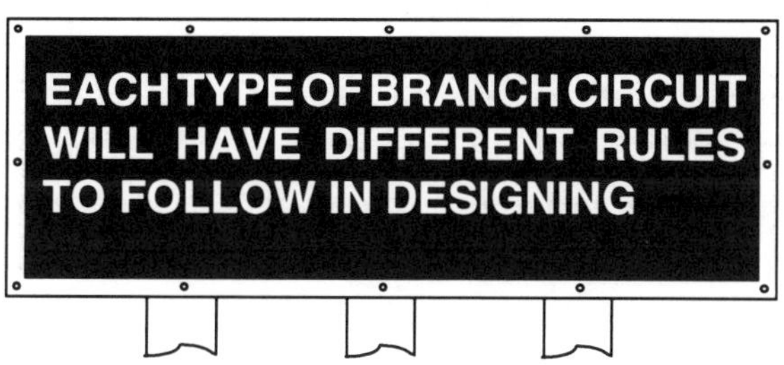

LIGHTING BRANCH CIRCUIT

A branch circuit supplying light fixtures is an every day installation by the electrician. The trouble starts by assuming a #12 wire for a 20 amp branch circuit. The first thing that must be determined is the load that will be on this branch circuit.

In the branch circuit shown below, the total ampere rating of each light is 2 amps. The branch circuit has a total of 8 lights.

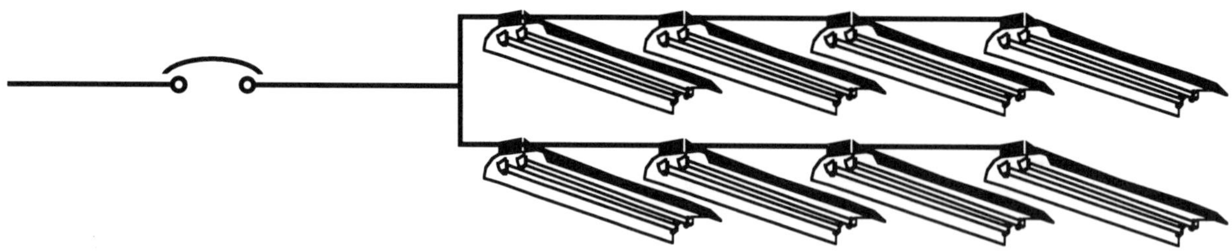

The first step in designing is to calculate the total load on the circuit. 8 lights x 2 amps = 16 amps.

At first glance one might say a 20 amp rated circuit with a #12 wire that has an ampacity of 20 amperes would be sufficient.

But, for proper designing and to follow the Code minimum requirements, let's examine the possible variables that could apply.

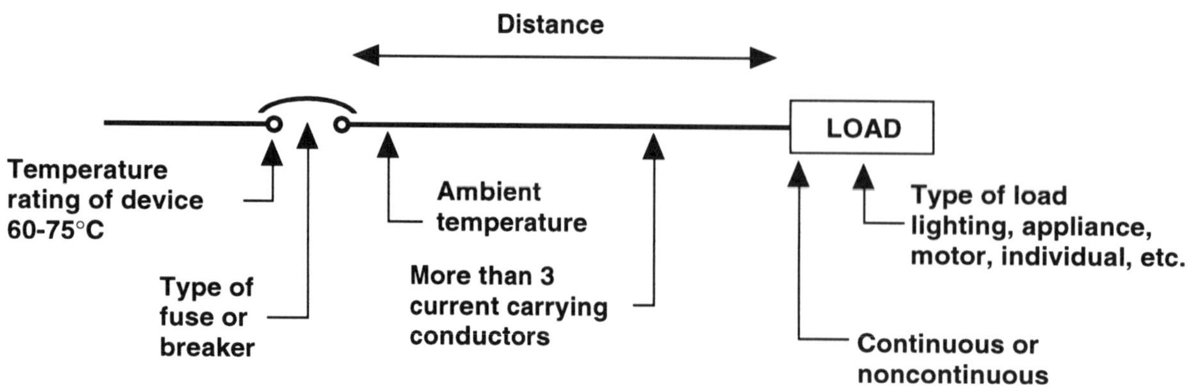

RECEPTACLE BRANCH CIRCUIT

The everyday branch circuit the electrician installs is the 20 amp receptacle branch circuit.

It's not merely selecting a #12 wire and installing the circuit.

The first thing that must be determined is the *load* that will be on this branch circuit.

YOKE IS THE MOUNTING STRAP

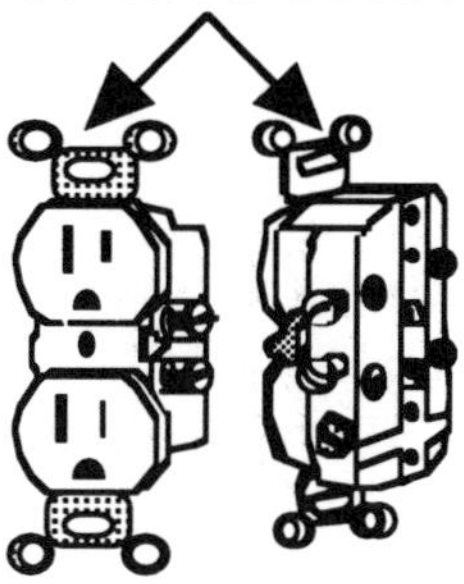

For buildings other than a house, the Code requires a *minimum* of 180va per yoke to be calculated for receptacles.

The *maximum* number of receptacles on a 20 amp branch circuit would calculate:

$$20 \text{ amp} \times 120\text{v} = \frac{2400\text{va}}{180\text{v}} = 13 \text{ receptacles}$$

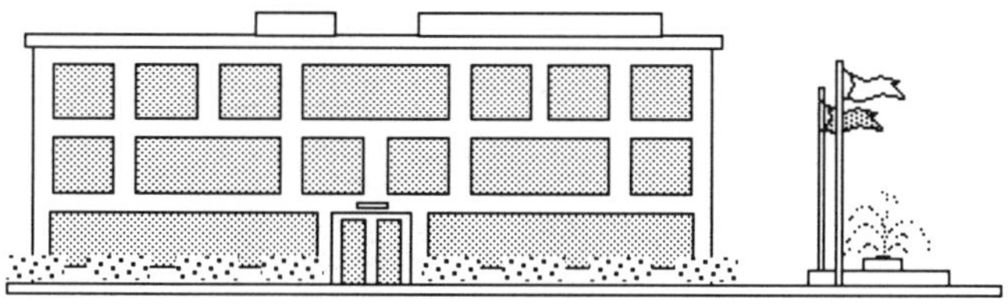

OFFICE BUILDING

Let's say we are to install this receptacle branch circuit in an office building. The first step is to find the load. These receptacles are convience receptacles so you don't know what load will be plugged into them. The Code *minimum* requires each yoke (in this case each duplex) to be calculated at 180va/120v = 1.5 amps. The circuit would be designed to carry 20 amps.

The Code recommendation is to limit voltage drop on a branch circuit to no more than 3%. For a 120 volt branch circuit the voltage drop allowed is 120v x 3% = 3.6 volts.

How far does the Code permit a 20 amp load on a branch circuit to be run?

$$\text{Distance} = \frac{\text{CM} \times \text{VD}}{2 \times \text{K} \times \text{I}} = \frac{6530\text{cm} \times 3.6\text{v}}{2 \times 12.9294 \times 20\text{a}} = \textit{45.4 feet}$$

The total length of this circuit can be no more than 45.4 feet! This means if 13 receptacles are installed, starting at the breaker, a receptacle would be installed each *3 1/2* feet.

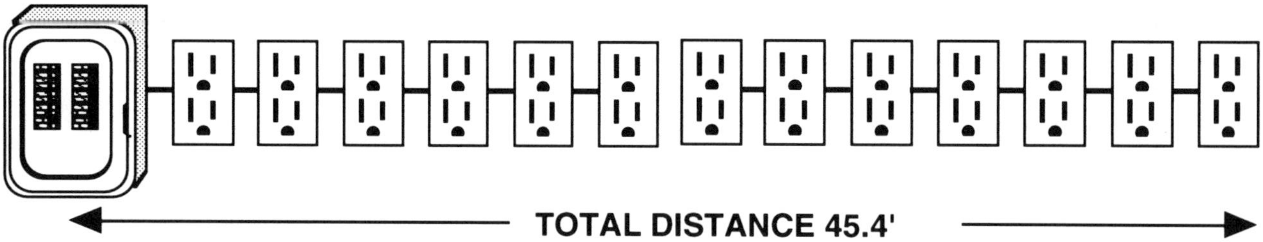

3% voltage drop is very poor designing. Let's calculate this same branch circuit at 2% voltage drop. 120v x 2% = 2.4 volts. Now the circuit can only be run 30.3 feet.

$$\text{Distance} = \frac{\text{CM x VD}}{\text{2 x K x I}} = \frac{\text{6530cm x 2.4v}}{\text{2 x 12.9294 x 20a}} = \textit{30.3 feet}$$

As mentioned earlier, some equipment will not operate properly at a lower voltage and good designing would limit voltage drop to 1or 2%. 120v x 1% = 1.2 volts. Now the circuit can only be installed 15 feet total distance to maintain a 1% voltage drop with a 20 amp load on a #12 wire.

$$\text{Distance} = \frac{\text{CM x VD}}{\text{2 x K x I}} = \frac{\text{6530cm x 1.2v}}{\text{2 x 12.9294 x 20a}} = \textit{15.1 feet}$$

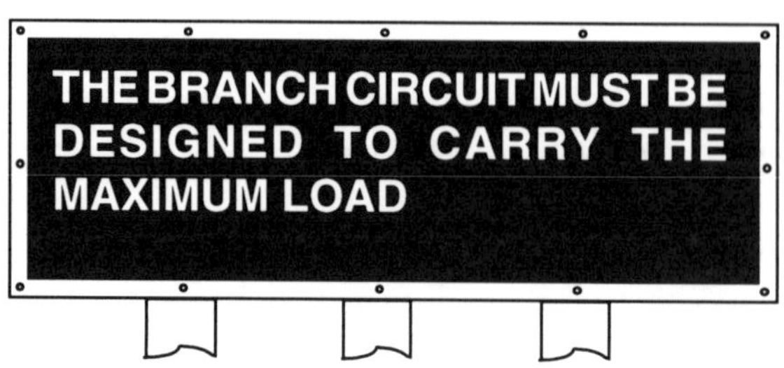

USING #12 WIRE

How far does the Code permit a 20 amp load on a branch circuit to be run?

$$\text{Distance} = \frac{CM \times VD}{2 \times K \times I} = \frac{6530cm \times 3.6v}{2 \times 12.9294 \times 20a} = \textit{45.4 feet}$$

The distance permitted at 2% voltage drop would be:

$$\text{Distance} = \frac{CM \times VD}{2 \times K \times I} = \frac{6530cm \times 2.4v}{2 \times 12.9294 \times 20a} = \textit{30.3 feet}$$

The distance permitted at 1% voltage drop would be:

$$\text{Distance} = \frac{CM \times VD}{2 \times K \times I} = \frac{6530cm \times 1.2v}{2 \times 12.9294 \times 20a} = \textit{15.1 feet}$$

USING #10 WIRE

How far does the Code permit a 20 amp load on a branch circuit to be run?

$$\text{Distance} = \frac{CM \times VD}{2 \times K \times I} = \frac{10380cm \times 3.6v}{2 \times 12.8712 \times 20a} = \textit{72.5 feet}$$

The distance permitted at 2% voltage drop would be:

$$\text{Distance} = \frac{CM \times VD}{2 \times K \times I} = \frac{10380cm \times 2.4v}{2 \times 12.8712 \times 20a} = \textit{48.3 feet}$$

The distance permitted at 1% voltage drop would be:

$$\text{Distance} = \frac{CM \times VD}{2 \times K \times I} = \frac{10380cm \times 1.2v}{2 \times 12.8712 \times 20a} = \textit{24.1 feet}$$

Now for proper designing let's take a close look at a very important variable; the cost factor.

Designing to the Code would permit a #12 wire feeding receptacles to be installed a maximum distance of 45.4 feet resulting in a 3% voltage drop.

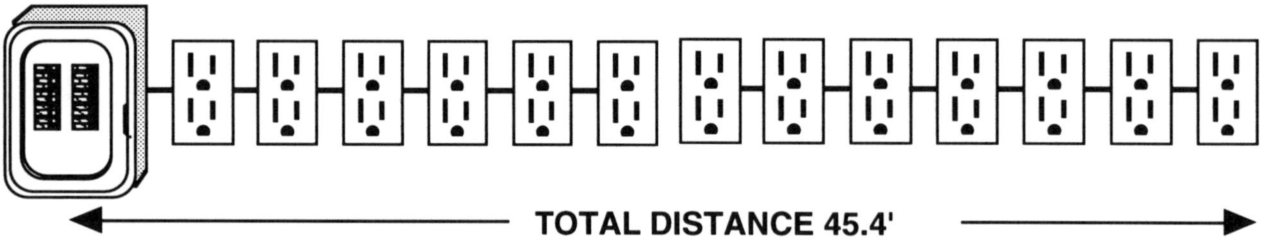

Power Loss = VD x I 3.6v x 20a = 72 watts

$$\text{Cost} = \frac{\text{72w x 8 hours used per day x .08 cents per kwh}}{1000} = \text{.046 cents per day}$$

Based on a 30 day month = .046 cents x 30 days = $1.38 cents per month wasted electricity for one 20 amp branch circuit feeding receptacles.

By installing a larger #10 wire, the same branch circuit could be installed 45.4 feet with a 2.25 voltage drop.

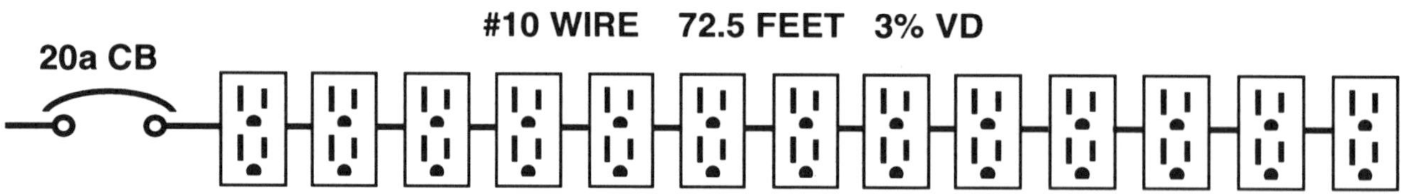

$$VD = \frac{2 \times K \times D \times I}{CM} \qquad \frac{2 \times 12.8712 \times 45.4' \times 20a}{10380 \text{ cm}} = \text{2.25 volts dropped}$$

Power Loss = VD x I 2.25v x 20a = 45 watts

$$\text{Cost} = \frac{\text{45w x 8 hours used per day x .08 cents per kwh}}{1000} = \text{.028 cents per day}$$

Based on a 30 day month = .028 cents x 30 days = .84 cents per month wasted electricity for one 20 amp branch circuit feeding receptacles. A savings of 54¢ a month for one 20 amp circuit.

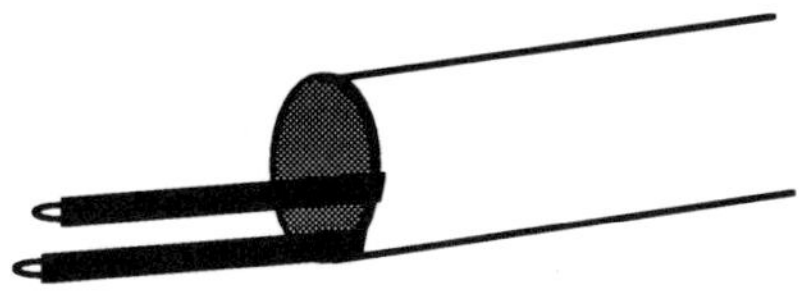

YOU CAN INSTALL 9 - #12 THHN or 5 - #10 THHN CONDUCTORS IN A 1/2" EMT CONDUIT

You can use the same size 1/2" EMT conduit for either a #12 THHN or #10 THHN so the price for the installation of the conduit would not change.

The price of a #12 THHN is approximately .04 cents a foot x 202' = $8.08.

The price of a #10 THHN is approximately .07 cents a foot x 202" = $14.14.

The #10 wire for the same branch circuit installed 45.4 feet in distance would cost $6.06 more for the larger wire.

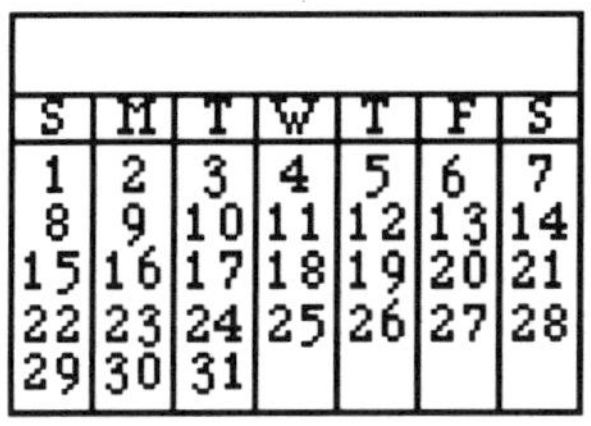

The cost for the power loss in the #12 wire was $1.38 a month compared to .84 cents for the larger #10 wire. This amounts to a savings of .54 cents a month by installing the #10 wire.

Since the cost of a #10 wire is $6.06 in approximately 12 months, the extra cost of the #10 wire would be paid for by the savings of power loss.

WHAT SIZE WIRE?

What size wire would be required for a 20 amp load @ 120v for a distance of 100 feet with 2% VD?

$$\text{Wire Size} = \frac{2 \times K \times D \times I}{\text{VD permitted}} \qquad \frac{2 \times 12.9 \times 100' \times 20a}{2.4 \text{ volts}} = \quad 21{,}500 \text{ cm}$$

A #6 wire has a cma of 26240, so by installing a #6 wire it would maintain a 2% drop.

What size wire would be required for a 20 amp load for a distance of 100 feet with 1% VD?

$$\text{Wire Size} = \frac{2 \times K \times D \times I}{\text{VD permitted}} \qquad \frac{2 \times 12.9 \times 100' \times 20a}{1.2 \text{ volts}} = \quad 43{,}000 \text{ cm}$$

A #4 wire has a cma of 41,740 so a #4 is not large enough. It would require a #3 wire with a cma of 52,620.

Receptacles are sometimes assumed to be lightly loaded. In the case of a house this is most likely true and that's why the Code has no limit on the number of receptacles that can be installed on a branch circuit in a house. But, for other than a house, like commercial or industrial installations of branch circuits it's easy to see why the Code has limited the number of receptacles to a branch circuit. When you calculate the heavier loads and distances it becomes easy to violate the minimum recommendation of the Code.

Proper calculations and designing of branch circuits is a must in todays overloaded world.

You ask how can you overload a branch circuit? Just look around your office or work station and look at the items plugged into the receptacles and this will quickly answer your question.

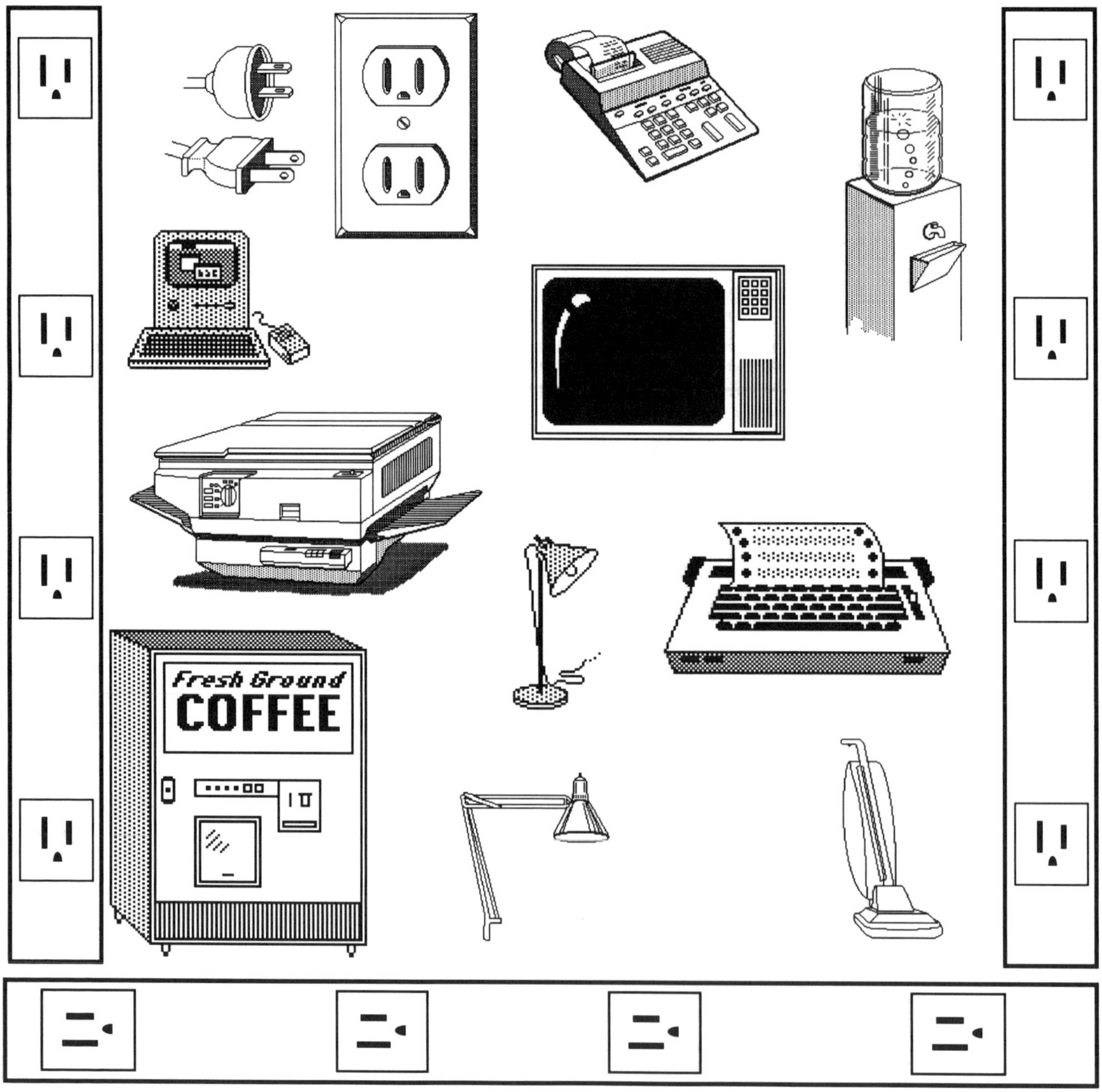

COST OF WATTAGE

A string of Christmas lights uses low wattage bulbs which represents a very small load. A standard 25' string of lights used for outline lighting for the roof of your house would be a C7 bulb @ 5 watts each. A string with 25 lights has a 7 amp fuse and #20 gauge wire. A maximum of 4 strings can be connected together.

5w x 25 bulbs = 125 watts per string x 4 strings = 500w/120v = 4.16 amps.

If 8 strings of lights were used for the roof decorating, 125w x 8 strings = 1000w/120v = 8.3 amps.

Only 4 strings can be connected together. In most cases an extension cord with a cube tap is used to plug in both of the 4 string sets.

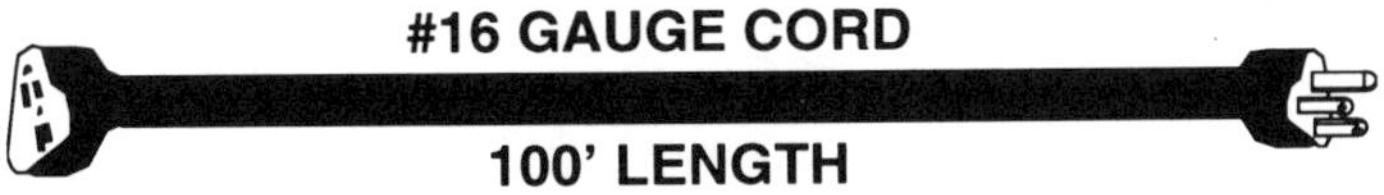

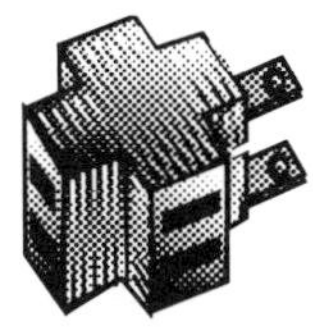

CUBE TAP

Often the cord used is a #16 gauge 100 feet in length.

The voltage drop would be: $\dfrac{2 \times 12.8742 \times 100' \times 8.3 \text{ amps}}{2580 \text{ cm}}$ = *8.28 volts dropped*

The power (watts) loss = VD x I   8.28v x 8.3 amps = 68.7 watts wasted

As an example, assume the roof lights are on 30 evenings from 6 pm to 10 pm during the Christmas season.

The cost of the watts wasted would be: $\dfrac{68.7 \text{ watts} \times 4 \text{ hrs} \times .08 \text{ per kwh}}{1000}$ = 2¢ per day

2¢ x 30 days = 60¢ wasted electricity

Using a #12 gauge extension cord = $\dfrac{2 \times 12.9294 \times 100' \times 8.3 \text{ amps}}{6530 \text{ cm}}$ = *3.28 volts dropped*

The power (watts) loss = VD x I   3.28v x 8.3 amps = 27.2 watts wasted

The cost of the watts wasted would be: $\dfrac{27.2 \text{ watts} \times 4 \text{ hrs} \times .08 \text{ per kwh}}{1000}$ = .008¢ per day

.008¢ x 30 days = 24¢ wasted electricity, a savings of 36¢, the pennies don't seem like much of a savings.

The next example is a large Christmas tree using a 33' string with 50 lights at .86w per bulb. 6 strings maximum can be connected using a 3 amp fuse.

50 lights x .86w per bulb = 43w per string x 6 strings = 258w/120v = 2.15 amps.

2.15 amps for 300 lights is not a heavy load. I have 3 Christmas trees in my front lawn, each tree has 900 lights. This is three circuits of 6 strings. 3 x 2.15 amps = 6.45 amps per tree.

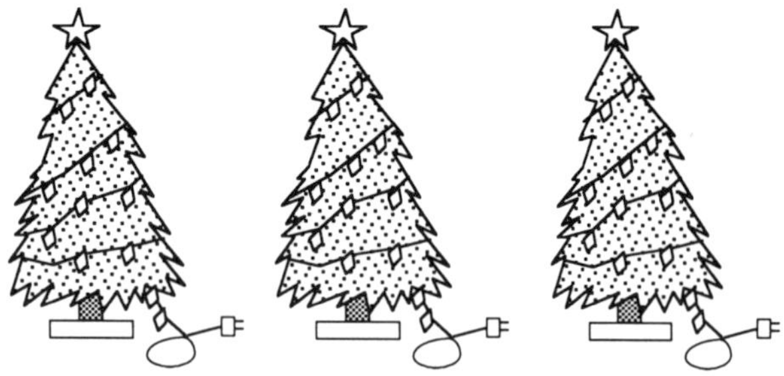

The 3 trees would have a total of 6.45a x 3 = 19.35 amps for 2700 Christmas lights, or 2700 lights x .86w = 2322 watts or 2.322 kwh. At a cost of 8¢ per kwh it costs me 18 1/2¢ per hour.

18.5¢ x 4 hours per evening = 74¢ per day or $22.20 for the 30 day Christmas season.

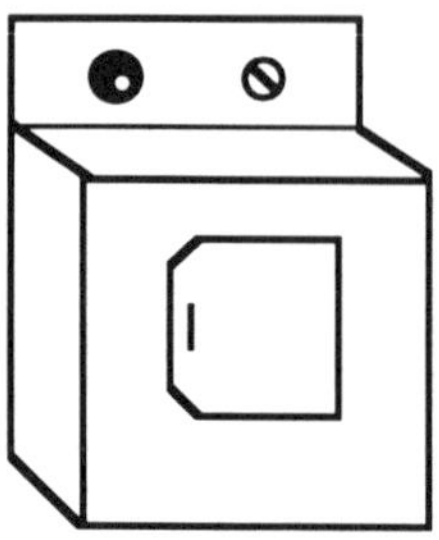

A 4000 watt dryer would cost you 32¢ per hour to operate, whereas, the 3 Christmas trees cost 18 1/2¢ per hour.

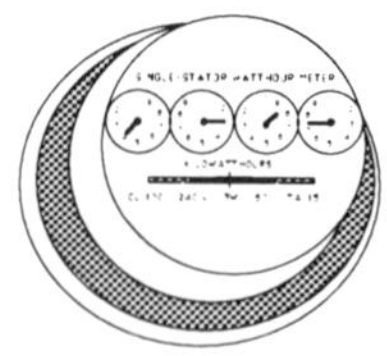

NON-MOTOR APPLIANCE BRANCH CIRCUIT

A branch circuit supplying energy to one or more outlets to which appliances are to be connected; such circuits shall have *no* permanently connected lighting fixtures which are not part of an appliance. Appliance is utilization equipment, generally other than industrial. Industrial machinery is covered in Article 670 of the Code.

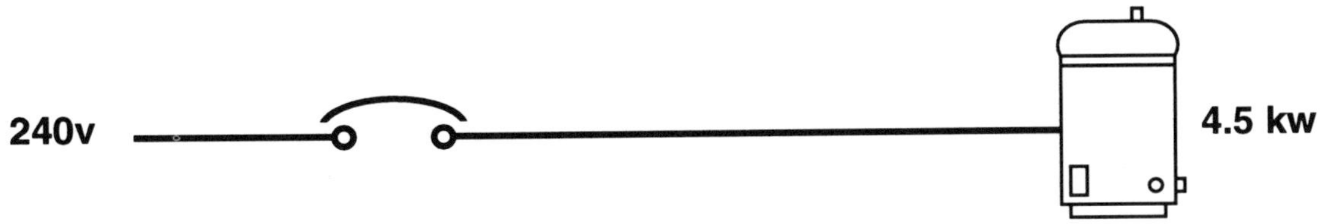

4.5 kw *water heater* 40 gallon 240v single-phase

Section 422-14b. A branch circuit supplying a fixed storage-type water heater having a capacity of 120 gallons *or less* shall have a *rating* not less than 125% of the nameplate rating of the water heater.

$$\frac{4500\text{w}}{240\text{v}} = 18.75 \text{ amps} \times 125\% = 23.43 \text{ amps}$$

Section 422-28e. If the branch circuit supplies a single *nonmotor*-operated appliance, the overcurrent device rating shall not exceed a protective device rating marked on the appliance, or if there is no marking, the value is specified as follows:
Appliance rating greater than 13.3 amperes = *150% of appliance rating.*

18.75 amps x 150% = 28.125 amps 422-28e ex. Use 30 amp circuit breaker #10 TW copper conductor.

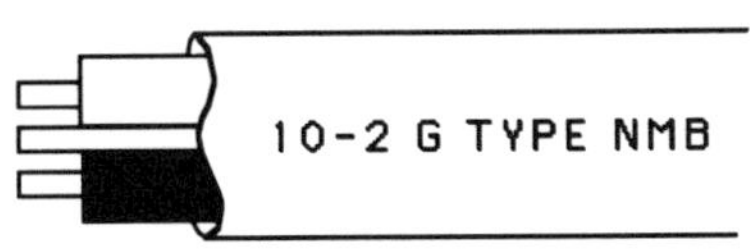

#10 ROMEX HAS AN *AMPACITY* OF 30

•Romex (NMB) has a 90°C *insulation*, but the *AMPACITY* is that of a 60°C conductor per Code section 336-26.

MOTOR APPLIANCE BRANCH CIRCUIT

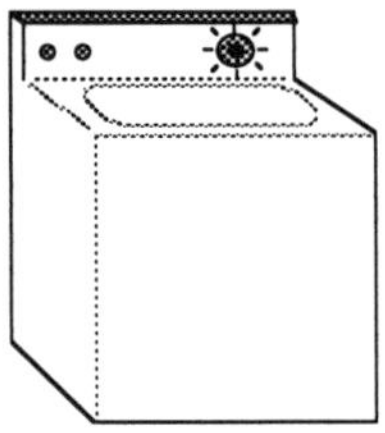

A clothes washing machine is an example of a motor operated *appliance.*

210-52f. At least one receptacle outlet shall be installed for the laundry in a dwelling.

220-4c. On a 20 amp branch circuit.

Definition of receptacle outlet: An outlet where one or more receptacles are installed.

A duplex receptacle with an iron plugged in is permitted by the Code on the *laundry* circuit.

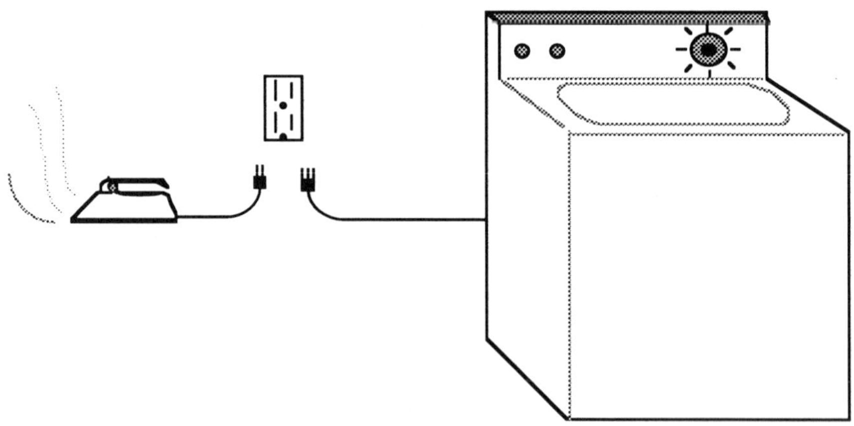

422-30 and 422-32a. Appliances consisting of motors and other loads shall be marked *volts and amps* or *volts and watts* to specify the minimum supply circuit conductor ampacity and the maximum fuse or circuit breaker.

BRANCH CIRCUIT SIZING FIXED SPACE-HEATING

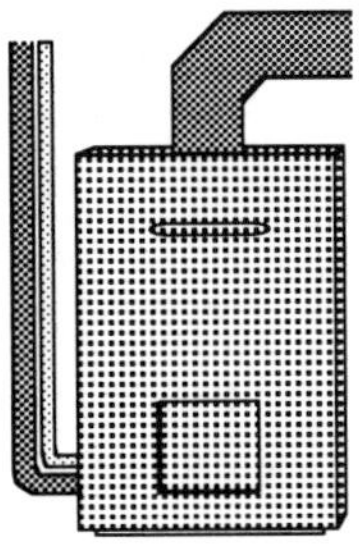

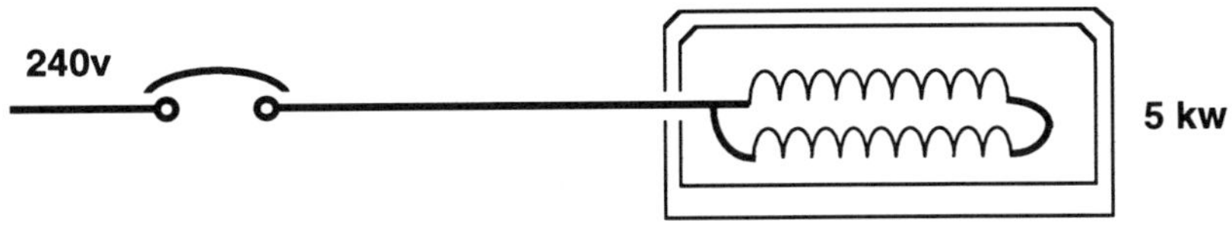

The 5 kw space heat unit would calculate:

$\frac{5000w}{240v}$ = 20.83 amps x 125% = 26 amps. Use 30 amp CB and #10 TW wire.

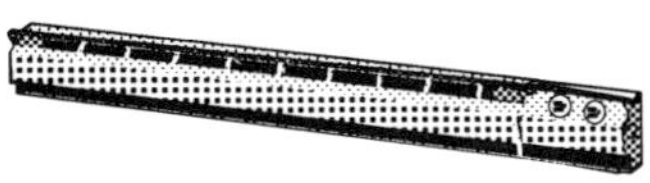

•Section 424-3a. *Individual* branch circuits shall be permitted to supply *any size* fixed electric space heating equipment.

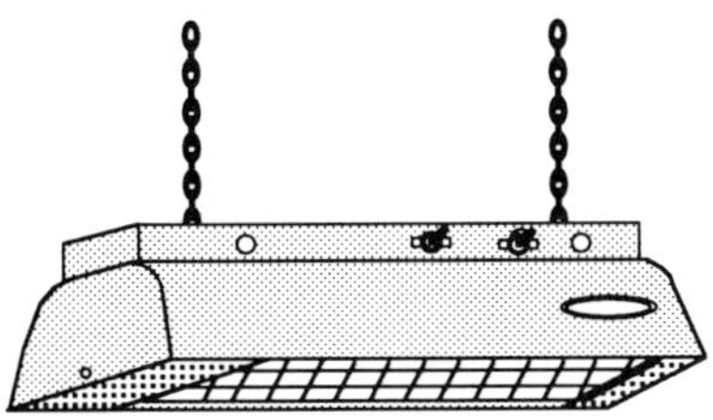

Branch circuits supplying *two or more* outlets for fixed electric space heating equipment shall be rated 15, 20, or 30 amperes. Exception: In other than residential occupancies, fixed *infrared* heating equipment shall be permitted to be supplied from branch circuits rated not over 50 amperes.

•Section 424-3b. The ampacity of the branch circuit conductors and the rating or setting of overcurrent protective devices supplying fixed electric space heating equipment consisting of resistance elements with or without a motor shall not be less than 125% of the total load of the motors and the heaters. Where the rating does not correspond to a standard size overcurrent device, the next higher standard rating shall be permitted per 240-3b.

424-22a. Other than motor operated, electric heat shall be permitted to be protected against overcurrent where supplied by one of the branch circuits in Article 210.

424-24b. Resistive elements can not have an overcurrent device larger than 60 amps.

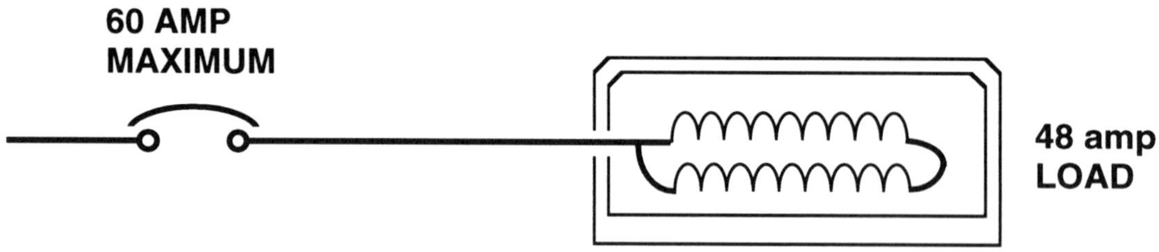

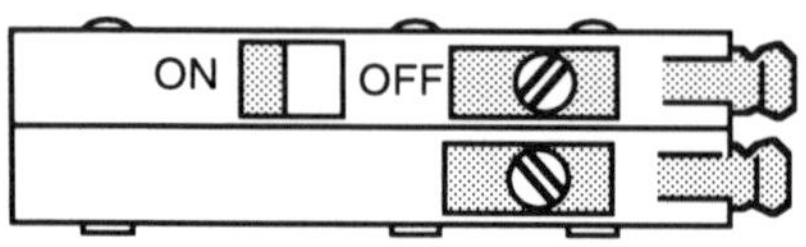

60 amp CB x 80% = 48 amps

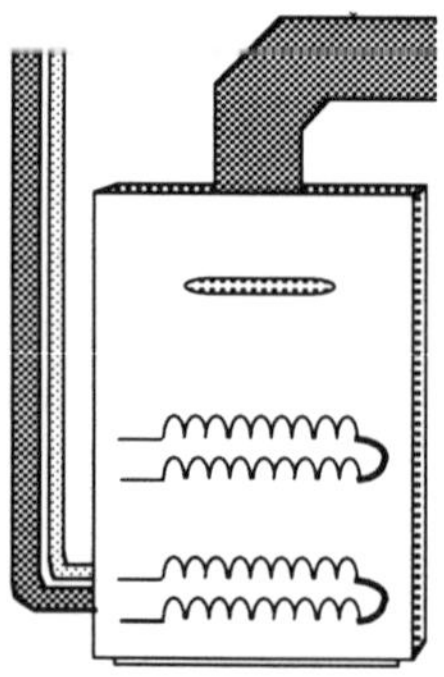

Equipment rated more than 48 amps shall have the elements subdivided, and each load shall not exceed 48 amps.

424-24d. The wire supplying the supplementary fuses shall be considered branch circuit conductors.

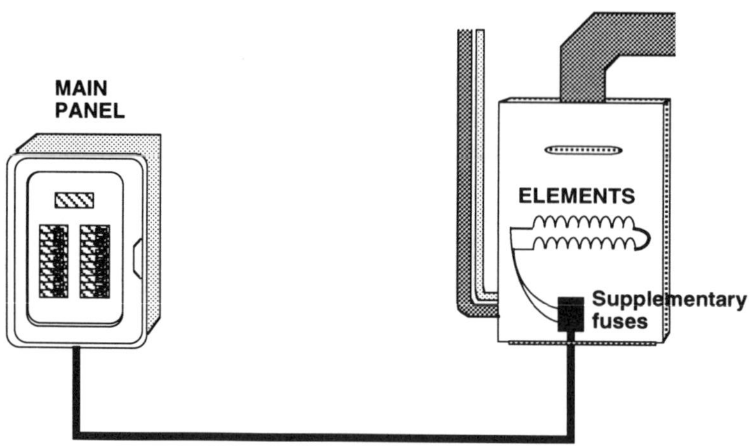

110-3b states all equipment shall be installed as listed.

10 KW @ 240 volts 1ø

If the heating unit has a nameplate of 10 kw at 240 volts, it must be wired for no less than 10 kw.

10,000w/240v = 41.6 amps. Sometimes the unit is connected to 208 volts which would only have an ouput of 7511 watts. 7511w/208v = 36.1 amps. But, per 110-3b the unit must be wired for 41.6 amps.

BOILERS

A boiler that has resistance-type immersion heating elements contained in an ASME rated and stamped vessel shall have the heating elements protected at not more than 150 amps. Such a boiler rated more than 120 amps shall have the heating elements subdivided into loads not exceeding 120 amps.

Where a subdivided load is less than 120 amps, the rating of the overcurrent device shall not be less than 125% of the total load of the element.

A boiler employing resistance-type heating elements *not* contained in an ASME rated and stamped vessel shall have the heating elements protected at not more than 60 amps. Such a boiler rated more than 48 amps shall have the heating elements subdivided into loads not exceeding 48 amps.

Where a subdivided load is less than 48 amps, the rating of the overcurrent device shall not be less than 125% of the total load of the element.

An electrode-type boiler in which heat is generated by the passage of current between electrodes through the liquid being heated has different design rules from the resistance-type boilers.

The branch circuit conductors and overcurrent protective device shall be calculated on the basis of 125% of the total load (not motors).

BRANCH CIRCUIT FOR AIR CONDITIONERS

Before sizing the branch circuit for air conditioning or refrigeration equipment you must first determine if the motor-compressor is hermetically sealed or is the compressor belt driven from the motor.

The rules of Article 440 apply to hermetically sealed motor-compressors. Whereas the rules of Article 430 apply where the compressor is belt driven.

Most equipment used today is the hermetic type which includes drinking fountains, household refrigerators, freezers, room air conditioners, etc.

Hermetic systems have the motor operating in the refrigerant whereas ordinary motors are cooled by the surrounding air. Hermetically sealed motors can safely consume more current and deliver more horsepower than an ordinary motor of the same size cooled by the ambient air.

While the hermetic motor-compressor is running, the refrigerant is continuously entering the case in a gaseous state that is a cooler temperature than the surrounding air.

After the hermetic motor-compressor sits idle for a period of time the refrigerant reaches a temperature of that of the ambient and provides poor cooling for a short period of time on start up. Consequently the motor heats up faster than an ordinary motor. This is the reason for the different rules of Articles 430 and 440.

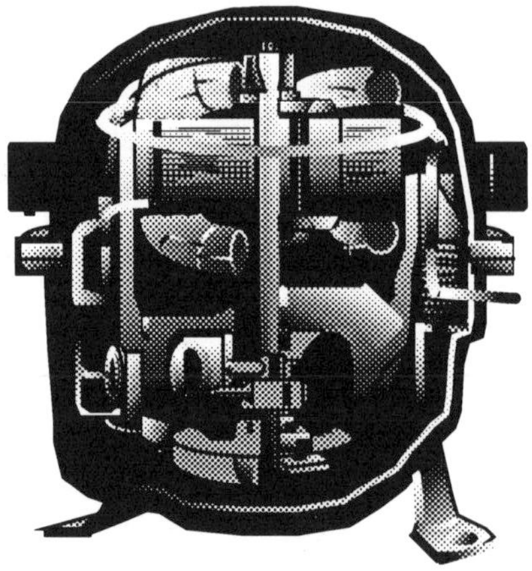

•Section 440-2. Definition of Hermetic refrigerant motor-compressor: A combination consisting of a compressor and motor, *both* of which are enclosed in the same housing with no external shaft or shaft seals, *the motor operating in the refrigerant.*

•Section 440-3b. Examples of equipment which are *not* hermetic, would be refrigeration compressors driven by conventional motors, furnaces with air conditioning evaporator coils installed, fan-coil units, remote forced air-cooled condensers, remote commercial refrigerators, etc.

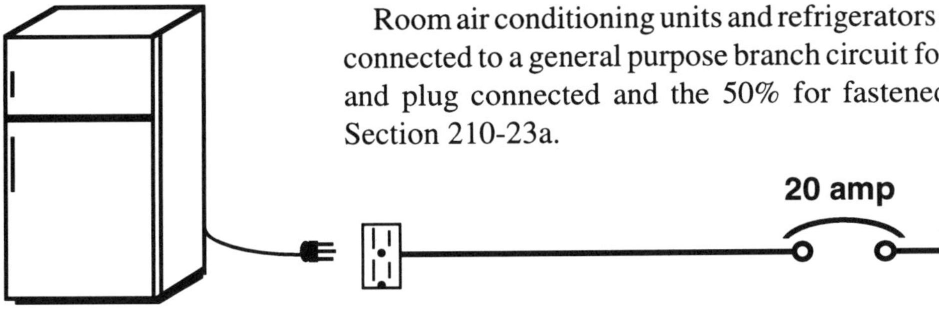

Room air conditioning units and refrigerators are usually cord and plug connected to a general purpose branch circuit following the 80% for cord and plug connected and the 50% for fastened in place appliances in Section 210-23a.

210-23a. The rating of any one *cord and plug connected* appliance shall not exceed 80% the branch circuit rating. The branch circuit rating is 20 amps x 80% = 16 amps. The load of the refrigerator could not exceed 16 amps, if it did, it would require an individual circuit.

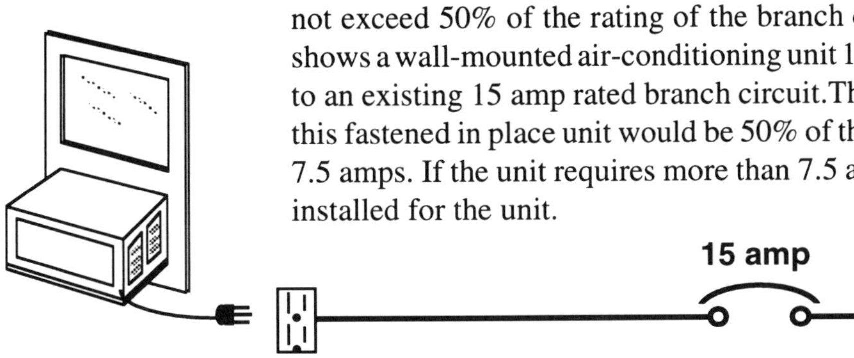

The total ampere rating of fastened in place utilization equipment shall not exceed 50% of the rating of the branch circuit. The following example shows a wall-mounted air-conditioning unit 120 volt cord and plug connected to an existing 15 amp rated branch circuit.The maximum permitted load for this fastened in place unit would be 50% of the 15 amp rated branch circuit = 7.5 amps. If the unit requires more than 7.5 amps then a new circuit must be installed for the unit.

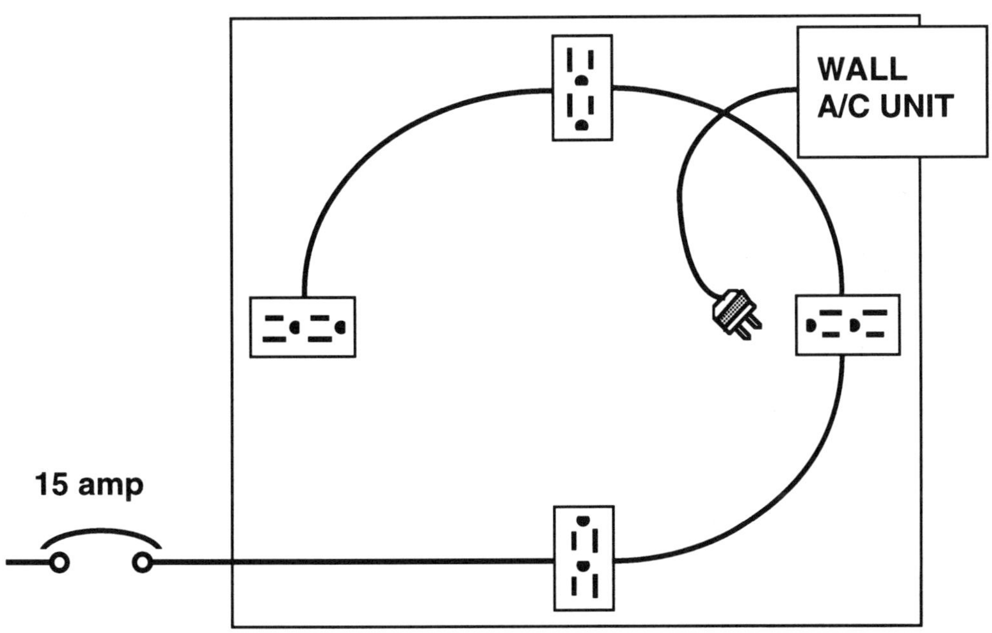

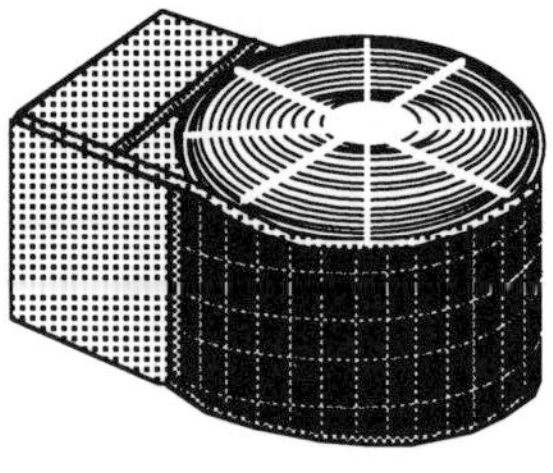

25 amps 240 volt 1ø
FULL LOAD CURRENT

An every day example of poor electrical designing is the branch circuit to a 3-ton A/C unit for a residence.

The installer would grab a box of #10 romex and proceed with the installation. The Code states a #10 conductor 60°C will carry 30 amperes. The A/C unit only has a full load of 25 amps, so #10 must be the proper wire size. WRONG!

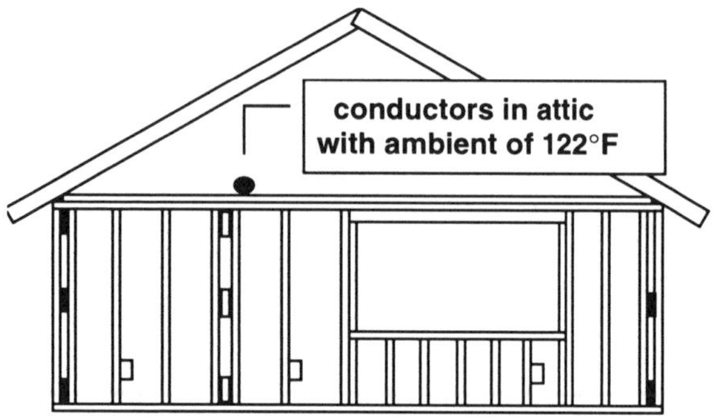

For this branch circuit let's say the conductors are installed in the attic where the ambient reaches 122°F. This is not unrealistic.

Now apply VARIABLE 3.

WIRE SIZE = LOAD/CORRECTION FACTOR

The correction factor for a 60°C wire at 122°F is .58.

25 amp load/.58 = 43.1 is the required ampacity for a 60°C conductor.

Table 310-16 requires a *minimum #6 copper* conductor which has an ampacity of 55 for this branch circuit to a 3-ton A/C unit or you have *violated* the Code minimum safety design standards.

As you can see a #10 wire is not the correct size for this A/C branch circuit, and this is looking at only one variable, there could be more variables that would apply in the circuit.

Let's take a look at VARIABLE 5 - COST for this A/C branch circuit.

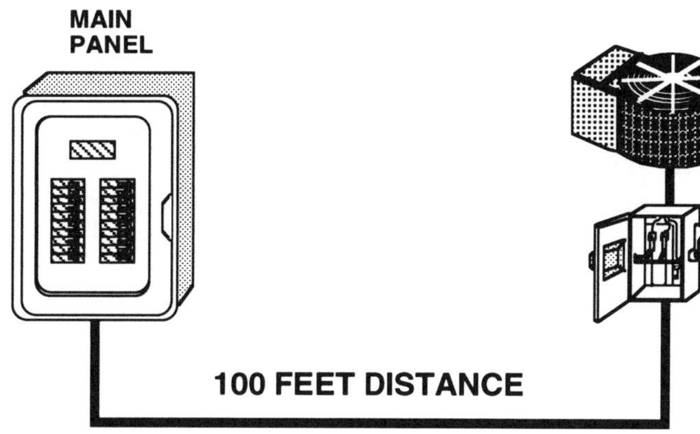

Calculate the difference in voltage drop using a #10 conductor and a #6 conductor based on 100 feet distance.

$$VD = \frac{2 \times 12.5598 \times 100' \times 25a}{10380 \text{ cm}} = 6.05 \text{ volts}$$

6.05 volts dropped using #10 wire.

By using a #6 wire the calculation would be:

$$VD = \frac{2 \times 12.88384 \times 100' \times 25a}{26240 \text{ cm}} = 2.455 \text{ volts dropped using the \#6 wire.}$$

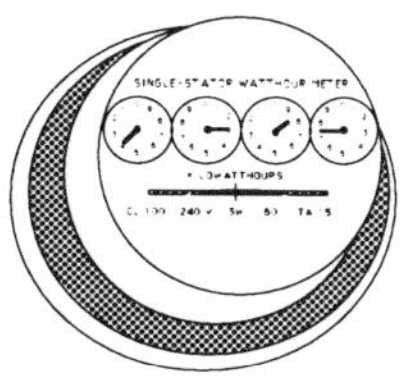

Power Loss = VD x I

$$COST = \frac{\text{Watts x Hours Used x Rate Per Hour}}{1000}$$

Based on 8 hours use per day, 8 cents per kwh.

#10 wire power loss = 6.05v x 25a = 151.25 power loss.
#6 wire power loss = 2.455v x 25a = 61.375 power loss

$$\text{\#10 wire cost} = \frac{151.25\text{w x 8 hours x .08 per kwh}}{1000} = .0968 \text{ or 10 cents per day}$$

$$\text{\#6 wire cost} = \frac{61.375\text{w x 8 hours x .08 cents per kwh}}{1000} = .03928 \text{ or 4 cents per day}$$

Based on a 30 day month the #10 wire would waste $3 in heat loss, the #6 only $1.20. For a year the #10 wire would waste $36 compared to $14.40 for the #6 wire. A yearly savings of $21.60, it wouldn't take to long to pay for the larger wire with these savings!

Now let's design the branch circuit to a three phase, 208volt, 10-ton air conditioner with a nameplate of 37.2 amps on the roof of a commercial building. The total distance of the branch circuit is 65 feet. All of the equipment is 75°C listed. Figure a minimum ambient from the solar sun of 113°F.

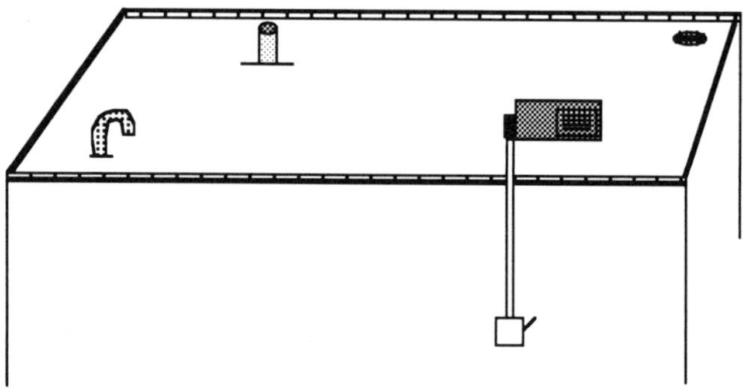

What size THHN wire should you select is now the question?

VARIABLE 2

$$\text{Voltage drop wire size} = \frac{1.732 \times 12.9 \times 65' \times 37.2a}{6.24 \text{ vd permitted}} = 8658 \text{ cm required Table 8} = \#10$$

A #10 conductor with 10380 cm would meet the VARIABLE 2 requirement.

440-42 requires the branch circuit conductors to have an ampacity of 125% of the load. 37.2a x 125% = 46.5 required ampacity. Table 310-16 lists a 75°C conductor with an ampacity of 35.

A #8 THHN has an ampacity of 55, now apply VARIABLE 3. Table 310-16 lists a correction factor of .87 for 113°F for a 90°C conductor.

VARIABLE 3

A #8 THHN has an ampacity of 55 x .87 = 47.85 ampacity, checking the 75°C column a #8 shows an ampacity of 50. So, in this condition a #8 THHN can be safely loaded to 47.85 amps.

To show how the ambient temperature can affect conductor sizing let's raise the ambient *one* degree from 113°F to 114°F and size the THHN conductor.

TYPICAL AMBIENT TEMPERATURES

LOCATION	TEMPERATURE	Minimum Rating of Required Conductor Insulation
Well ventilated, normally heated buildings	30°C (86°F)	* (see note below)
Buildings with such major heat sources as power stations or industrial processes	40°C (104°F)	75°C (167°F)
Poorly ventilated spaces such as attics	45°C (113°F)	75°C (167°F)
Furnaces and boiler rooms (minimum) (maximum)	 40°C (104°F) 60°C (140°F)	 75°C (167°F) 90°C (194°F)
Outdoors in the shade	40°C (104°F)	75°C (167°F)
In thermal insulation	45°C (113°F)	75°C (167°F)
Direct solar exposure	45°C (113°F)	75°C (167°F)
Places above 60°C (140°F)	-----	110°C (230°F)

*Note: 60°C for up to and including #8 copper and up to and including #6 aluminum. 75°C for over #8 copper and #6 aluminum.

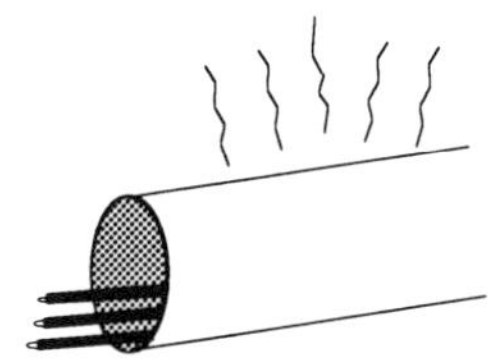

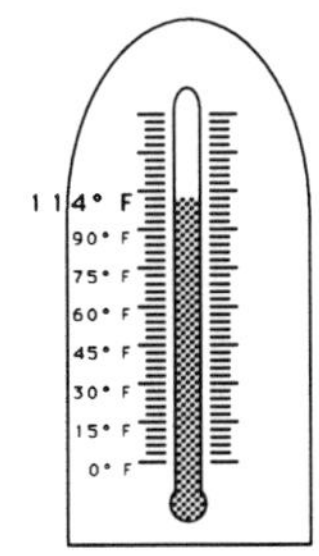

A #8 THHN has a normal ampacity of 55. From Table 310-16 the correction factor for 114°F for a 90°C wire is .82. The normal ampacity, 55, must be lowered x .82 = 45.1 is the new derated ampacity.

440-42 requires an ampacity of 37.2a x 125% = 46.5. A #8 THHN is a violation at 114°F.

A #6 THHN has a normal ampacity of 75 x .82 = 61.5 ampacity.

An increase of temperature of *one degree* changes the size of this branch circuit.

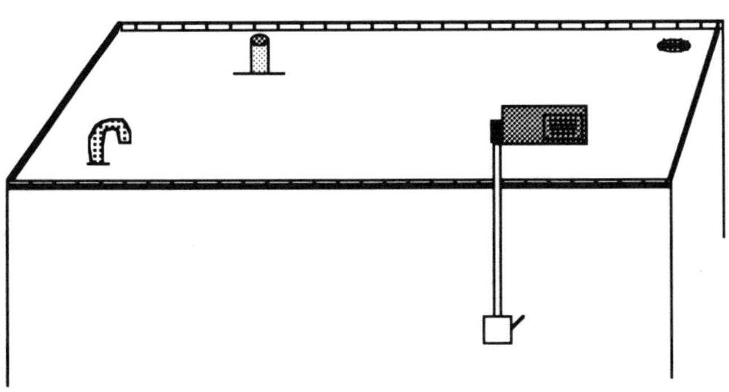

The Table shown above at one time was in the Code showing approximate ambients in different areas. Remember the wire is to be sized to the worst *heat* condition it would ever encounter. When in doubt use a higher ambient.

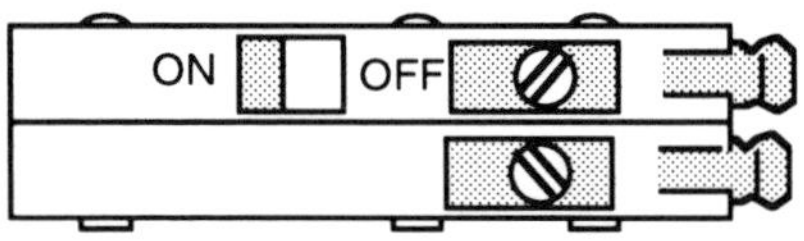

60 AMP MINIMUM
80 AMP MAXIMUM

440-22a requires the O.C.P. to not exceed 175% of the load unless it is not sufficient to start the motor, then the O.C.P. can be sized not to exceed 225% of the load.

37.2a x 175% = 65.1 a   MINIMUM 60 AMP
37.2a x 225% = 83.7a   MAXIMUM 80 AMP

440-12a1 requires the disconnecting means ampere rating to be at least 115% of the load. 37.2a x 115% = 42.78a.

Use a 60 amp rated disconnect for this branch circuit.

STANDARD DISCONNECT SIZES

30 - 60 - 100
200 - 400 - 600
800 - 1200 amperes

OVERLOAD

Generally the overload is built into the equipment (integral).

440-52a states that each motor-compressor shall be protected against overload and failures by one of the four following means:

(1) A separate overload relay selected to trip at not more than 140% of the motor-compressor load.

(2) An integral (built-in) thermal protector.

(3) A fuse or circuit breaker rated at not more than 125% of the motor-compressor load.

(4) A protective system separate from the motor-compressor shall be arranged so the opening of the control circuit will interrupt the current to the motor-compressor.

The *overload relay* is the heart of motor protection. Like the dual element fuse, the overload relay holds during the starting period (inrush current), yet provides protection on small overloads above the full load current of the motor when it's running. Unlike the dual element fuse, the overload relay can withstand repeated trip and reset cycles without having to be replaced. The overload relay does *not* provide short circuit or ground fault protection, that is the function of a breaker or fuse, *not* the overload relay.

The *overload relay* consists of a current sensing unit connected in the conductor to the motor, plus a mechanism, actuated by the sensing unit, which serves directly or indirectly to break open the circuit. In a manual motor starter an overload trips a mechanical latch causing the power contacts to the motor to open. In a magnetic (coil) starter an overload opens a set of contacts within the overload relay itself. These contacts are wired in *series* with the coil of the motor starter. When the coil de-energizes, it opens the power contacts to the motor.

A/C DISCONNECT

440-14 states a disconnect shall be located within sight from and *readily accessible* from A/C or refrigerating equipment. The disconnect shall be permitted to be installed on or *within* the A/C or refrigerating equipment.

Does a rooftop air-conditioning unit require a disconnect on the roof on or within the equipment even though the the disconnect on the side of the building is capable of being locked in the open position?

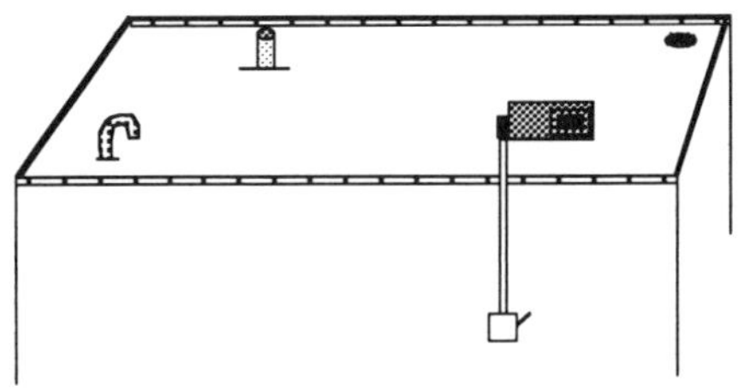

The answer is *yes.*

"Within sight from" is defined as equipment *that is visible* and *not more than 50 feet* from the other equipment.

440-12 does *not* provide an exception for the *air conditioning* disconnect that is capable of being locked in the open position.

430-102b provides an exception for a *motor* that is capable of being in the locked position.

Readily accessible is defined as capable of being reached quickly for operation, renewal, or inspections, without requiring those to whom ready access is necessary to climb over or remove obstacles or to resort to *portable* ladders, chairs, etc.

The term *"readily accessible"* is not crystal clear in all situations. Many inspectors do not accept equipment as being readily accessible where tools are needed to gain access to the equipment.

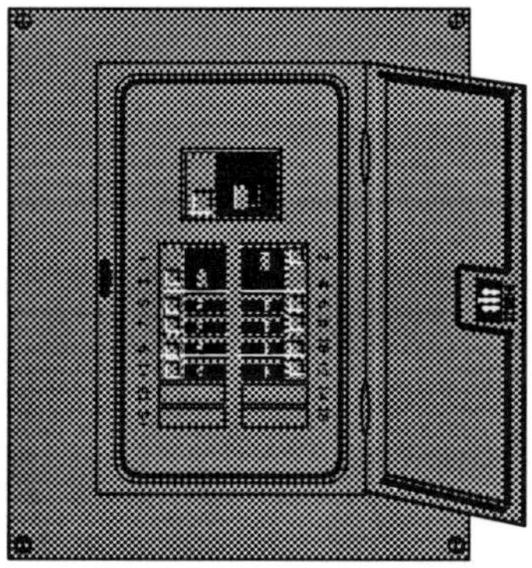

240-24 requires overcurrent devices (fuses or breakers) to be readily accessible.

However, overcurrent devices and disconnects behind hinged and latched doors or covers secured by wing nuts or thumb screws are generally considered as meeting the intent of the definition of readily accessible.

Except for minor changes in punctuation, the definitions of "accessible" and "readily accessible" (as applied to equipment) have remained unchanged in the Code for more than 40 years.

When applied to equipment, which of the following would you think would be more restrictive; "accessible" or "readily accessible"?

Some believe that "readily accessible" is a stiffer requirement than "accessible". However, the definition of "readily accessible" says nothing about not guarded by locked doors.

The definition of "accessible" includes *not guarded by locked doors*. The definition of "readily accessible" does *not* mention not guarded by locked doors. Accessible is more restrictive than readily accessible in reading the definition.

I feel the question you should ask is "readily accessible" to *whom*?

Do you want a high-voltage primary switch in a substation, 480 volt switchgear, a disconnect that could disconnect life-support equipment *readily accessible*? So that any person by throwing a switch could pose a threat to safety by maliciously throwing the disconnect.

This equipment needs to be "readily accessible" only to *qualified* persons that have the knowledge of what equipment they are disconnecting.

In some cases I want the disconnect behind a panel where bolts or screws have to be removed to prevent the passer-by from shutting off the equipment thus causing a disorderly shutdown. I want the disconnect "readily accessible" to the person it should be; the *qualified* person.

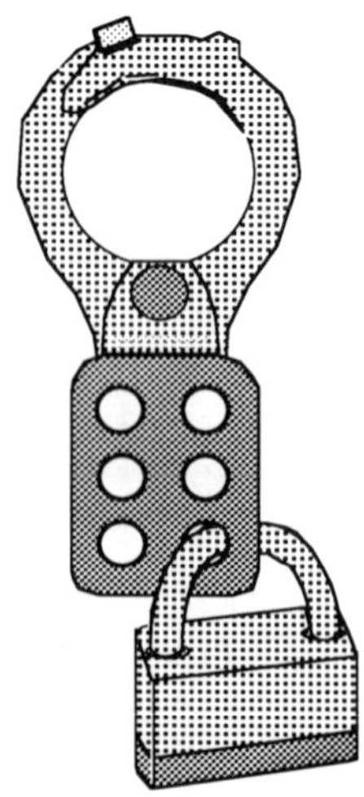

In industrial buildings the primary function of the disconnect is for lock-out purposes when working on the equipment. Quick de-energization of the motor and equipment in an emergency situation is accomplished by pushing the emergency stop button, not pulling the disconnect.

Disconnects are isolating switches, not load break rated, they could blow up if opened under a heavy load. Disconnects are usually 480 volts whereas the emergency stop button is 120 volt control. Let the machine operator handle shutting off the machine with the 120 volt stop and leave the disconnecting of the heavier loads and higher voltages up to the *qualified* person.

380-8a readily accessible *place***.** This section of the Code states that all switches and circuit breakers used as switches shall be so located that they may be operated from a readily accessible place. They shall be so installed that the center of the grip of the operating handle, when in its highest position, will not be more than 6' 7" above the floor or working platform.

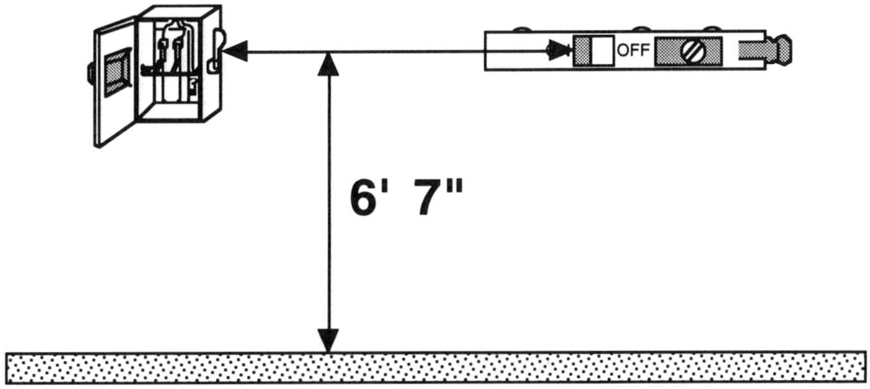

The safety factor requires a circuit to be shut off as quickly as possible in some cases. Portable ladders are not permitted as a means to reach the switch, as they can be removed from the location. Permanent ladders or stairs are permitted. Hook sticks shall be permitted for greater heights for operating isolation switches.

PERMANENT LADDER OKAY

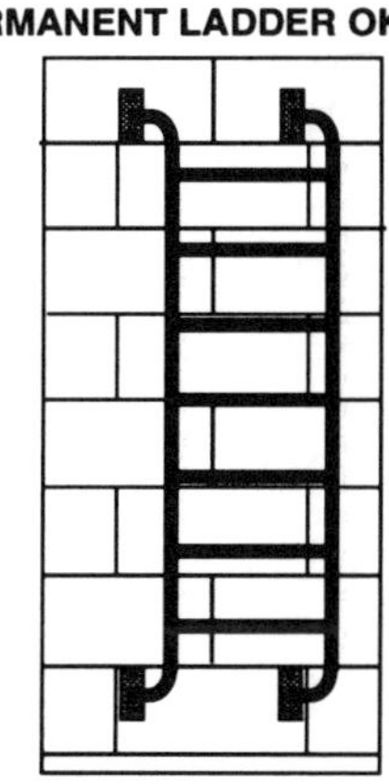

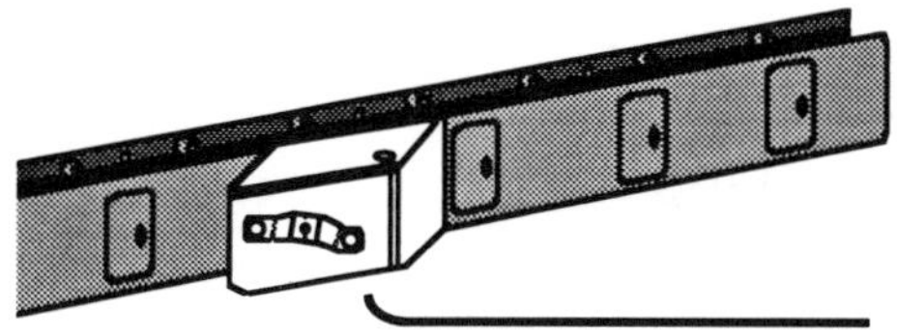

HACR TYPE CIRCUIT BREAKERS

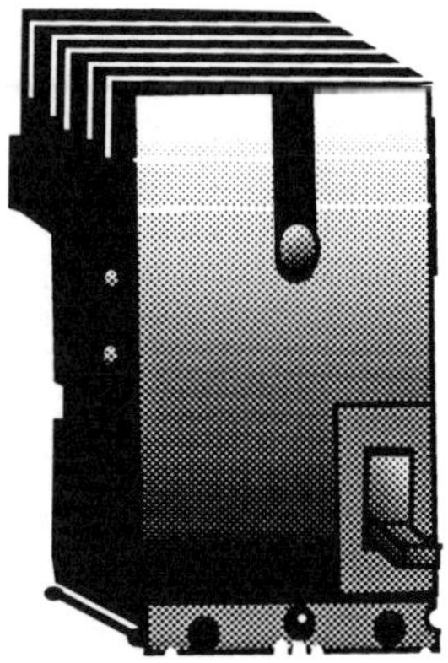

HACR TYPE CIRCUIT BREAKER

Circuit breakers suitable for heating, air conditioning and refrigeration equipment comprised of *multimotor* or *combination* loads are labeled HACR type.

HACR type circuit breakers are the same as other type inverse-time circuit breakers, except they have been additionally investigated by UL and determined to be suitable for motor branch circuit protection on *group installations*. The additional investigation includes tests to determine if the circuit breaker affords an appropriate level of protection for the circuit components. UL testing is performed on circuit breakers and equipment intended for use with HACR circuit breakers.

HACR rated circuit breakers are tested on heating, air conditioning and refrigeration equipment. This equipment usually incorporates more than one motor load, typically a compressor motor and fan motor, and is designed as a *group* motor load. Use of HACR circuit breakers with heating, air conditioning and refrigeration equipment *is limited to installations where equipment is marked.*

Code section 430-53c3 states that for group motor installations, "Each circuit breaker is one of the inverse-time type and listed for group installation."

AIR CONDITIONER MARKINGS

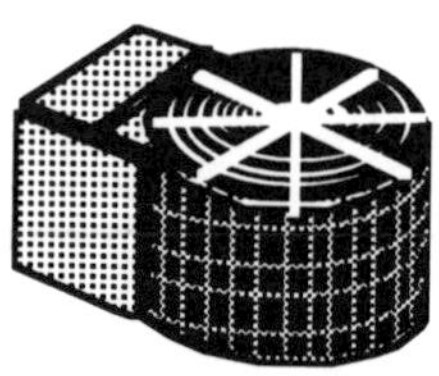

"Maximum rating of overcurrent protective device"

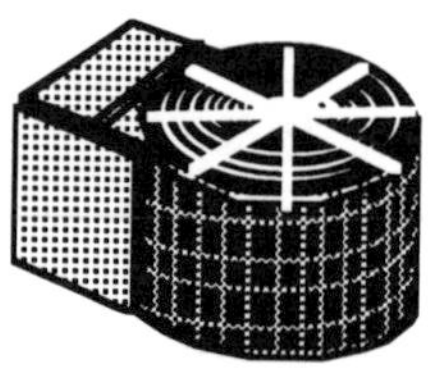

"HACR circuit breaker"

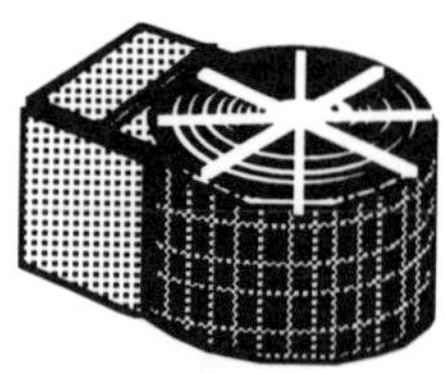

"Maximum fuse or HACR circuit breaker size"

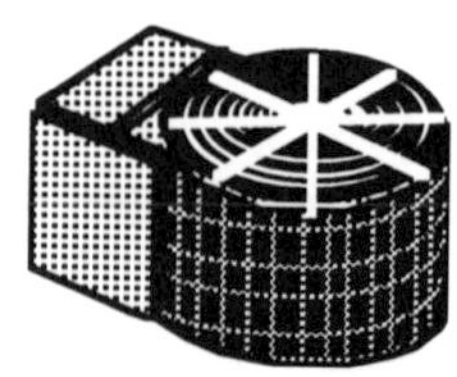

"Maximum fuse size"

SIGN BRANCH CIRCUIT

The question that often arises on outdoor electric signs is, "Doesn't the Code require electric signs to be listed?"

The answer is yes, no question about it, 600-3 states they shall be listed.
But, you can use signs that are *not* listed, if given *special permission* per 600-3.

Special permission is defined as the *written* consent of the authority having jurisdiction (generally the electrical inspector).

Therefore, you can install any sign that is not listed by a recognized testing laboratory only if you have received permission, *in writing*, from the inspector prior to the installation.

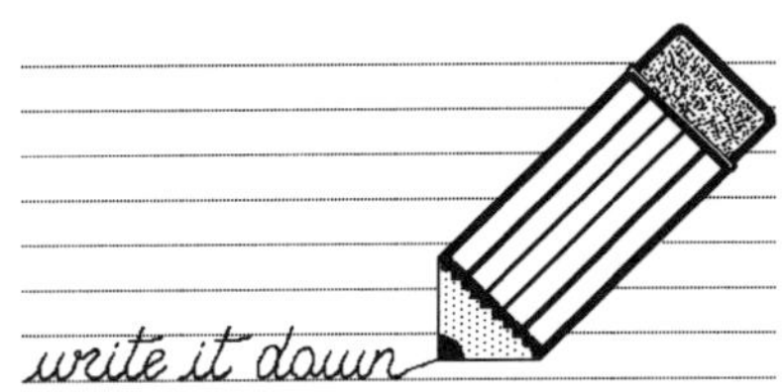

Each permanently installed sign (not portable) and each outline lighting installation shall be controlled by an externally operated switch or breaker that will open all the hot wires.

The disconnect shall be within sight of the sign.

The branch circuit rating for circuits supplying lamps, ballasts, and transformers, or combinations shall not exceed 20 amps.

Branch circuits limited to electric-discharge lighting transformers shall not be over 30 amps in rating.

Each commercial building accessible to pedestrians shall have at least one 20 amp sign outlet accessible outside the entrance to *each* tenant space.

The argument that arises from this is, some zoning ordinances do not permit *electric* illuminated signs on the exterior of buildings. The contractor feels this is a waste of money to install a 20 amp outdoor branch circuit that will never be used.

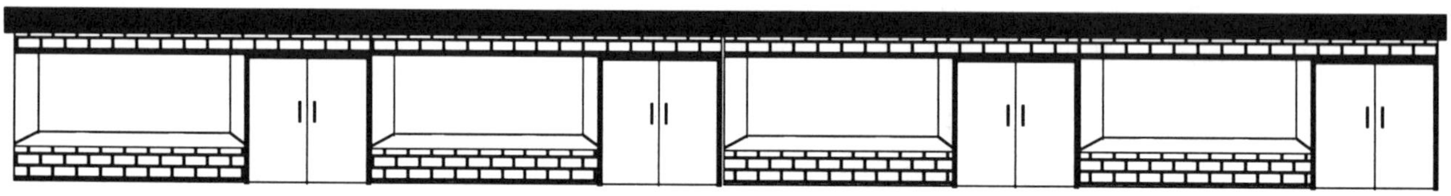

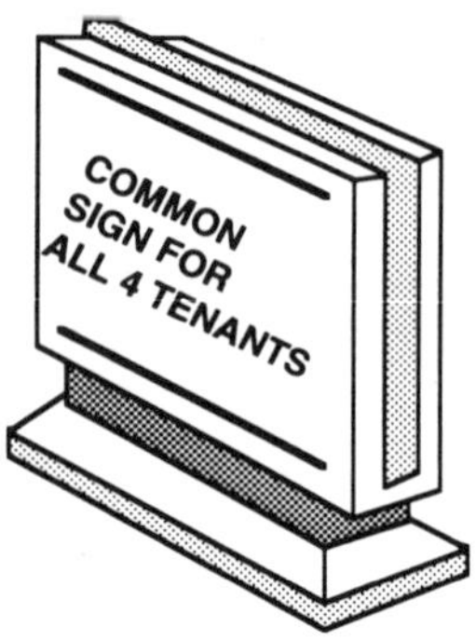

Or the design question is "if I install one 20 amp circuit to a common sign which lists all four of the tenant business, do I still need a 20 amp circuit sign outlet exterior at each tenant unit"?

The intent of the Code is, yes. No exception is made for a common sign or zoning ordinances that prohibit electric signs. An outlet is required at *each* tenant space.

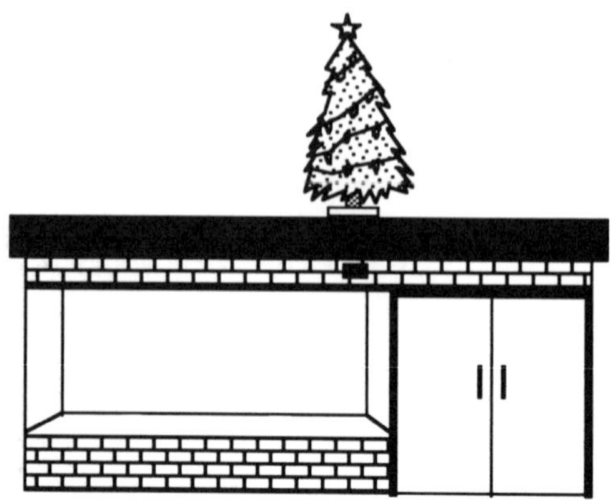

Myself, I've never seen an outdoor receptacle that was *never* used! I agree with the Code feelings that even though there may be areas which prohibit signs, the required 20 amp circuit will be very useful at Christmas time for plugging in the outdoor lighting, etc.

PORTABLE SIGNS

A portable sign shall have a weatherproof receptacle and attachment plug with one pole for grounding.

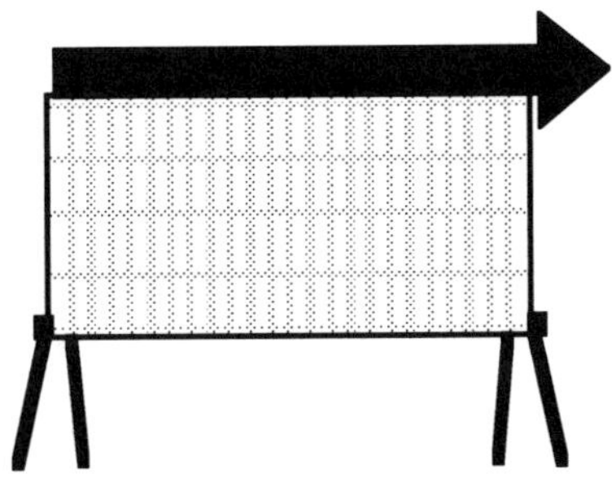

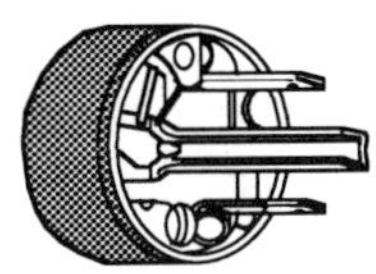

Cords shall be of the hard service or junior hard service type 3-conductor with one conductor grounded.

Outdoor portable signs shall be provided with *factory-installed* ground-fault circuit-interrupter for the protection of personnel. It shall be an integral part of the attachment plug or located in the supply cord within 12" of the attachment plug.

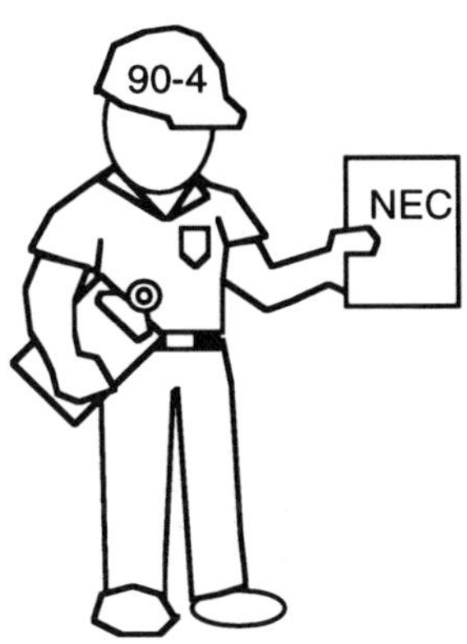

Code section 90-7 states that factory-installed internal wiring or the construction of the equipment does *not need to be inspected* at the time of installation of the equipment, except to inspect for alterations or damage. If the sign has been listed by a qualified electrical testing laboratory it meets the Code requirements.

MOTOR BRANCH CIRCUIT

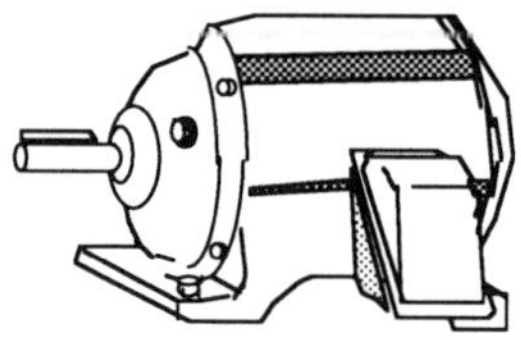

Designing motor circuits are rather complicated so first we need to understand how a motor functions.

A properly protected motor will last for many years.

Some loads that are driven by motors require more starting torque such as water pumps or compressors.

Motors deliver more torque while starting than when running. Up to 6 times more in some applications. So naturally the current is much higher during the starting period.

If the motor is delayed too long during the starting period it will overheat before ever reaching full run speed.

Proper design in selecting the right type and size motor for the load is very important.

Rewinding a motor is very expensive so always design so the motor is never continuously overloaded.

Selecting a motor too large for the load is not good designing. Motors are designed to operate at the best efficiency when delivering their rated horsepower.

When a motor is overloaded or even underloaded it is not efficient.

Almost any motor can deliver 1 1/2 to 2 times its rated horsepower for a short time. A 1/2 hp motor that is driving a water pump is using less than the 1/2 hp with no water in the water tank. But as the pressure in the tank builds up the horsepower gradually increases.

It costs more to run a 1 hp motor at half load than it does to run a 1/2 hp at full load.

SERVICE FACTOR 1.00 to 1.40

If a motor has a service factor of 1.00 it means if the motor is installed in a location where the ambient temperature (surrounding temperature) is not over 40°C or 104°F, it can deliver its rated horsepower continuously without damage to the motor windings.

If a motor has a 1.15 SF it means it can deliver 115% its rated horsepower without damage. It's best not to use a motor above its rated horsepower. But a higher service factor provides *temporary* extra power when needed.

Code Tables 430-148 and 430-150 for motor full load currents show voltages of 115 - 230 whereas in design we are using 120 - 240 volts.

Motors are designed to operate as much as 10% over their rated voltage. Although we may have 240 volts at the source of supply it is doubtful we would still have 240 volts at the motor due to the voltage drop in the conductors. 200 volt motors are available for use on 208 volt systems.

Each motor nameplate must identify the motor time rating of 5 - 15 - 30 or 60 minutes or continuous duty. Almost all motors are *continuous duty rated* unless used in a special application like a hoist or elevator motor.

Full load current (F.L.C.) means when the motor reaches full speed, delivering its rated horsepower, at rated voltage.

The first step is to start the motor, this is when the heavy currents appear and the branch circuit must be designed to protect the motor and the wire.

When a motor is first started the current is much higher than after the motor reaches speed. The ratio of the starting current to the running current varies with the type and size of motor. It could be 600% or more. It also depends on what type of machinery the motor is starting to how high the current will reach on starting.

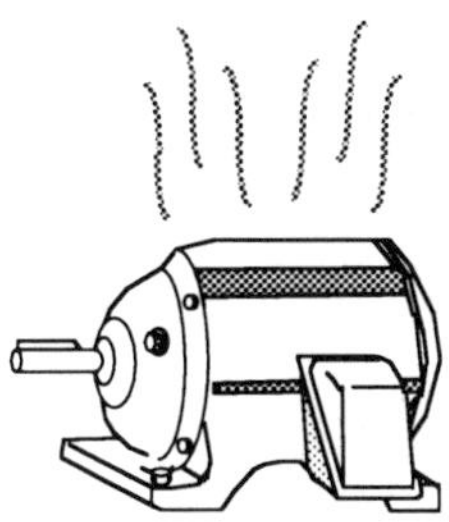

Any motor 1/2 hp or larger has on its nameplate a "code letter" that indicates the approximate kva consumed per hp with the motor in a locked-rotor condition.

Locked rotor means the motor is locked so the rotor can't turn. When the switch is first turned on, during start, the motor sometimes is locked (stalled).

Generally speaking, the smaller the motor, the higher the locked-rotor current per horsepower.

Code Table 430-7b lists the code letters and kva per horsepower with locked-rotor.

CODE LETTER	kva per hp with locked-rotor	CODE LETTER	kva per hp with locked-rotor
A	0 - 3.14	L	9.0 - 9.99
B	3.15 - 3.54	M	10.0 - 11.19
C	3.55 - 3.99	N	11.2 - 12.49
D	4.0 - 4.49	P	12.5 - 13.99
E	4.5 - 4.99	R	14.0 - 15.99
F	5.0 - 5.59	S	16.0 - 17.99
G	5.6 - 6.29	T	18.0 - 19.99
H	6.3 - 7.09	U	20.0 - 22.39
J	7.1 - 7.99	V	22.4 - and up
K	8.0 - 8.99		

The nameplate on the motor is required to show the code letter. Refer to Table 430-7b which lists for code letter "K" 8.0 to 8.99. An average would be 8.5. Multiply the horsepower times 8.5.

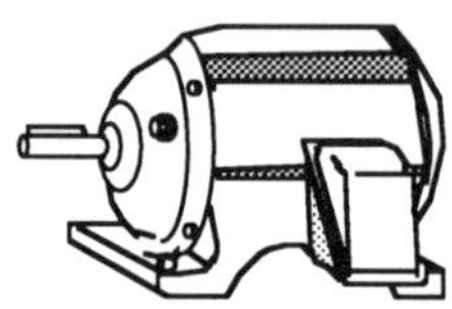

MOTOR NAMEPLATE
5 hp 1ø 240v
CODE LETTER K

5 hp x 8.5kva = 42.5 kva or 42,500va. I = VA/E 42,500va/240v = 177 amps. When this motor is first started it will draw approximately 177 amps and then level off to approximately 28 amps the full load current when it has reached speed.

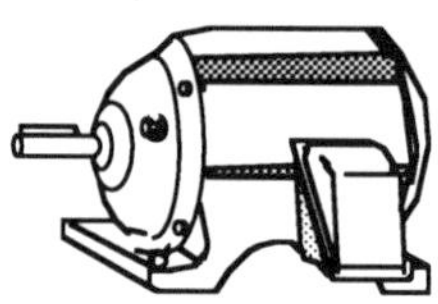

MOTOR NAMEPLATE
10 hp 3ø 208v
CODE LETTER C

The nameplate on this motor lists a code letter "C" which from Table 430-7b lists from 3.55 - 3.99. An average would be 3.75 kva per horsepower.

10 hp x 3.75kva = 37.5 kva or 37,500va. 37500va/208v x 1.732 = 104 amps. When this motor is started it will draw approximately 104 amps and then level off to approximately 30 amps the full load current when it has reached speed.

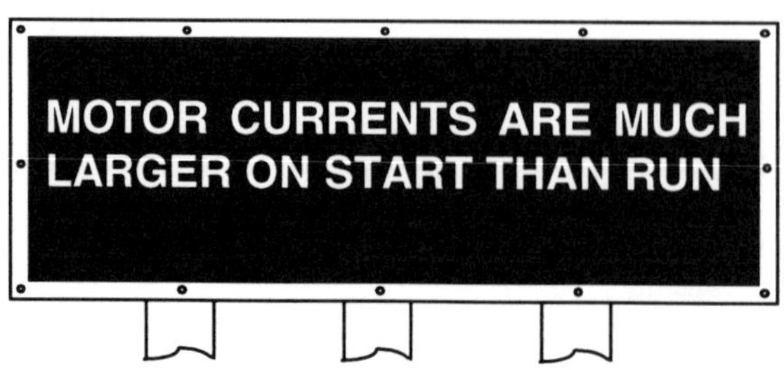

Code section 240-3 states the conductors shall be protected against overcurrent in accordance with their ampacities, *unless otherwise permitted.*

For example, lighting branch circuit conductors would have to be protected within their ampacities.

But for a motor branch circuit, the rules are different per 240-3f, which states motor circuits are sized per Article 430. The reason for the different rules is because a motor circuit will see a higher current on start than a lighting circuit, etc.

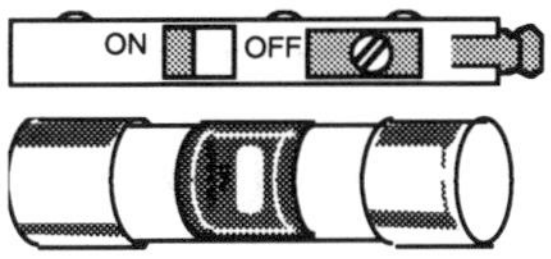

Code section 430-52 states the fuse or breaker shall be capable of carrying the starting current. The rating or setting value is calculated using the values given in Table 430-152.

The *type* of overcurrent protection can be nontime delay fuse, time delay fuse, instantaneous trip breaker, or the regular breaker the inverse time.

The *type* of motor lists single-phase all types.

All three-phase squirrel cage motors.

Wound rotor has a slip ring with brushes, whereas the squirrel cage motor has no brushes.

Table 430-152. Maximum Rating or Setting of Motor Branch Circuit Short-Circuit and Ground-Fault Protective Devices

	Percent of Full-Load Current			
Type of Motor	**Nontime Delay Fuse**	**Dual Element (Time-Delay) Fuse**	**Instantaneous Trip Breaker**	*** Inverse Time Breaker**
Single-phase motors	300	175	800	250
AC polyphase motors other than wound-rotor				
Squirrel-cage:				
Other than Design E	300	175	800	250
Design E	300	175	1100	250
Synchronous†	300	175	800	250
Wound-rotor	150	150	800	150
Direct-current (constant voltage)	150	150	250	150

Type of Motor	Nontime Delay Fuse	Dual Element (Time-Delay) Fuse	Instantaneous Trip Breaker	* Inverse Time Breaker
	Percent of Full-Load Current			
Single-phase motors	300	175	800	250
AC polyphase motors other than wound-rotor				
Squirrel-cage:				
Other than Design E	300	175	800	250
Design E	300	175	1100	250
Synchronous†	300	175	800	250
Wound-rotor	150	150	800	150
Direct-current (constant voltage)	150	150	250	150

The number shown in each column are percentages of the full load current.

Example, the full load current of a single phase 230 volt, 1 1/2 hp motor from Table 430-148 is 10 amps.

The size nontime delay fuse would be 10a x 300% = 30 amp fuse.

The next column lists 175% for a dual element *time delay* fuse. 10a x 175% = 17.5 amps.

But, is 17.5 a standard size fuse? Code section 240-6a lists standard sizes.

15, 20, 25, 30, 35, 40, 45, 50, 60, 70, 80, 90, 100,110, 125, 150, 175, 200, 225, 250, 300, 350, 400, 450, 500, 600,700, 800, 1000, 1200, 1600, 2000, 2500, 3000, 4000, 5000, and 6000 amperes.

Exception: Additional standard ratings for FUSES shall be considered 1, 3, 6, 10 and 601.

17.5 amps is *not* listed as a standard size fuse.

Now for the $64 dollar question. Do you select a 15 amp or a 20 amp fuse?

Code section 430-52c exception 1 states if you do not correspond to a standard size the next higher size is permitted.

430-52c exception 2b permits the next higher size time-delay fuse in exception 2b where Table 430-152 is not sufficient for the *starting of the motor*. 10a x 225% = 22.5 amps. You can use a 20 amp time delay fuse to start the motor.

The next column in Table 430-152 lists an instantaneous trip breaker at 800%. F.L.C. 10a x 800% = 80 amp circuit breaker for a 10 amp motor load is permitted. An instantaneous trip breaker shall be used only if adjustable, and if part of a listed combination controller having coordinated motor overload and short-circuit and ground-fault protection in each conductor.

The last column lists the everyday circuit breaker, the inverse time breaker. Inverse means the higher the current the faster the breaker opens, the lower the current the slower the breaker opens. 10 a x 250% = 25 amp inverse time breaker.

	Percent of Full-Load Current			
Type of Motor	Nontime Delay Fuse	Dual Element (Time-Delay) Fuse	Instantaneous Trip Breaker	* Inverse Time Breaker
Single-phase motors	300	175	800	250
AC polyphase motors other than wound-rotor				
Squirrel-cage:				
Other than Design E	300	175	800	250
Design E	300	175	1100	250
Synchronous†	300	175	800	250
Wound-rotor	150	150	800	150
Direct-current (constant voltage)	150	150	250	150

What size inverse time breaker is required for this motor that has a code letter "K"?

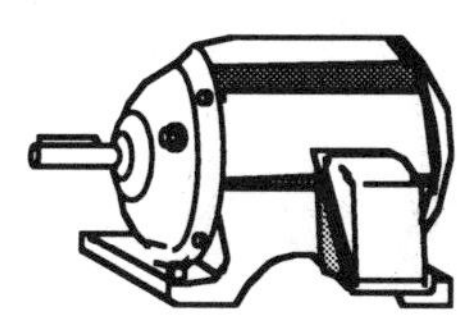

MOTOR NAMEPLATE
5 hp 1ø 240v
CODE LETTER K

Table 430-148 lists the F.L.C. at 28 amps x 250% = 70 amp circuit breaker.

We learned earlier the locked-rotor current could reach approximately 177 amps on start. The 70 amp circuit breaker will allow this higher current for a short period of time during starting.

Section 430-52 exception 2c permits an inverse time breaker of 100 amps or less to be sized up to 400% if the 250% is not sufficient to *start the motor*. Use a 70 amp per 240-6.

When multiplying the F.L.C. of a motor by the percent of value from Table 430-152 it calculates to a figure that is not a standard fuse or breaker size per 240-6 and you are faced with selecting the next lower size or the next higher size the Code allows. If it is my motor, I would first try the lower size fuse or breaker as it would provide the best protection.

Some select the next higher size where the Code allows it. Their thinking is there is less chance of the device tripping and this means no call backs for the service man.

Keep in mind when selecting fuses, time-delay dual element is the best choice.

Code section 430-57 requires when using *nontime* delay fuses the fuseholder must be capable of holding the *largest* fuse permitted. This may require using a fuse adapter to permit a 60 amp fuse to be used in a 70 amp fuseholder. 430-57 exception does not require time-delay fuses to meet this requirement.

PLUG FUSE
30 AMP OR LESS

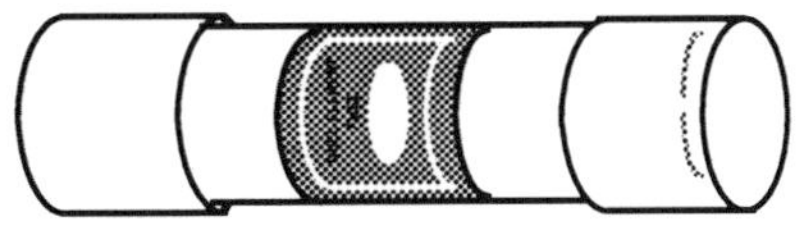

FERRULE-CONTACT TYPE
SIZES UP TO 60 AMPS

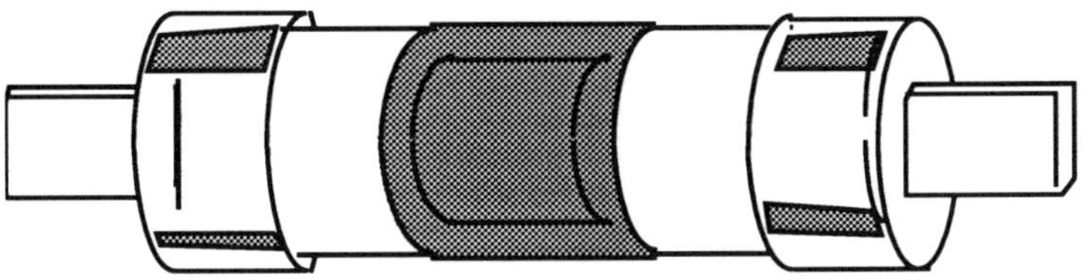

KNIFE-BLADE CONTACT TYPE
SIZES OVER 60 AMPS

Example: Sizing nontime delay fuses for the 3 hp motor would calculate 18.7a x 300% = 56.1 amps. But, if a 50 or 60 amp nontime delay fuse won't allow the motor to start:

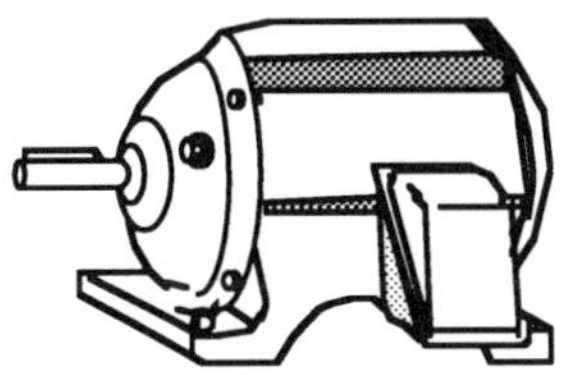

3 hp 1ø 208v
F.L.C. 18.7 amps

Section 430-52a exception 2a permits using 400% in the case where the motor won't start with the 300%. 18.7a x 400% = 74.8 amps. A 70 amp nontime delay fuse is permitted.

The designer must keep in mind the fuseholders must be sized to accommodate the largest size fuse permitted, in some case this will require different fuseholders as the cartridge fuse is up to 60 amps and above 60 amps is the knife-blade fuseholder.

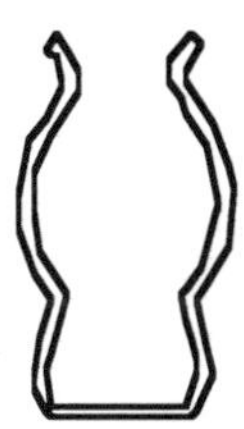

CARTRIDGE

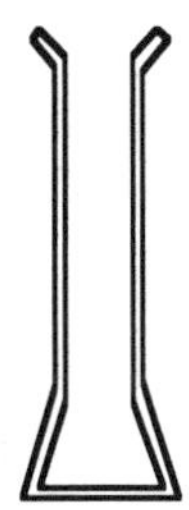

KNIFE-BLADE

OVERLOAD

The Code requires thermal overload protection for most motor branch circuits. As discussed on the prior pages, the overcurrent protection (fuse or breaker) protects the motor and branch circuit against short-circuits and ground-faults. But the overcurrent device does *not* protect the motor or circuit against current *higher* than the normal running.

These higher than normal currents are caused by overloading the motor or by failure of the motor to start when turned on.

The 25 amp breaker or fuse is selected to handle the high inrush currents on the starting of a 10 amp motor.

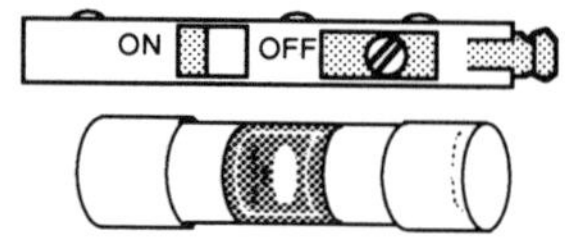

Once the motor has reached running speed and the higher currents drop down to the F.L.C. of 10 amps the 25 amp overcurrent device provides no protection for an overload on the motor.

A motor is designed to allow for a temporary overload when the motor is required to deliver more than its rated horsepower only for a *short* period of time. If this overload would continue, then the overload which senses this increase of current (heat) will open thus protecting not only the motor but also the wiring.

The words are similar "over*current*" and "over*load*" but they have entirely different functions and this is very important to understand when designing motor circuits.

OVERCURRENT =

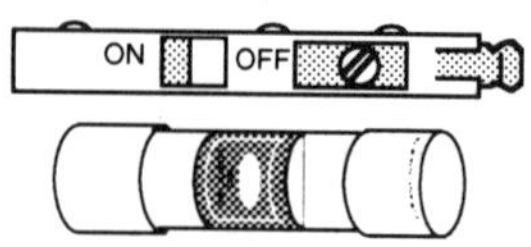

OVERLOAD =

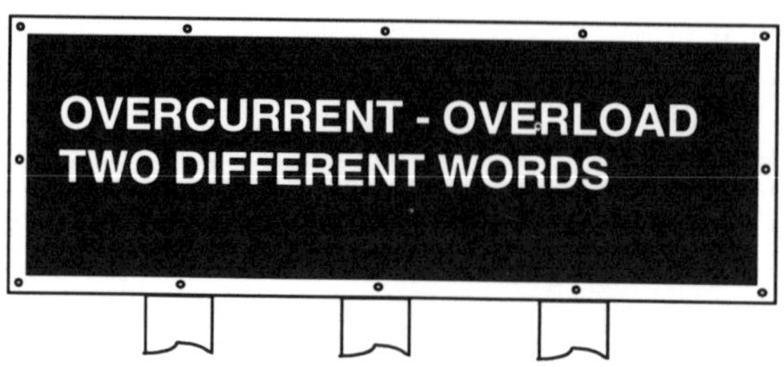

Over*loads* can be electrical as well as mechanical, in origin. Low voltage or single phasing of a three phase system are examples of electrical overloads. Mechanical overloads can come from a high head pressure on a pump, a bearing on a shaft freezing up, etc.

The effect of an *overload* is a rise in temperature in the motor windings. The larger the overload, the more quickly the temperature will increase to the point of damaging the insulation and lubrication of the motor bearings. It has an inverse effect, the higher the current in a given time, the shorter the time before motor damage occurs.

All *overloads* shorten the life of a motor by deteriorating the motor insulation. The intensive heat concentration generated by excessive current in the motor windings is what causes the insulation to fail and burn up the motor.

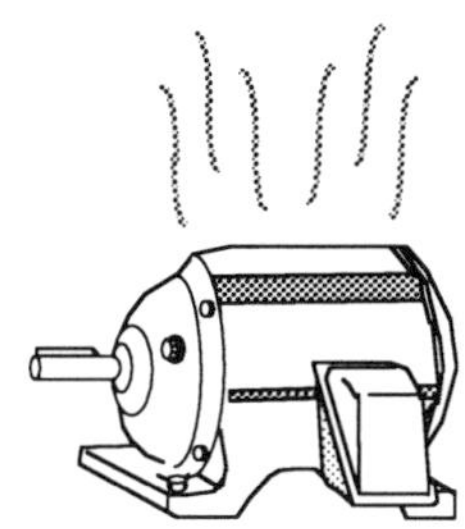

It has been estimated that for every 1°C rise over *normal* ambient temperatures, it can reduce the life of the motor one year for each degree above the insulation rating. The *normal* rating for most motors is 40°C which is 104°F. Ambient temperature is the temperature surrounding the motor and the environment where the motor is placed.

Fuses are not designed to provide over*load* protection. Their basic function is to provide protection against short circuits and ground faults, the over*current* protection. A single element fuse chosen on the basis of motor full load current would blow every time the motor is started. On the other hand, if a fuse were chosen large enough to pass the starting (inrush) current, it would not protect the motor from overloads. A dual element or time delay fuse can provide motor overload protection, but the disadvantange is, if the fuse blows it must be replaced.

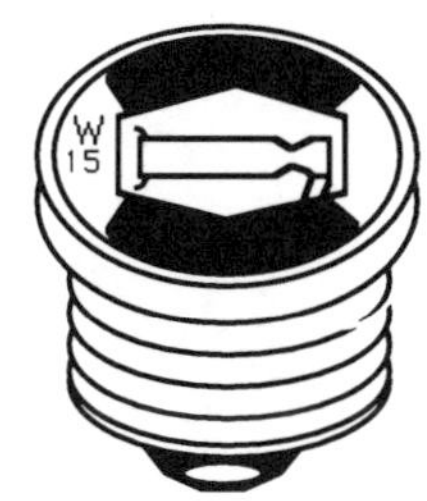

The *overload relay* is the heart of motor protection. Like the dual element fuse, the overload relay holds during the starting period (inrush current), yet provides protection on small overloads above the full load current of the motor when it's running. Unlike the dual element fuse, the overload relay can withstand repeated trip and reset cycles without having to be replaced. The overload relay does *not* provide short circuit or ground fault protection, that is, the function of a breaker or fuse, *not* the overload relay.

The *overload relay* consists of a current sensing unit connected in the conductor to the motor, plus a mechanism, actuated by the sensing unit, which serves directly or indirectly to break open the circuit. In a manual motor starter an overload trips a mechanical latch causing the power contacts to the motor to open. In a magnetic (coil) starter an overload opens a set of contacts within the overload relay itself. These contacts are wired in *series* with the coil of the motor starter. When the coil de-energizes it opens the power contacts to the motor.

Some motors have built-in (integral) overload protection, making the need for separate overload protection unnecessary.

Overload relays can be classified as being either *thermal* or *magnetic*. Magnetic overload relays react only to excess current and are not affected by temperature. The device operates through the use of a *current coil*. At a specified overcurrent value, the current coil acts as a solenoid, causing a set of normally closed contacts to open thus disconnecting the motor from the circuit.

An advantage of the magnetic overload relay is that it is extremely quick on reset, since it does not require a cooling off period before it can be reset. The disadvantage of the magnetic overload relay is that it is more expensive that the thermal overload relay. The thermal overload relay is the one most often used.

Thermal overload relays rely on the rising temperatures caused by the excessive current and trip the overload mechanism when the maximum temperature is reached. Thermal overload relays can be subdivided into *two* types, the *melting alloy* (solder pot) and *bi-metallic*.

The *melting alloy* thermal overload relay is also referred to as the "solder pot relay". With this overload relay the motor current passes through a small heater winding (heater coil). Under overload conditions the heat *causes a special solder to melt*, allowing a ratchet wheel to spin free, opening the interlock contacts that are wired in series with the coil in a magnetic motor starter. When this occurs, the overload relay is said to have "tripped". The heat transfer characteristic and the accuracy of the heater coil cannot be accidentally changed. Thermal overload heater coils are rated in amperes and are selected on the basis of the motor full load current, not the horsepower of the motor. To obtain appropriate tripping current for motors of different sizes, or different full load currents, a range of thermal units (heaters) are available.

Heater separate

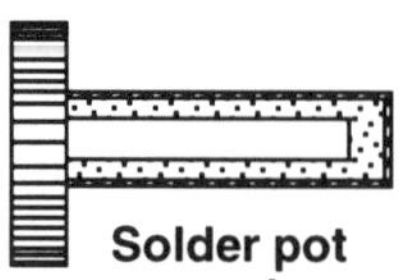
Solder pot separate

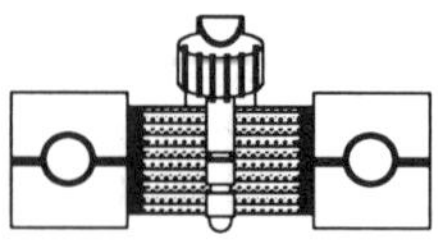
Heater coil and solder pot are combined into one piece

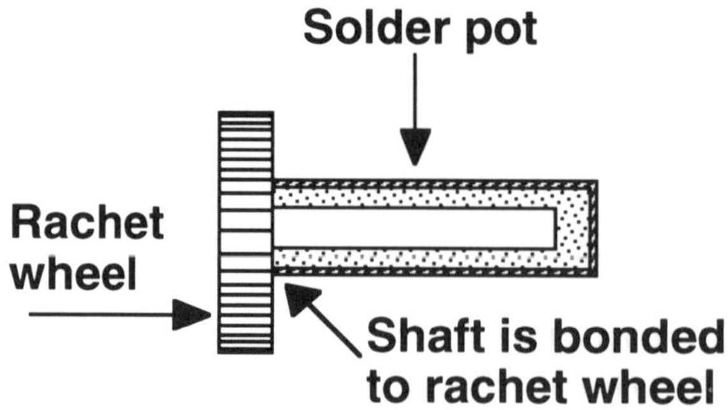

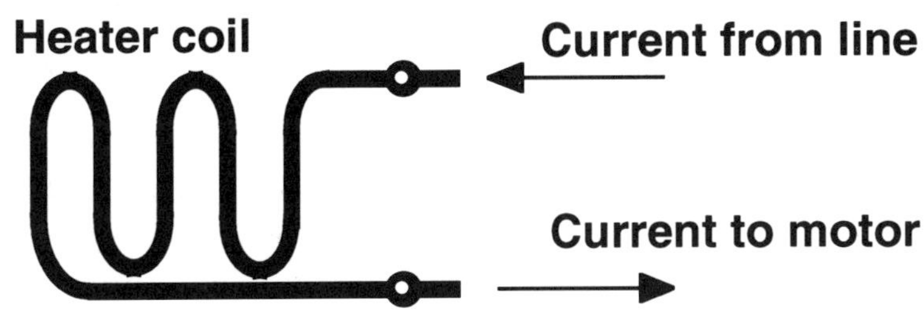

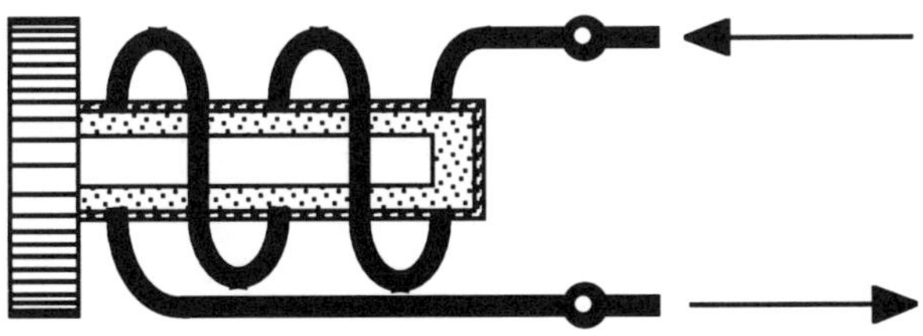

Heater coil slides over the shaft

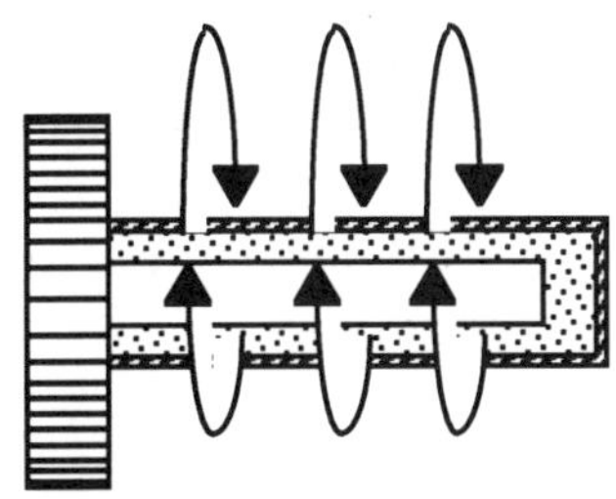

The heat from the current flowing through the motor is monitored in the shaft

Shown below is a sketch of the current flowing from the line through the heater coil and on to the motor. The solder pot is monitoring the amount of heat from the current flow. This application would be for a *manual starter* as the *line* current would be interrupted in tripped condition.

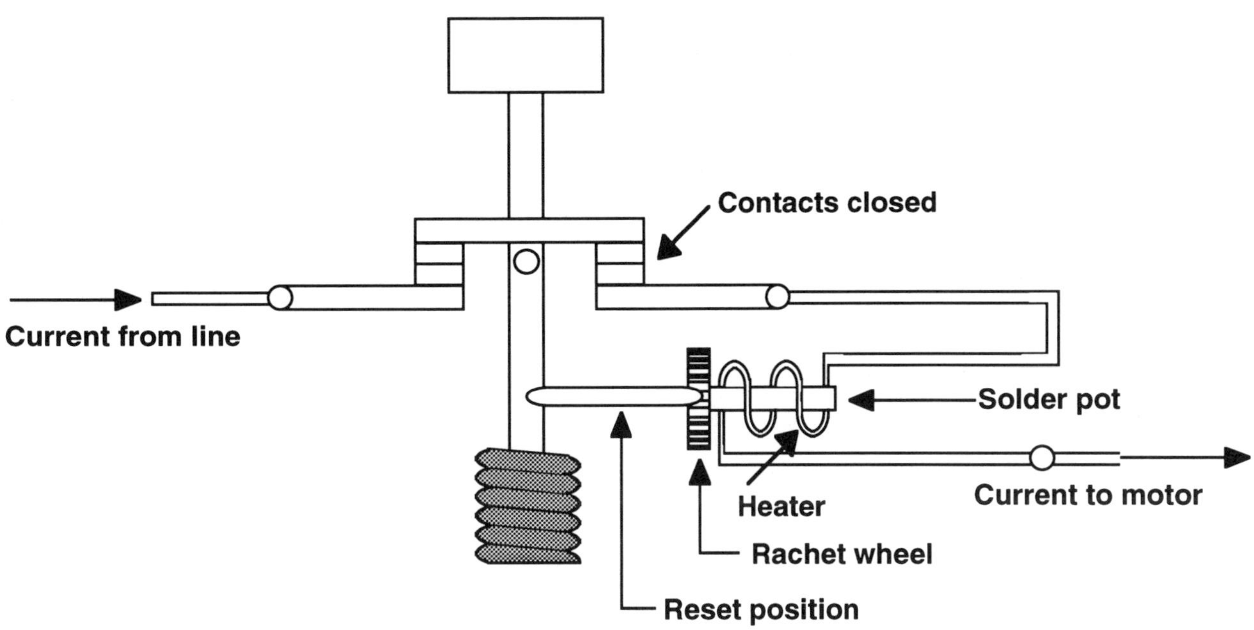

The other type of overload relay is the *bimetallic.* The bimetallic thermal overload relays employ a U-shape bimetal strip, associated with a current carrying heater element. When an overload occurs, the heat will cause the bimetal to warp and open a contact.

The bimetallic overload relay can be hand reset or automatic reset. However, automatic reset overload relays should not normally be used with 2-wire control circuits. The reason is, when the overload relay contacts reclose after an overload relay trip, the motor will restart, and unless the cause of the overload has been removed, the overload will trip again. This cycle of tripping and resetting will continue until eventually the motor will burn up. More important is the danger of restarting when personnel are involved as mentioned earlier. The unexpected restarting of a machine may find the operator in a hazardous situation.

Shown below is the motor circuit in normal position and when an overload has occurred with a bimetallic overload relay.

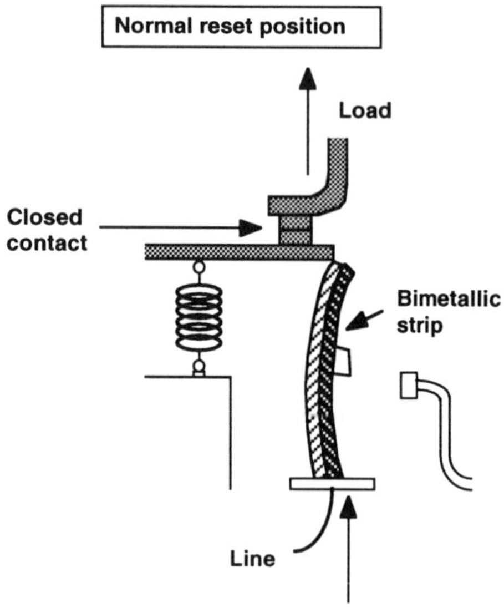

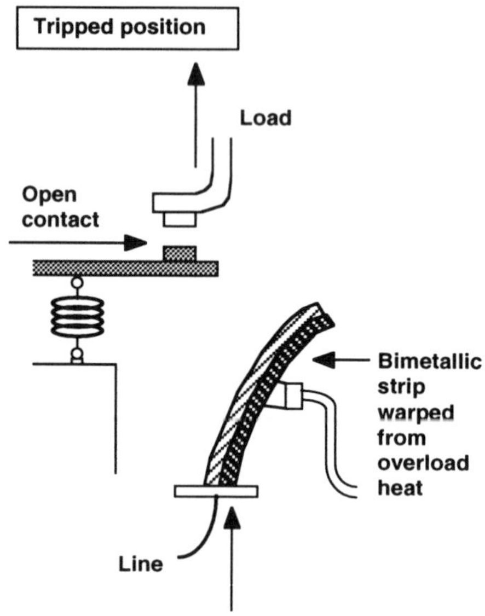

Ambient-compensated bimetallic overload relays were designed for one particular situation; when the motor is at a constant temperature and the motor controller is located separately in a varying temperature. In this case, if a standard thermal overload relay were used, it would not trip consistently at the same level of the motor current if the motor controller temperature changed. This thermal overload relay is always affected by the surrounding temperature. To compensate for the temperature variations the motor controller may see, an ambient temperature-compensated overload relay is applied. Its trip point is *not* affected by temperature and it performs consistently at the same value of current.

Melting alloy (solder pot) and bimetallic overload relays are designed to approximate the heat actually generated in the motor.

Heater coils will not detect high ambient temperatures at the motor location because they are monitoring line current. The high ambient temperatures at the motor location can burn the motor up without being detected.

As you can see by now, heat plays a very important role in motor protection. Both motor controllers and motors are subject to a 40°C (104°F) ambient temperature *limit*. This is the temperature *outside* the motor, not inside.

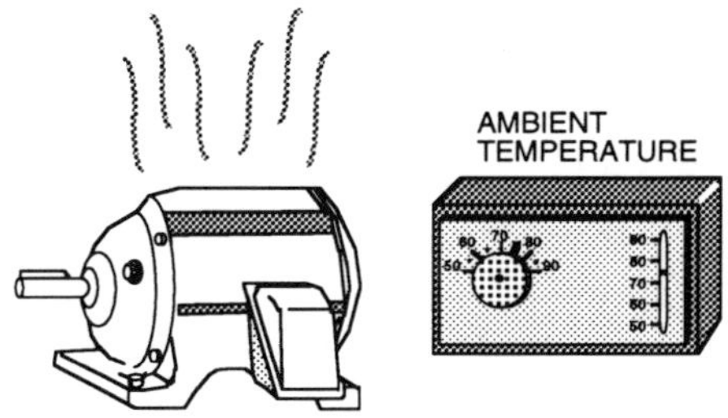

Current passing through the windings of a motor results in an increase in the motor temperature. The difference between the winding temperature of the motor when running and the ambient temperature is called the *temperature rise*.

The temperature rise produced at full load is not harmful provided the motor ambient temperature does not exceed 40°C (104°F).

Higher temperatures caused by increased current or higher ambient temperatures produces a deteriorating effect on the motor insulation and lubrication. An old "rule of thumb" states that for each increase of 10°C above rated temperature the motor life is cut in half.

When a Celsius thermometer reads 30°C, a Fahrenheit thermometer will read 86°F or when a Celsius reads 40°C, a Fahrenheit will read 104°F.

40° Celsius = 104° Fahrenheit

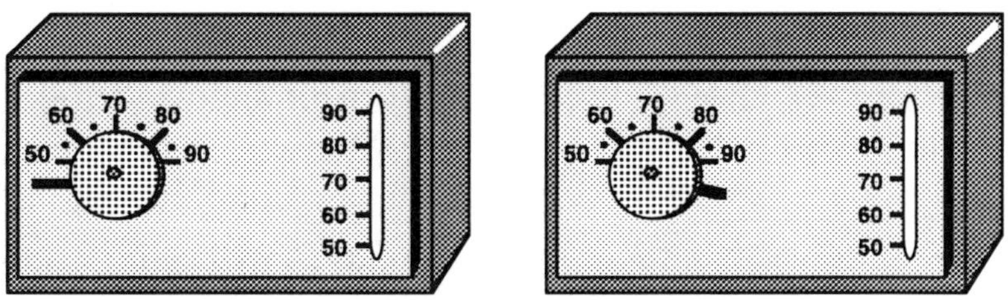

Temperature *RISE* is over and above the ambient temperature while the motor is delivering its rated horsepower. An ordinary motor is based on a temperature *rise* of 40°C which is a *rise* of 72°F. Don't confuse a *change* in the reading of the two different thermometers with their *actual* readings. When a Celsius changes from 40°C to 80°C (a 40°C rise), the Fahrenheit changes from 104°F to 176°F (a 72°F rise). One degree *rise* on the Celsius scale is the equivalent of 1.8° on the Fahrenheit scale.

Example, a 30°C (86°F) *rises* to 40°C (104°F) a 10°C rise. 10°C x 1.8 factor = 18°F.
86°F + 18°F = 104°F.

A 40°C *rise* is a 72°F rise. 104°F ambient + 72°F rise = 176°F actual heat. In a pump house or other closed in areas where the ambient is expected to exceed 104°F, a 50°C rise motor should be selected.

If the motor manufacturer has given the motor a *service factor*, it means that the motor can be allowed to develop *more* than its rated or nameplate horsepower, without causing undue deterioration of the insulation. Example; a 10 hp motor has a service factor of 1.15. The motor can be allowed to develop 11.5 horsepower (10 hp x 1.15 service factor = 11.5 hp). The service factor depends on the motor design.

Over the years, better insulation on the wires used to wind the motors has been developed so the motor will not be damaged by the temperatures that would have destroyed older motors. This has reduced the physical size of todays motors. Todays motors run hotter without damaging the motor.

Remember, a 10 hp motor with a service factor can develop 11.5 hp without insulation damage providing the ambient temperature does not exceed 40°C or 104°F. A motor will operate longer and more efficiently if not used over its *rated* horsepower. The service factor provides *temporary* extra power if needed.

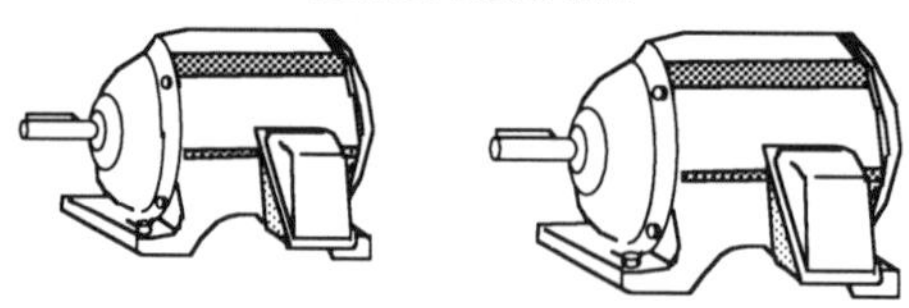

When selecting the size of overload relay heaters, the full load current of the motor, type of motor, and the possible difference in ambient temperature between the motor controller and the motor must all be taken into account. Always refer to the *nameplate* rating on the motor for the full load current.

The full load current is determined from the Code Tables 430-148 for single-phase motors and Table 430-150 for three-phase motors, unless the *nameplate* rating is given per section 430-6a.

To size the heaters, Code section 430-32 is for continuous-duty motors more than one horsepower.

Heater

Section 430-32a1. A separate overload device that is responsive to motor current. This device shall be selected to trip or rated at no more than the following percent of the motor nameplate full load current rating.

Motors with a marked service factor not less than 1.15 ...... 125%
Motors with a marked temperature rise not over 40°C 125%
All other motors 115%

Example; a standard squirrel cage motor with no service factor or temperature rise markings that has a nameplate full load current rating of 12 amps would have the overload heater sized at 12 amps x 115% = 13.8 amps.

With the heater sized to 13.8 amps with the motor full load current of 12 amps, a 1.8 amp increase in current will trip the overload.

At first when you see a 28 amp motor on a circuit that has a 70 amp circuit breaker you feel the circuit is not protected properly. But the truth is, it's just the opposite.

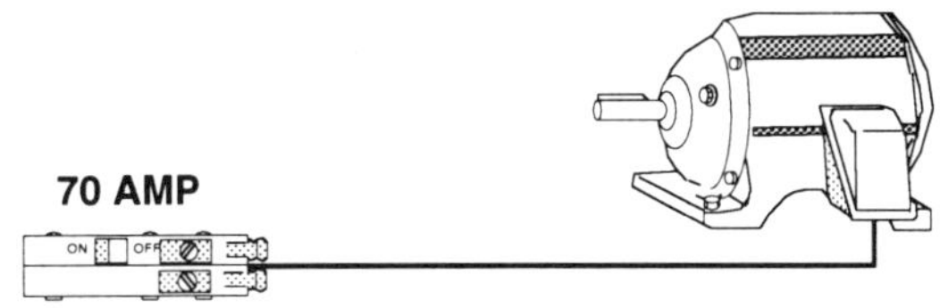

With thermal overload protection the motor is the best protected circuit.

A motor with a full load current of 28 amps would require a #10 THW conductor. The overload protection would be 28 amps x 115% = 32.2. The overload trips out at 32.2 amps, a #10 THW has an ampacity of 35 amps. The 70 amp circuit breaker is to allow the heavy inrush currents when starting and to protect against short circuits and ground faults.

The overload protection is the difference between a motor circuit and a lighting, appliance or receptacle circuit.

Example; the table lamp in the living room of your home generally has a #16 gauge cord wire or smaller. This lamp could be plugged into a 20 amp rated branch circuit.

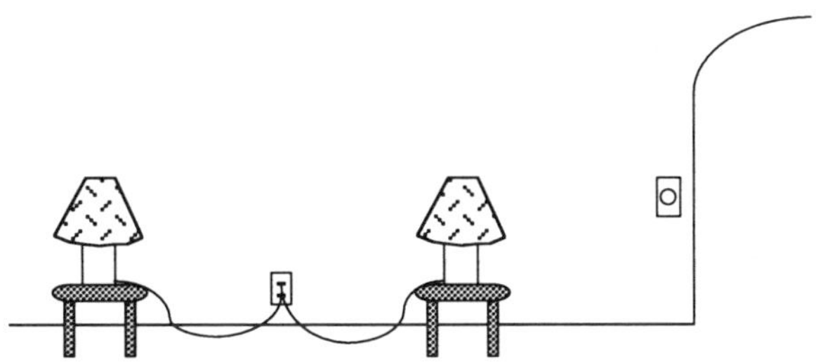

The living room is not a heavy loaded circuit. With the television and a couple of lamps turned on the current flowing would be around 4 amps. The lamp cord would have to increase 18 to 20 amps before the circuit breaker would trip or the fuse would blow. The overcurrent device has to do *both*, protect against short circuits, ground faults *and overloads*.

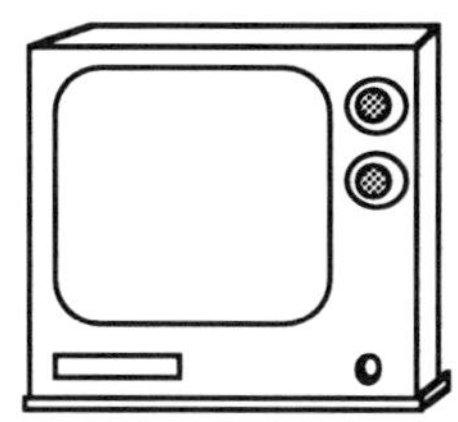

Two functions - overcurrent and overload protection

With a motor circuit, the breaker provides overcurrent protection and the heater provides overload protection.

Overcurrent

Overload

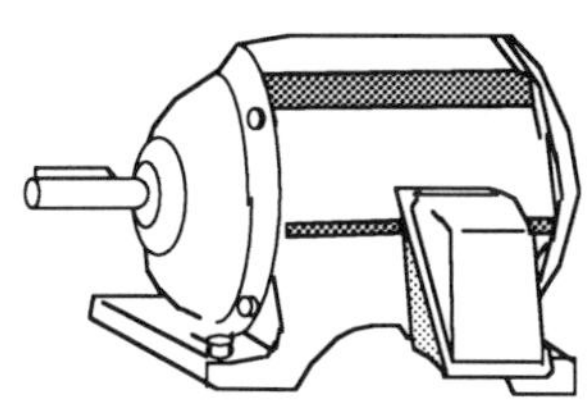

The Code requires the motor to be controlled. By definition, a motor controller is any switch or device normally used to start and stop a motor.

Not only must the controller be capable of starting and stopping the motor it controls, it is also required to be able to interrupt the stalled-rotor current of the motor.

A manual starter is a motor controller whose contact mechanism is operated by a mechanical linkage from a toggle handle or pushbutton, which is operated by hand. A thermal overload relay and *direct acting overload mechanism* provides motor overload protection and will interrupt the line contact. Basically, a manual starter is an "on-off" switch with overload relays.

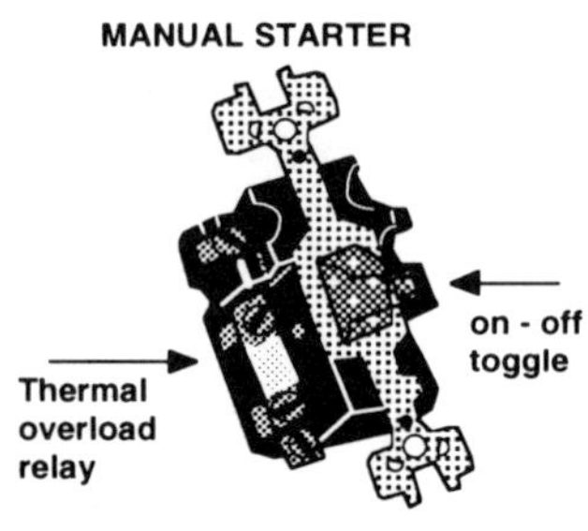

Manual starters usually provide across the line starting. They are generally used on fans, blowers, pumps, compressors, small machine tools, etc.

Manual starters are the lowest in cost of all motor starters. They have a simple mechanism action and provide quiet operation with no A.C. magnet humming noise. By moving a toggle or pushing a "start" button, the contacts close and remain closed until the toggle is moved to "off" or the "stop" button is pushed, or the thermal overload relay trips.

The advantage of using a manual starter for exhaust fans, pumps, compressors, oil burners, etc. is if power fails, the contacts remain closed, and when power is restored the motor will start again without having to manually push the toggle or start button again. This is good to keep in mind with an industrial building that has several fan motors. Manual starters save the maintenance person from having to start each motor in the building when power is restored.

Keep in mind the *restarting* of a motor when power is restored can also be a *disadvantage*. A manual starter should *not* be used where a machine operator may have his fingers or hands involved in the operation of a machine. When power is restored it could cause an accident. The 3-wire magnetic coil starter would be used in this case requiring the operator to push a button to restart the machine.

Shown below is the *tripped* position of a manual starter. The contacts cannot be reclosed until the overload relay has been reset by moving the handle to the full "off " position, after allowing time for the thermal unit to cool.

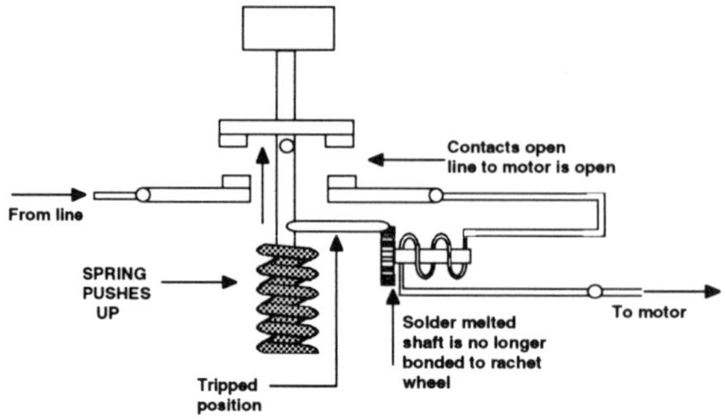

Most applications require the motor controller to be capable of operation from remote locations, or to provide *automatic* operation in response to signals from limit switches, pressure switches, float switches, thermostats, etc. Manual starters cannot provide this type of control, and therefore, *magnetic* motor starters are used.

The operating principle which distinguishes a *magnetic* from a *manual* starter is the use of an electromagnetic coil as used in the relay.

The basic difference between a magnetic motor starter and a relay or contactor is the motor starter provides thermal overload protection.

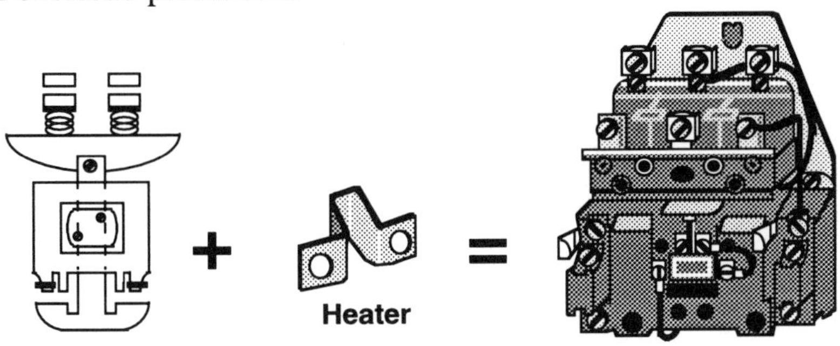

MAGNETIC RELAY **MAGNETIC MOTOR STARTER**

With the manual motor starter, the starter must be mounted so that it is easily accessible to the operator. With a magnetic motor starter, the limit switches, pressure switches, etc. can be mounted anywhere on the machine and connected by the control wiring to the coil of the magnetic motor starter.

The Code requires a thermal overload relay to be installed:

Single-phase motor, 2-wire, ungrounded----------------------one in either conductor

Single-phase motor, 2-wire, one conductor grounded-------one in the *ungrounded* conductor

Three-phase motor, 3-wire--------------------------------------one in *each* phase

On a schematic drawing, the thermal overload (heater) will be shown by the symbol —∞—.

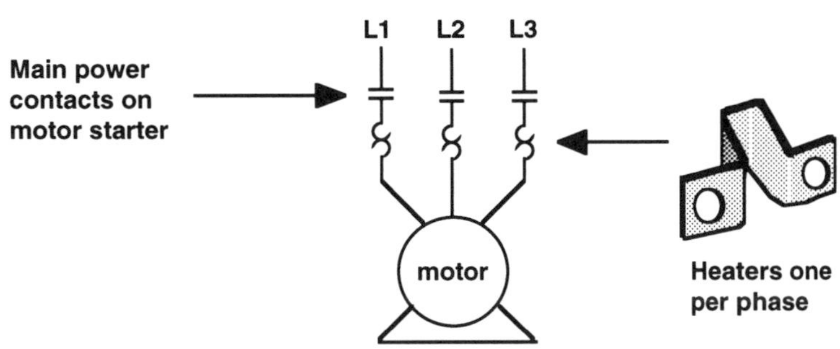

In order to properly read a schematic diagram of a magnetic motor starter, I feel you should understand how a motor starter is constructed and the function of the components.

A three-phase motor starter has 3 power contacts that switch the current to the motor. The magnetic starter contains a coil that when energized, closes the power contacts. The starter contains 3 thermal overloads, one per phase. Also the magnetic starter contains one normally open control contact which is used as a holding circuit for the coil.

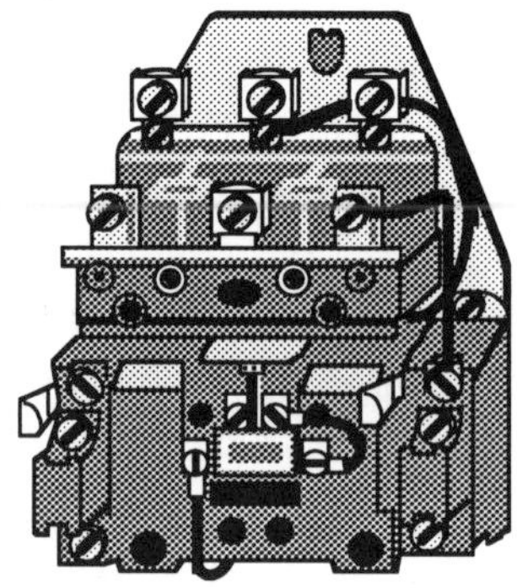

Smaller fractional horsepower motors used a manual starter in which the thermal overloads opened the power to the motor. This is not the case with the magnetic motor starter which normally is used to carry much larger currents. The current is too large for the overload to interrupt the line. A magnetic starter is used with the power contacts interrupting the line. The thermal overloads actually trip a normally closed control interlock that is wired in series with the coil. When an overload occurs, the interlock is opened which de-energizes the coil which opens the power contacts to the motor.

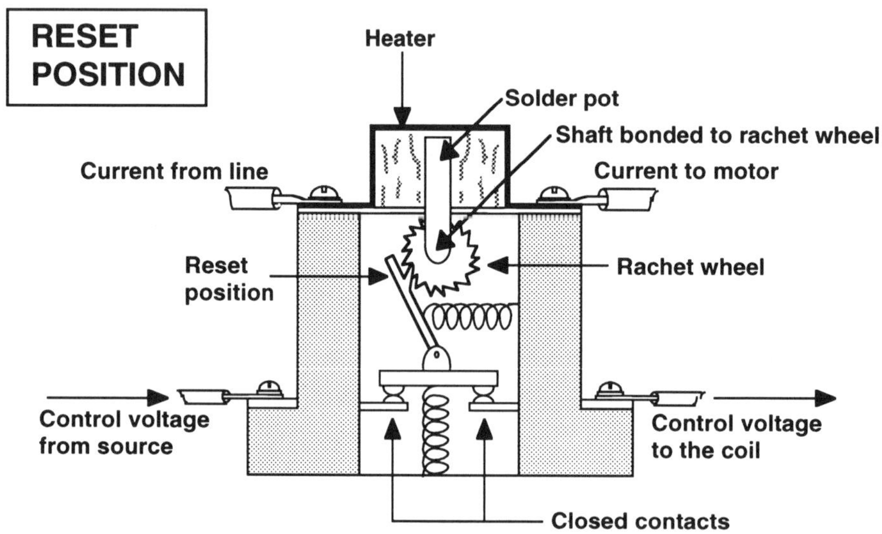

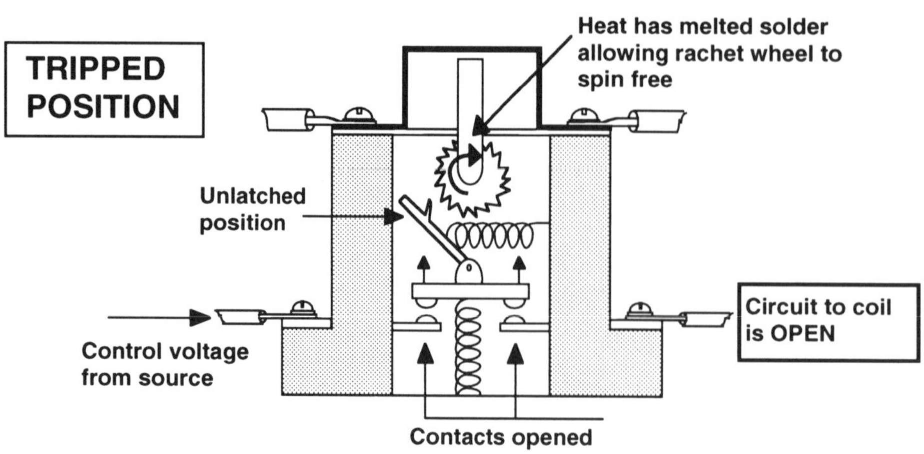

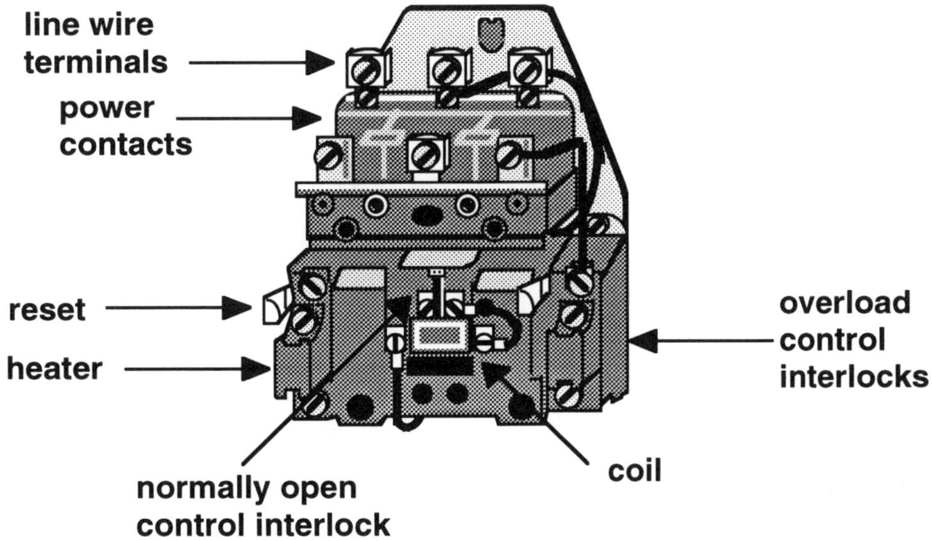

Shown below is a main wiring diagram of a three-phase magnetic motor starter with 3 thermal overloads. The starter is controlled by a start-stop push button control circuit using the holding circuit. The 3 overload interlocks are connected in series with the coil. The 3 thermal overloads are connected in series with the motor. The control wires are shown in a fine line, the power wires in a heavier line.

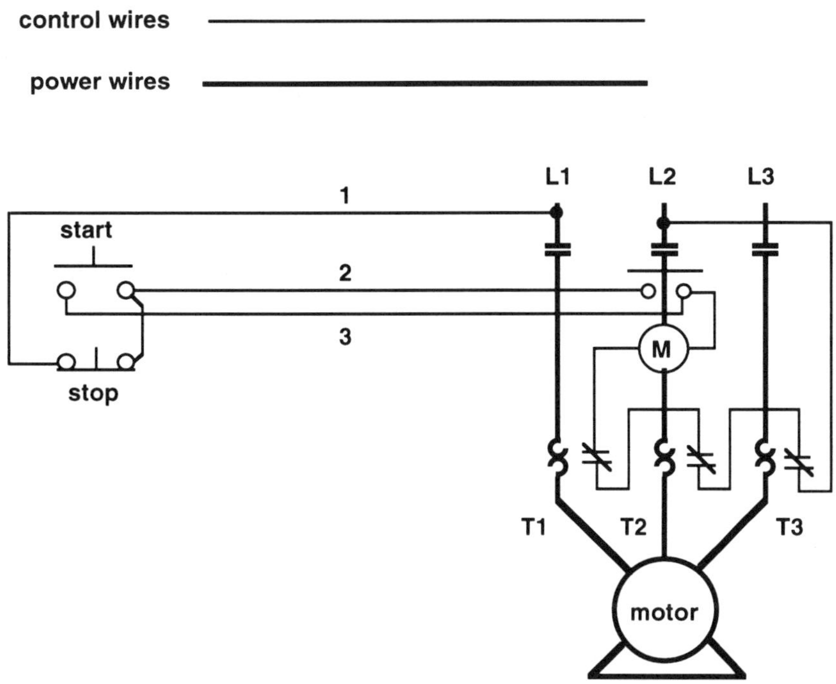

Code section 430-102a states a disconnecting means shall be located in sight from the controller location and shall disconnect the controller.

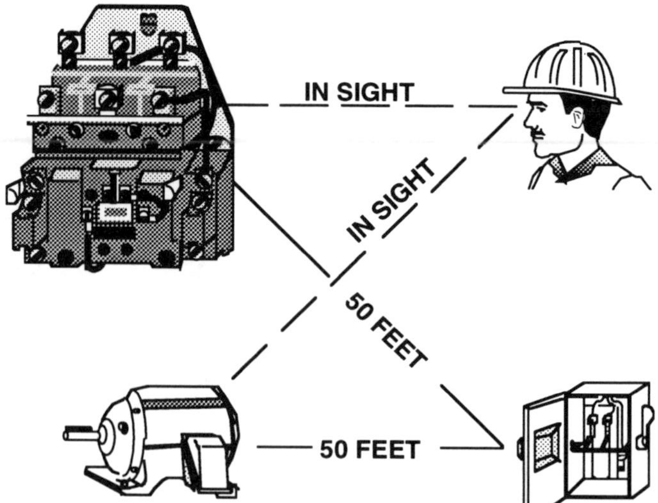

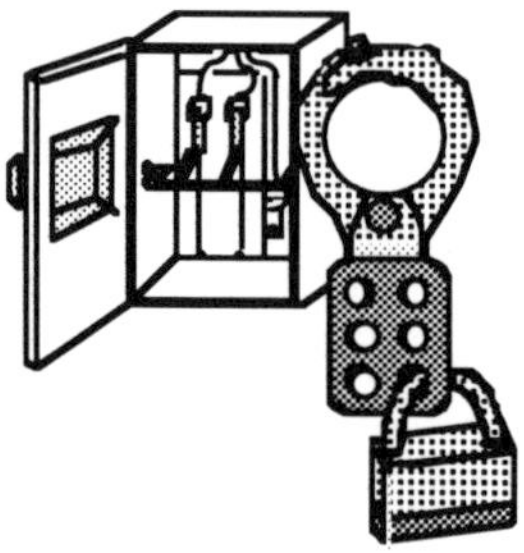

Section 430-102b requires that a disconnecting means be located in sight from the motor location and the driven machinery location. The exception states unless the disconnecting means is capable of being locked in the open position.

The disconnecting means shall be a motor-circuit switch rated in horsepower, a circuit breaker or a molded-case switch.

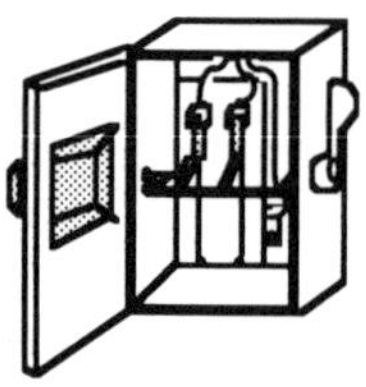

MOTOR DISCONNECT RATED AT LEAST 115% OF THE F.L.C.

Code section 430-110a requires the disconnect for a motor circuit to be *at least* 115% of the F.L.C. of the motor.

The required disconnect switch for this 5 hp motor is calculated 28a x 115% = 32.2 amps.

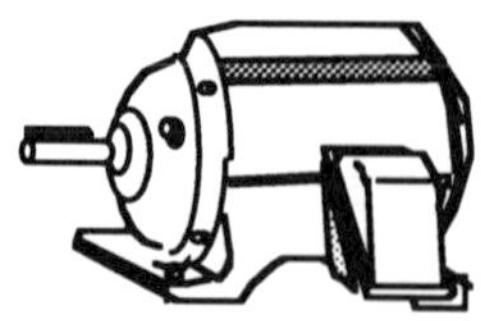

5 hp 230v 1ø
F.L.C. 28 amps

STANDARD DISCONNECT SIZES

30 - 60 - 100
200 - 400 - 600
800 - 1200 amperes

This disconnect switch is required to have a rating of *at least* 32.2 amps. The next standard rating is *60 amps.*

DISCONNECT SWITCH 60 AMP RATING

The Code requires the wires to a motor branch circuit to be 25% larger than the F.L.C. of the motor.

The overload device generally has a rating 25% higher than the F.L.C. of the motor so Code section 430-22 requires the wire to have an ampacity 25% higher.

WIRES TO MOTOR BRANCH CIRCUITS ARE REQUIRED TO HAVE AN AMPACITY 125% x F.L.C.

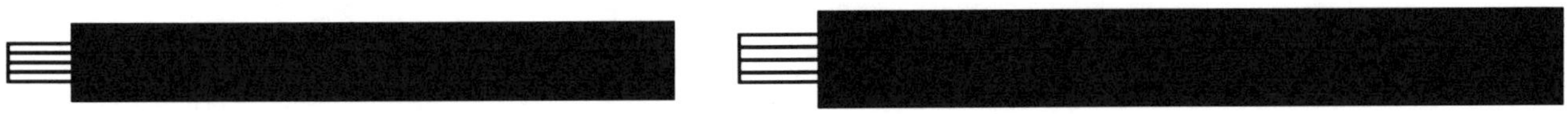

Section 430-22 does not take voltage drop into consideration in the 125% ampacity required for motor branch circuits.

As mentioned throughout this book voltage drop is an important VARIABLE in the proper sizing of wires.

It is not good design to permit voltage drop in excess of 2% in a motor branch circuit.

Voltage drop is merely wasted electricity by unnecessarily heating the wire which can become very costly over a period of time.

Excessive voltage drop means the motor will not develop full power, can stall under heavy loads, or slow to reach full running speed.

If you have designed your motor branch circuit to the Code allowed 3% drop, just remember, when the motor is trying to start and the motor consumes 6 times its normal current, the 3% drop in the circuit can become 18%. When starting heavier type loads this voltage drop can stall out the motor on starting.

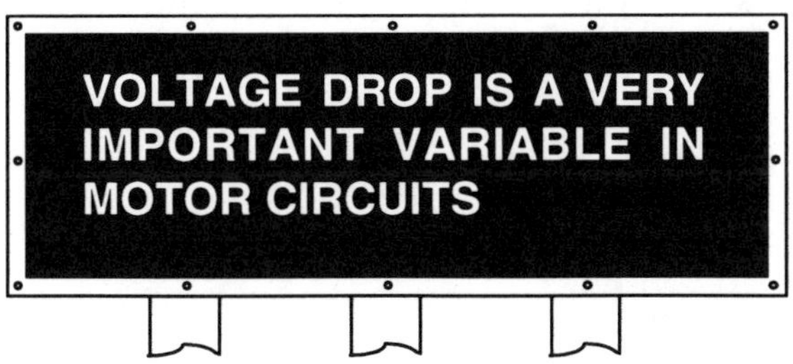

When selecting the conductor for a motor circuit, I have seen confusion when reading the footnotes at the bottom of Table 310-16, the ampacity table.

† *Unless otherwise specifically permitted elsewhere in this Code*, the *overcurrent protection* for conductor types marked with an obelisk (†) shall not exceed 15 amperes for #14 AWG, 20 amperes for #12 AWG, and 30 amperes for #10 AWG copper; or 15 amperes for #12 AWG and 25 amperes for #10 AWG aluminum and copper-clad aluminum after any correction factors for ambient temperature and number of conductors have been applied.

This footnote does *not* say ampacity, it says overcurrent protection.

OVERCURRENT PROTECTION =

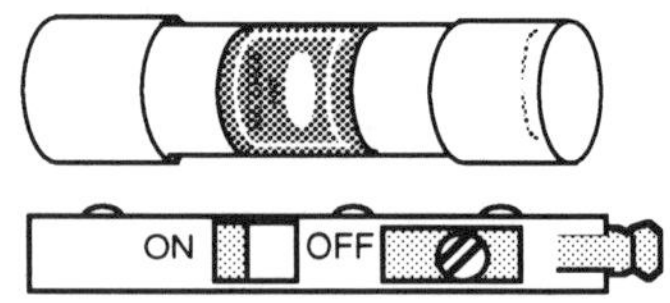

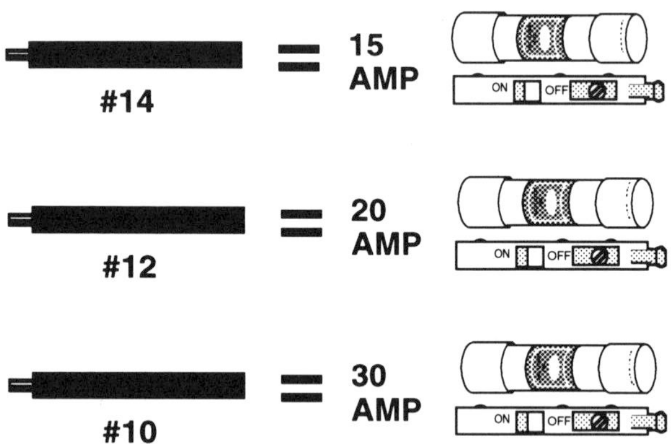

The † applies *unless otherwise specified in the Code.*

240-3f is where it is otherwise specified for motors, 240-3h is for air-conditioners.

Example: A 5 hp motor with a F.L.C. of 28 amps x 125% = 35 required ampacity. A #10 THW has an ampacity of 35. The breaker size would calculate 28a x 250% = 70 amp breaker.

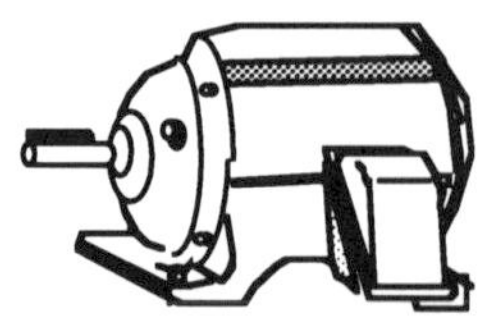

5 hp 230v 1ø
F.L.C. 28 amps

If you followed the † in the footnotes you are only allowed a 30 amp breaker on a #10 wire. A 30 amp breaker would not allow a motor with a full load current of 28 amps to start. This is the reason 240-3f was put into the Code. The difference between a motor circuit and another circuit such as a lighting circuit, is the motor circuit has a thermal overload. The overload not only protects the motor but it also protects the wire. You *can't* apply the † to a motor circuit. The motor would trip the overcurrent device on start.

† THE OBELISK APPLIES TO THE FUSE OR BREAKER, BUT NOT TO MOTOR CIRCUITS

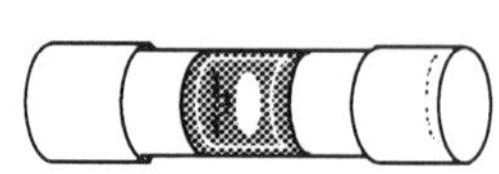

VOLTAGE DROP - MOTORS

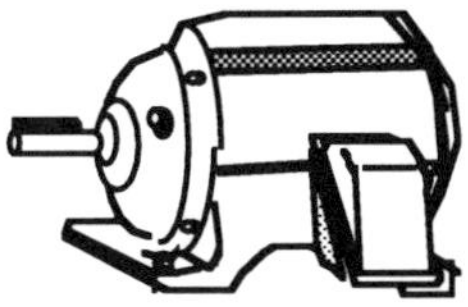

5 hp 230v 1ø
F.L.C. 28 amps

The branch circuit wire size is 28a x 125% = 35 required ampacity. Table 310-16 requires a #8 TW.

The Code recommendation is to limit the voltage drop on a branch circuit to not more than 3%. For a 230 volt branch circuit the allowed voltage drop would be 230v x 3% = 6.9 volts.

The distance this branch circuit could be run would be: $\frac{16510\text{cm} \times 6.9\text{vd}}{2 \times 12.84478 \times 28\text{a}} = 158'$

Watt loss = VD x I   6.9 vd x 28a = 193 watts

Watt loss cost (based on 12 hr. day, 8¢ per kwh) $\frac{193\text{w} \times 12 \text{ hrs.} \times .08}{1000}$ = 18.5¢ per day

18.5¢ per day x 30 day month = $5.55 per month wasted electricity.

The branch circuit looks like this:

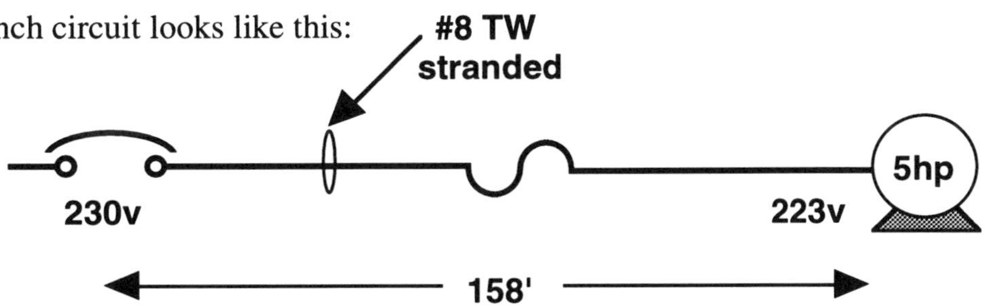

The approximate 7 volts dropping in the wire is a result of the power consumed in the heating of the wires because of the wire's resistance to the flow of current. This means there is *less* power to operate the motor.

Voltage drop not only leads to wasted power, but also seriously affects the operation of the equipment. A motor operated at 95% of its rated voltage will only deliver about 90% as much power as it would have at its rated 230 volts.

I^2R is heat. The resistance of 1000 feet of #8 stranded wire from Table 8 is .778Ω. At 158 feet distance x 2 (down and back) is 316 feet of wire x .778Ω = .245848Ω.

I^2R = 28a x 28a x .245848Ω = 192 watts of heat.

Suppose the full load current was doubled to 56 amps using the same #8 wire at the same voltage.

VD = I x R, 56a x .245848Ω = 13.76 volts. The voltage drop is *doubled* from the 6.9 volts. But, the power loss is 56a x 56a x .245828Ω = 770 watts, *four* times more than the 192 watts.

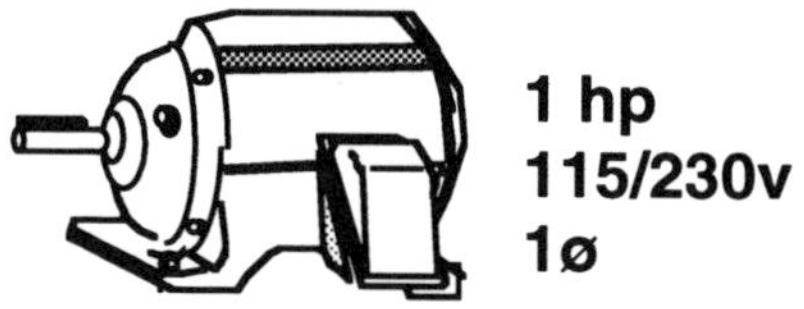

A single-phase 1 hp squirrel cage motor can be connected either 115 volts or 230 volts.

The F.L.C. is 16 amps @ 115v and 8 amps @ 230v. The branch circuit wire size would be calculated either 16a x 125% = 20 required ampacity or 8a x 125% = 10 required ampacity. In either case a #14 TW with an ampacity of 20 could be used.

At 115 volts the distance the branch circuit could be run is:

$$\text{Distance} = \frac{4110\text{cm} \times 3.45\text{vd}}{2 \times 12.9054 \times 16\text{a}} = \mathit{34.3\ feet}$$

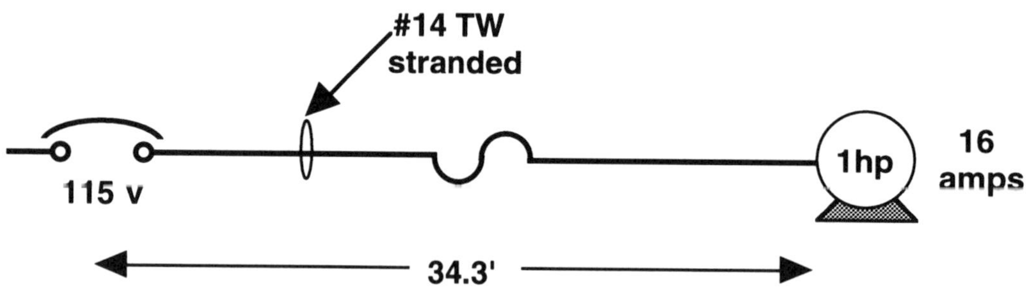

The inverse-time circuit breaker would calculate 16a x 250% = 40 amp circuit breaker.

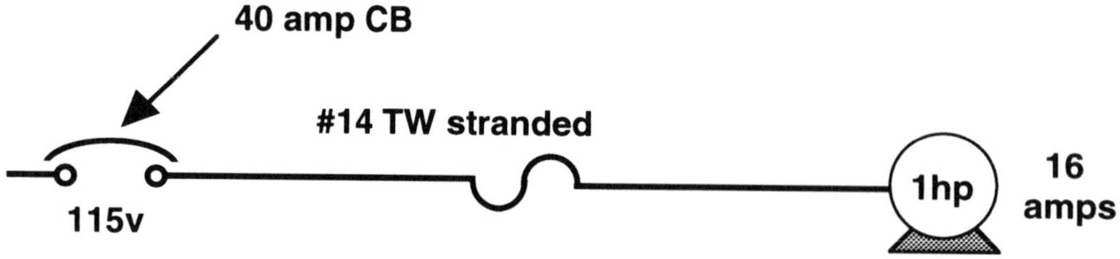

Power loss = 16a x 3.45vd = 55.2 watts

Cost of power loss = $\frac{55.2\text{w x 12 hr day x .08¢ kwh}}{1000}$ = .05¢ per day x 30 days = *$1.50* per month

If the wire size is increased we can reduce the voltage drop and cut the wasted power loss. Since the branch circuit has a 40 amp circuit breaker, let's calculate a #4 conductor.

VD = $\frac{2 \text{ x } 12.85592 \text{ x } 34.3' \text{ x } 16\text{a}}{41740\text{cm}}$ = .338 volts dropped

Power loss = 16a x .338vd = 5.4 watts

Cost of power loss = $\frac{5.4\text{w x 12 hr day x .08¢ kwh}}{1000}$ = .005¢ per day x 30 days = *.15¢* per month

By installing a #4 conductor rather than the required #14 conductor to this 1 hp motor branch circuit, *$1.35* per month is saved in wasted electricity. Also the motor will perform more efficiently and deliver more power.

If the conductor installed was a #14, as permitted by the Code, but the voltage is *230* rather than 115.

At 230 volts the distance the branch circuit could be run is:

Distance = $\frac{4110\text{cm x } 6.9\text{vd}}{2 \text{ x } 12.9054 \text{ x } 8\text{a}}$ = *137.3 feet*

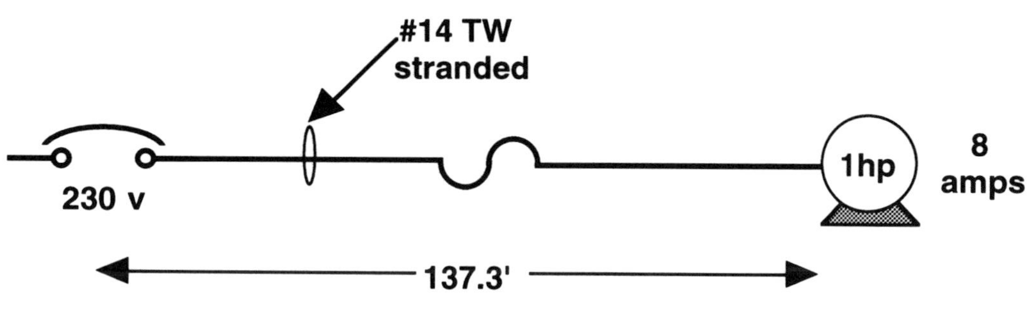

The #14 conductor can be run *four* times farther at 230 volts than at 115 volts.

Now the power loss cost is raised back to the \$1.50 per month.

Power loss = 8a x 6.9vd = 55.2 watts

Cost of power loss = $\frac{\text{55.2w x 12 hr day x .08¢ kwh}}{1000}$ = .05¢ per day x 30 days = *\$1.50* per month

By using a #4 conductor instead of a #14 and doubling the voltage from 115 to 230v we can run the branch circuit *four* times farther with the same power loss cost.

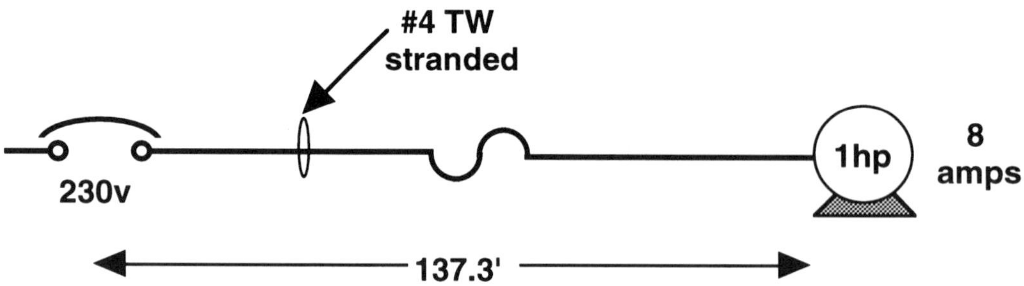

VD = $\frac{\text{2 x 12.85592 x 137.3' x 8a}}{\text{41740cm}}$ = .676 volts dropped

Power loss = 8a x .676vd = 5.4 watts

Cost of power loss = $\frac{\text{5.4w x 12 hr day x .08¢ kwh}}{1000}$ = .005¢ per day x 30 days = *.15¢* per month

The general rule for a single motor is to multiply the full-load current x 125% and size the branch circuit conductors.

There is an exception to the general rule that applies to conductors to a motor used for short-time, intermittent, periodic, or varying duty. The required ampacity for these motors is determined from Table 430-22a ex.

Any motor application shall be considered as *continuous duty* unless the nature of the apparatus it drives is such that the motor will not operate continuously with load under any condition of use.

The definition of *continuous duty* is operation at a substantially constant load for an indefinitely long time.

Table 430-22a ex. lists 5-minute, 15-minute, 30 & 60 minute, and continuous rated motors. The classification of service is short-time duty, intermittent duty, periodic duty and varying duty.

Short-time duty is operation at a substantially constant load for a short and definitely specified time. Examples are operating valves, raising or lowering rolls, etc.

Intermittent duty is operation for alternate intervals of (1) load and no load; or (2) load and rest; or (3) load, no load, and rest. Examples are freight and passenger elevators, tool heads, pumps, drawbridges, turntables, etc.

Periodic duty is intermittent operation in which the load conditions are regularly recurrent. Examples are rolls, ore and coal handling machines, etc.

Varying duty is operation at loads, and for intervals of time, both of which may be subject to wide variation.

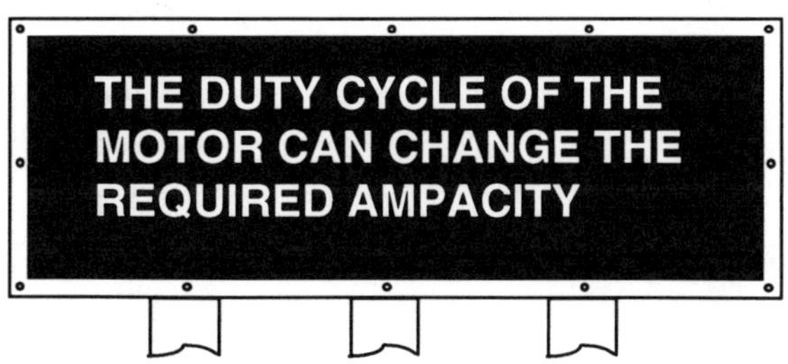

DUTY-CYCLE SERVICE

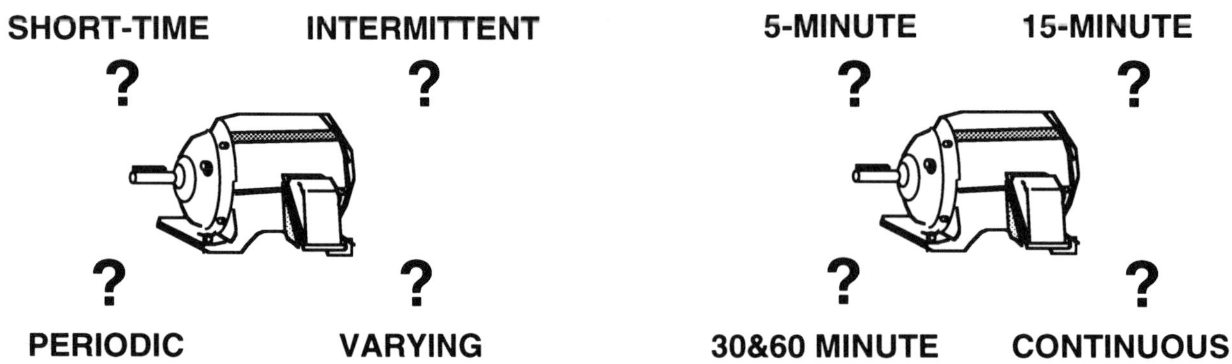

An elevator motor is classified intermittent, Table 430-22a ex. requires for a 5 or 15 minute rated motor for the ampacity of the conductors supplying the motor to be at least 85% of the motor nameplate current. For a 30 and 60 minute rated motor the ampacity of the motor conductors are required to be 90% of the motor nameplate rating. Elevators with continuous rated motors, the ampacity of the conductors is required to be 140% of the motor nameplate rating.

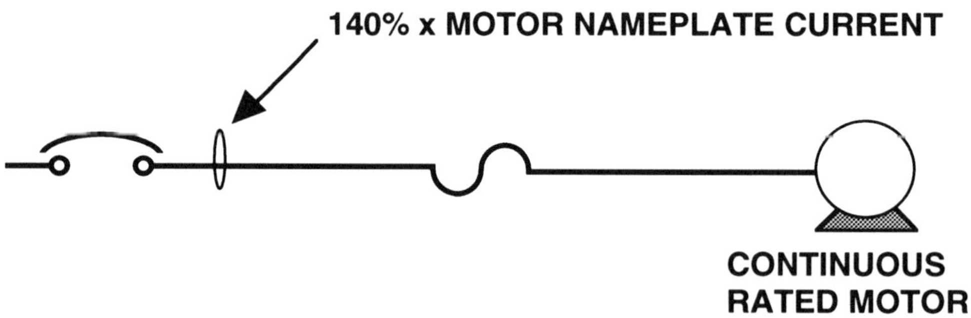

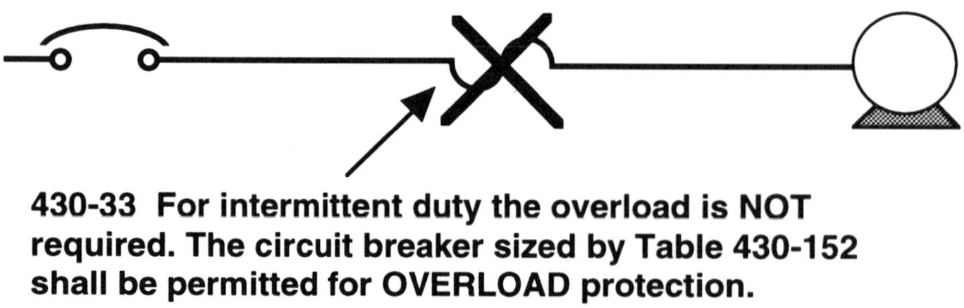

Ampacities for conductors used with short-time rated crane and hoist motors are found in Table 610-14a.

LIFT STATION PUMP

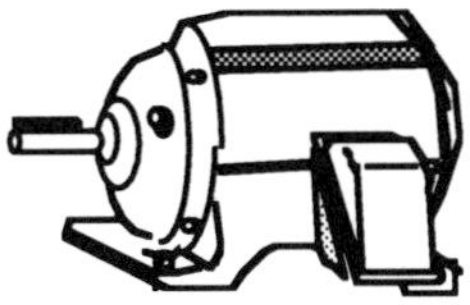

10 hp
208v
3ø

Lift station pumps in sewage treatment plants play important roles. They run when liquid levels reach specified levels, determined by a float switch, and tend to continue to run for long periods of time. The motors are rated for continuous duty.

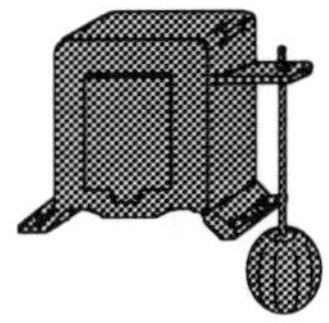

Code section 430-22a ex.1 states branch circuit conductors supplying a single motor used for short-time, intermittent, periodic, or varying duty shall have an ampacity not less than the percentages shown in Table 430-22(a), Exception.

Definitions Article 100 lists the definitions of short-time, intermittent, periodic, and varying duty. Table 430-22(a), Exception lists a pump as *intermittent duty*, and for a continuous motor the percentage is 140%.

Table 430-150 F.L.C. 30.8 x 140% = 43.12 required ampacity. Table 310-16 requires a minimum #6 TW conductor.

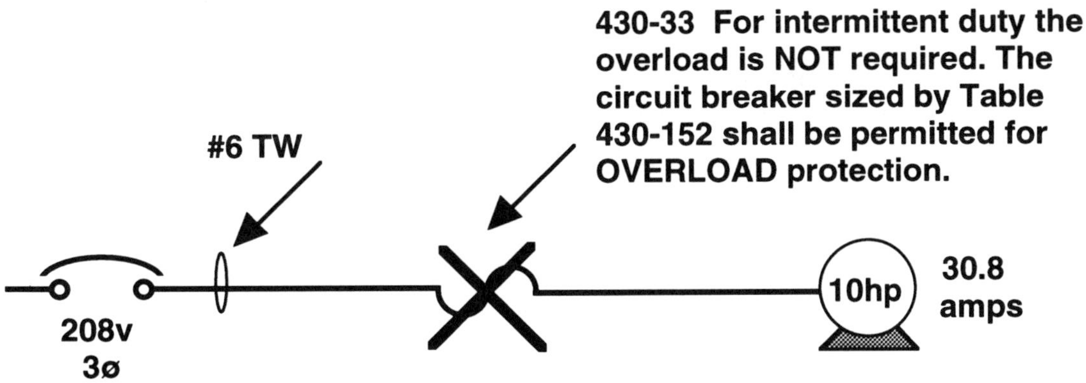

WELL PUMP BRANCH CIRCUIT

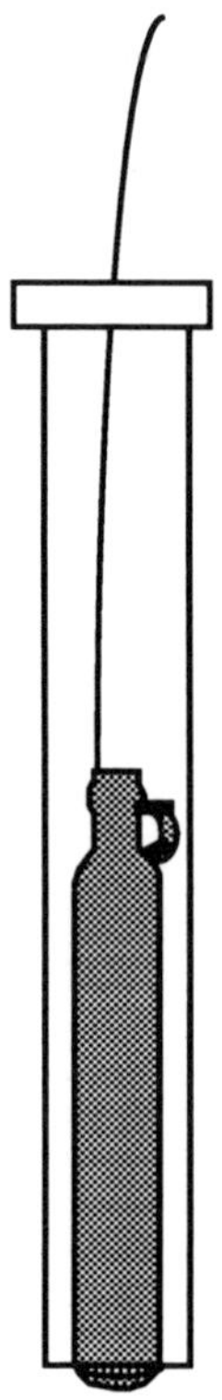

Submersible pumps can deliver more consistent water pressure than a jet pump and the electrical power consumption can run 50-75% less than a jet pump for the same amount of water delivered.

Three-wire pumps have the advantage of high *starting* torque motors. A high starting torque is sometimes needed to break free mineral deposits that may adhere to the pump's moving parts. Two-wire pumps have higher *running* torque and are more electrically efficient. Two-wire pumps should *not* be used where abrasives or heavy water scale is present.

Two-wire pumps have "in the motor" controls and wire directly to the water system pressure switch or power supply.

Three-wire pumps require an *above ground control box* between the pressure switch and the pump motor. The above ground control box includes the motor protector, current sensors, capacitors, and the start and run winding overloads. Most models are capacitor-start, capacitor-run for greater electrical efficiency and longer motor/pump life.

When sizing the conductors for the branch circuit, distance becomes an important factor in the design as the depth of the pump will vary. The NEC recommendation is a maximum of 3% voltage drop on a *branch circuit.*

The Code recommends a branch-circuit voltage drop of not more than 3% of the source voltage and a total voltage drop not to exceed 5%.

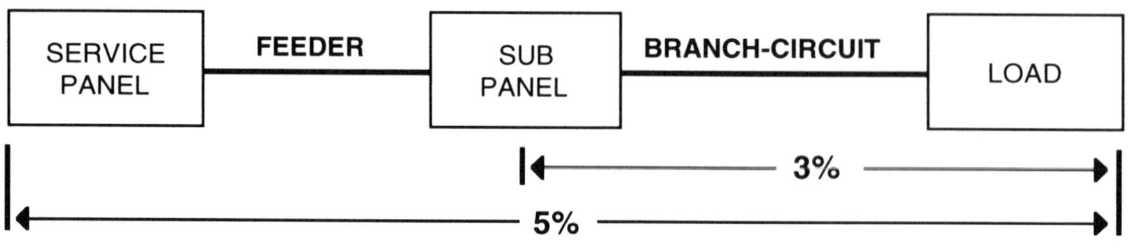

•Some manufacturer's provide tables that list the cable size for distances from the service to the pump. Some of these cable selection tables are based on 5% voltage drop which is permitted on the entire service, but not on a branch circuit.

BRANCH CIRCUIT SIZING INDUSTRIAL MACHINERY

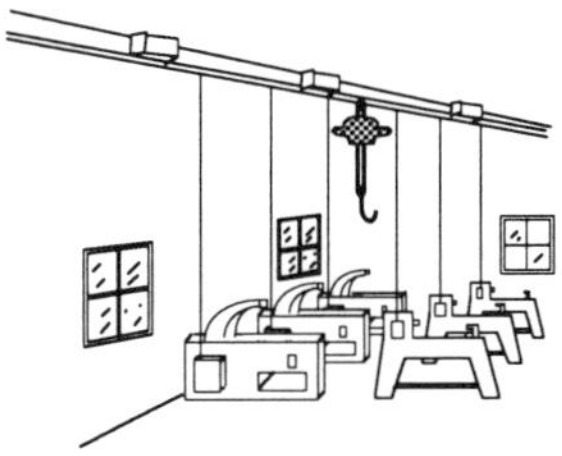

Article 670 of the Code covers industrial machinery. These machines or tools are *not portable* by hand and are used to shape or form metal or plastic. These machines are considered to be *individual* loads even though the machine may employ several motors or heaters.

An example is an industrial crane or hoist: An overhead crane generally has three motors: (1) a hoist motor for up and down (2) a motor to move the hook from side to side (3) a bridge motor to move the crane from end to end.

Code section 670-3 requires the machine to have a permanent nameplate attached to the machine where plainly visible.

VOLTAGE - PHASE - FREQUENCY
FULL LOAD CURRENT - MAXIMUM
FUSE OR BREAKER - AMPERE
RATING OF LOAD - INTERRUPTING
CAPACITY OF FUSE OR BREAKER

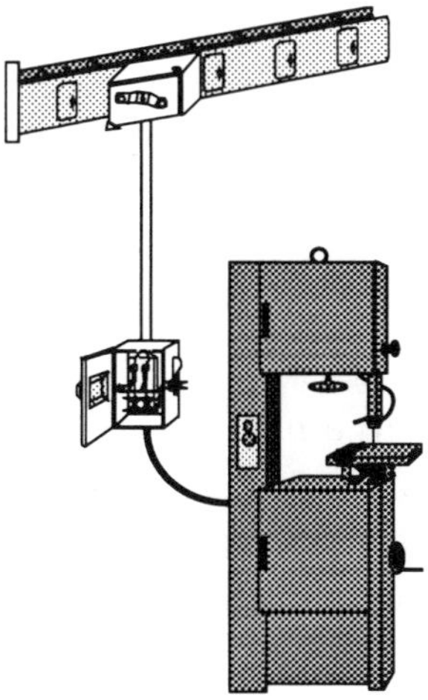

A machine shall be considered as an *individual* unit and have a disconnecting means. The disconnect does not have to have overcurrent protection.

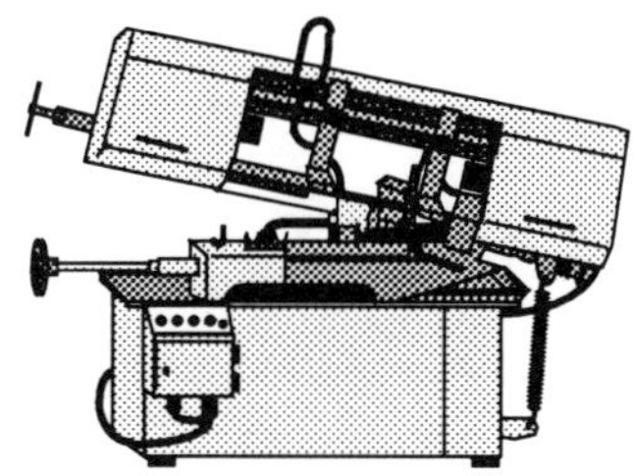

Overcurrent protection shall consist of a single circuit breaker or set of fuses.

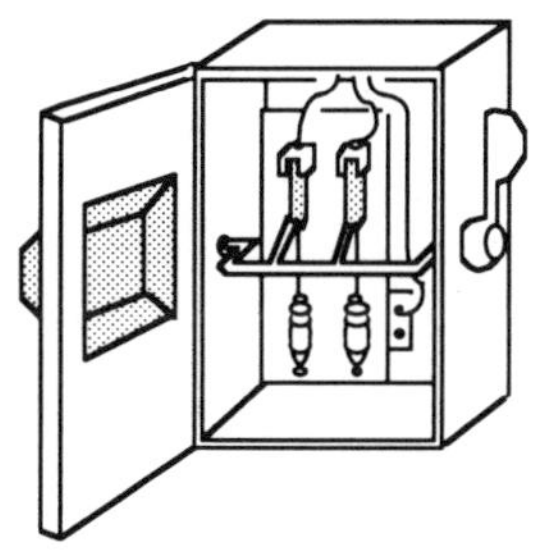

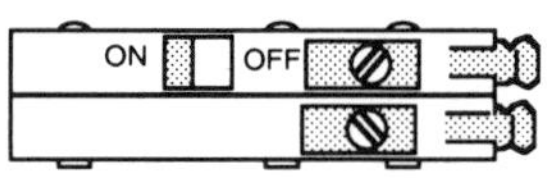

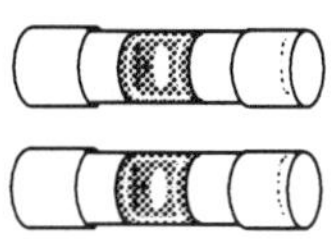

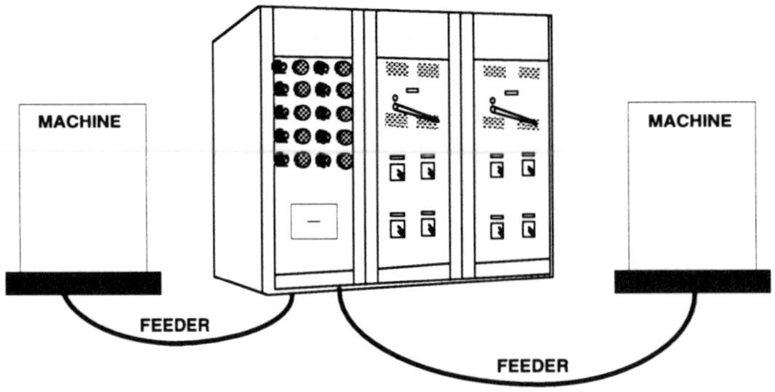

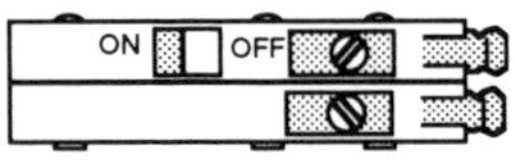

The rating of the overcurrent device for the circuit supplying the machine (feeder or tap) shall not be greater than the largest branch circuit overcurrent device provided with the machine plus 125% of the full load current of the resistive heat loads plus the sum of the full load currents of all other motors and apparatus that may be in operation at the same time.

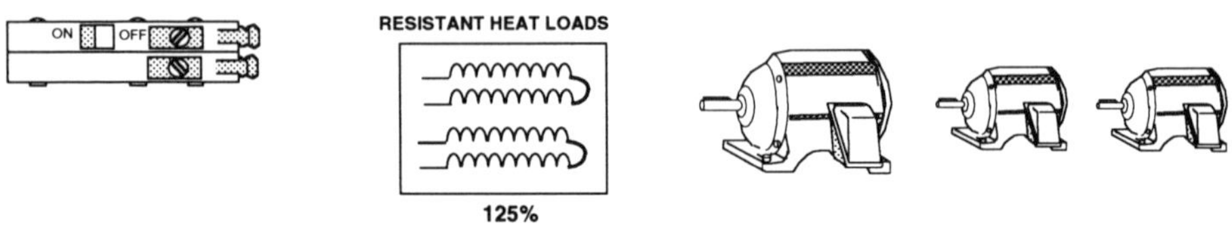

Where no branch circuit overcurrent device is provided with the machine, the rating of the overcurrent device shall be based on 430-52 and 430-53 as applicable.

The supply conductors shall have an ampacity not less than 125% of the full load current of all resistant heat loads plus 125% of the full load current of the highest rated motor plus 100% full load current of all other motors and apparatus that may be in operation at the same time.

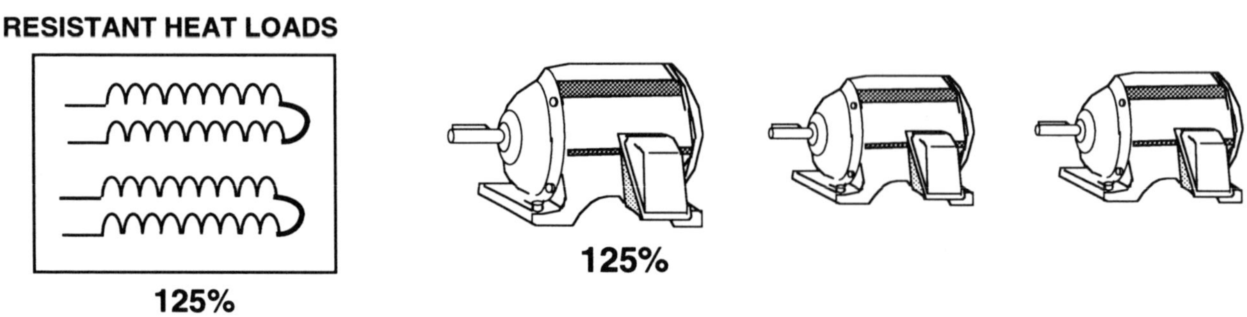

COST COMPARISON

Often the designer must determine the most economical circuit makeup which will meet the minimum standards required by the Code.

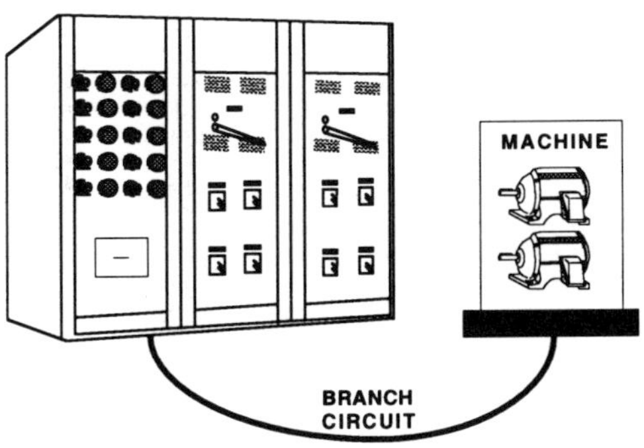

Example: A machine has two 15 hp, 230v, 3ø squirrel-cage induction motors. Each motor is fed by a *separate* branch circuit originating at a combination motor controller in a motor control center. Rigid steel conduit is to be used, assume ambient not over 86°F, and voltage drop is limited to 3%. 60°C equipment.

To calculate the load current, Table 430-150 shows the full load current for each motor is 42 amps. Section 430-22 requires each circuit to have an ampacity of not less than 125% x 42a = 53a.

There are various circuit arrangements that could be used to supply the two motors in the machine.

I. Two separate conduit runs to the machine. From the motor control center, the branch circuits to the two motors on the machine could be carried in separate conduits. This would require three #6 TW wires which have a 55 ampacity. Table C8 would require a 3/4" conduit for each branch circuit.

3/4" **3/4"** **#6 TW or THW INSULATION**

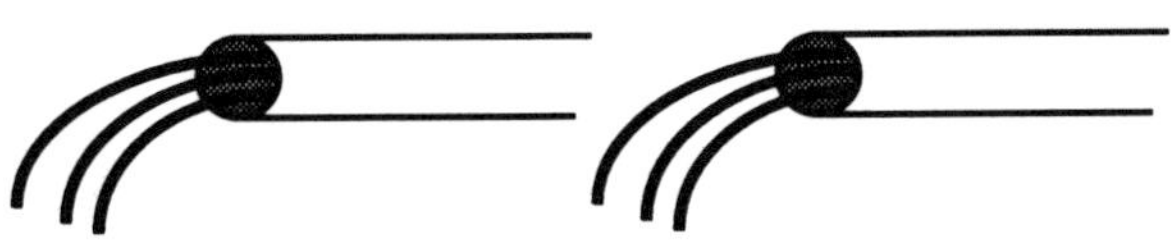

Or, two runs of 3/4" conduit using #6 THW which have a 65 ampacity.

By using two runs of #6 THHN with a 70 ampacity, Table C8 also allows a 3/4" conduit for each circuit. •The THHN must be used at the 60°C ampacity per Code section 110-14c.

3/4" **3/4"** **#6 THHN INSULATION**

II. Single conduit run to the machine. Both branch circuits to the two motors on the machine could be carried in a single conduit. The *variable* for more than three current carrying conductors would affect the wire size in this case.

Using TW insulation, the normal ampacity of 55a x 80% (Note 8) = 44 ampacity. The #6 TW would not meet the 53 amp required ampacity for the motor with six wires in a single conduit.

The next larger size is a #4 TW with a normal ampacity of 70a x 80% (Note 8) = 56 ampacity. A #4 TW would satisfy the Note 8 requirements of the Code.

Table C8 shows that six #4 TW wires require a minimum conduit size of 1 1/4".

Using a THW instead of TW insulation would also result in using a #4 with a 1 1/4" conduit.

1 1/4" #4 TW or THW INSULATION

Using THHN insulation, each #6 THHN conductor has a normal ampacity of 70a x 80% (Note 8) = 56 ampacity, which more than meets the 53 amp required ampacity.

110-14c states the 90°C THHN conductors cannot be loaded to more than the 60°C TW ampacity, which is 55 amps. The full load current is 42 amps.

With a 53 amp required ampacity the #6 THHN wires are being used within the 55 ampacity rating of a #6 TW and are therefore, suitable for connection to the terminals of the motor starter and other circuit components, which are recognized by UL for use with conductors rated not over 60°C.

Table C8 shows that six #6 THHN wires require a minimum conduit size of 1".

1" #6 THHN INSULATION

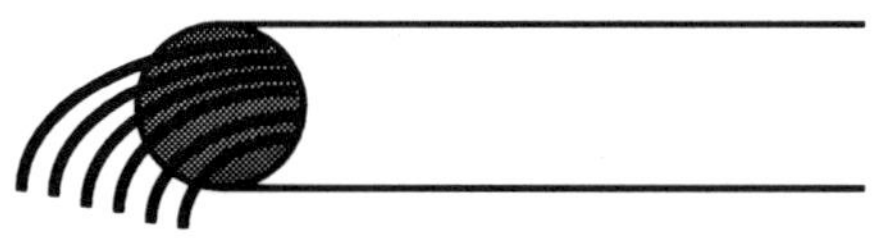

Now calculate the #6 THHN for the 3% limit on voltage drop.

VD = I x R x 1.732 Table 8 resistance is .491Ω per k/ft. x 50 feet distance = .491 x .050' = .02455Ω

42 amps x .02455Ω x 1.732 = 1.03 vd The Code permits 230v x 3% = 6.9 vd

Conclusions

After the material and labor costs have been compared, these comparisons show that the installation on THHN wires is less costly than installation of TW or THW, whether two conduits are run to the machine or whether both branch circuits are run in a single conduit. The installation of six THHN wires in a single 1" conduit is actually less costly than three THHN wires in each of the two 3/4" conduits.

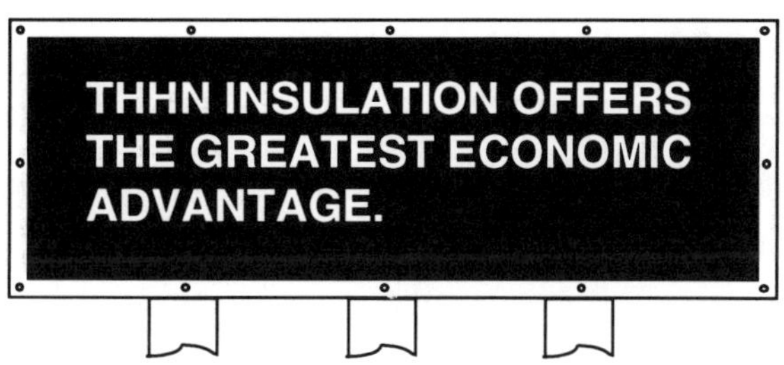

BRANCH CIRCUIT SIZING WELDER

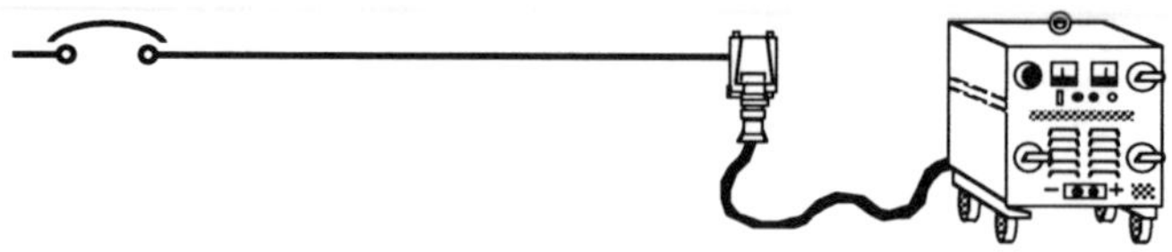

There are three types of welders and each has a duty cycle.

(1) AC Transformer and DC Rectifier Arc
(2) Motor - Generator Arc
(3) Resistance Welders

A duty cycle is the percentage of time the welder is actually drawing a current when it is welding.

Shown below and on the next page are three different types of welders, each with a different nameplate.

SINGLE AC TRANSFORMER DC RECTIFIER TYPE

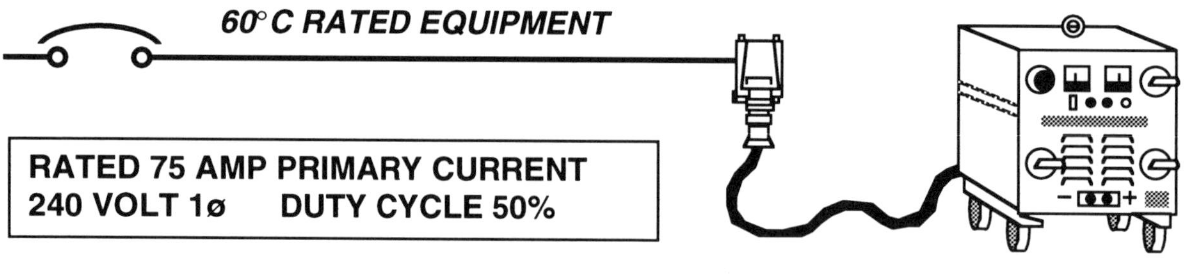

Wire size: 630-11a lists a multiplier of .71 for a 50% duty cycle. The rated primary current 75a x .71 = 53.25 required ampacity. Table 310-16 shows a #6 THHN has an ampacity of 75 amps, but connected to 60°C equipment a #6 THHN can be loaded to 55 amps. The required ampacity is 53.25 amps so a #6 THHN would meet the Code requirements.

Circuit breaker size: 630-12a states each welder shall have overcurrent protection rated or set at not more than 200% of the rated primary current of the welder.

75a x 200% = 150 amp circuit breaker

SINGLE MOTOR-GENERATOR ARC TYPE

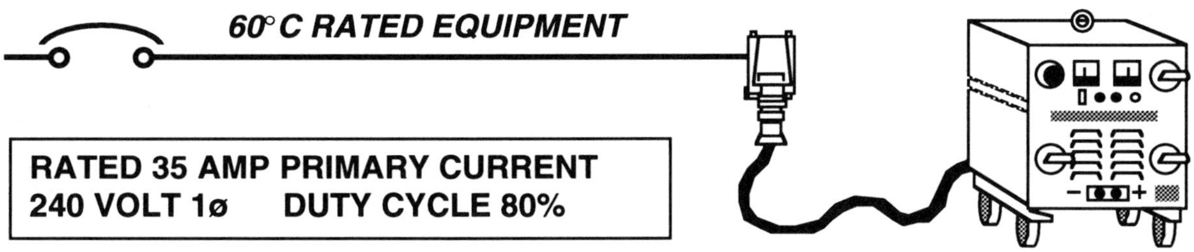

Wire size: 630-21a lists a multiplier of .91 for a 80% duty cycle. The rated primary current 35a x .91 = 31.85 required ampacity. Table 310-16 shows a #8 THHN has an ampacity of 55 amps, but connected to 60°C equipment a #8 THHN can be loaded to 40 amps. The required ampacity is 31.85 amps so a #8 THHN would meet the Code requirements.

Circuit breaker size: 630-22a states each welder shall have overcurrent protection rated or set at not more than 200% of the rated primary current of the welder.

35a x 200% = 70 amp circuit breaker

SINGLE RESISTANCE WELDER TYPE

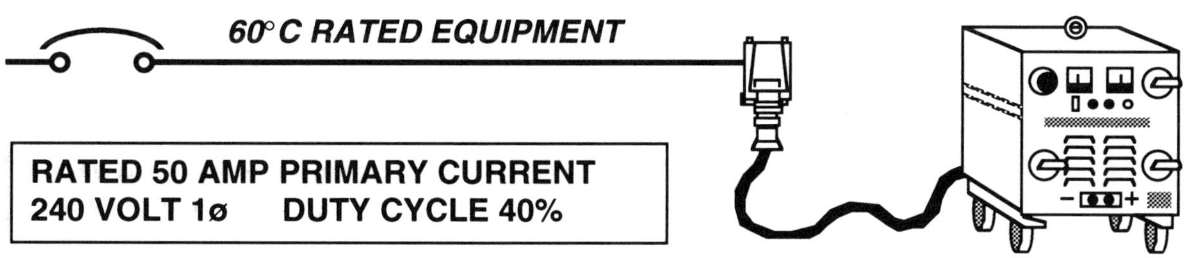

Wire size: 630-31a2 lists a multiplier of .63 for a 40% duty cycle. The rated primary current 50a x .63 = 31.5 required ampacity. Table 310-16 shows a #8 THHN has an ampacity of 55 amps, but connected to 60°C equipment a #8 THHN can be loaded to 40 amps. The required ampacity is 31.5 amps so a #8 THHN would meet the Code requirements.

Circuit breaker size: 630-32a states each welder shall have overcurrent protection rated or set at not more than 300% of the rated primary current of the welder.

50a x 300% = 150 amp circuit breaker

Designing the branch circuit to a group of welders.

ALL WELDERS ARE MOTOR-GENERATOR ARC TYPE

75° C RATED EQUIPMENT

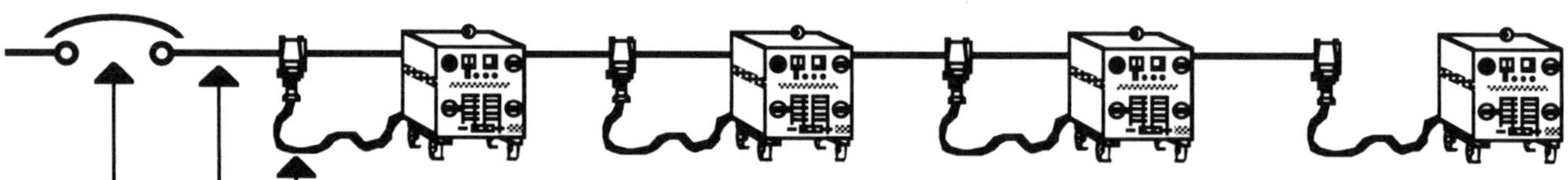

ALL THE WELDERS HAVE THE SAME NAMEPLATE

RATED 75 AMP PRIMARY CURRENT
240 VOLT 1ø DUTY CYCLE 50%

630-21a. **75a x .75 = 56.25 required ampacity = #6 THHN**

630-21b FPN. **Welder #1 = 75a x .75 x 100% = 56.25a**
Welder #2 = 75a x .75 x 100% = 56.25a
Welder #3 = 75a x .75 x 85% = 47.81a
Welder #4 = 75a x .75 x 70% = 39.38a
199.69a
199.69 required ampacity = #3/0 THHN

630-22b. **#3/0 THHN conductor rating @ 75°C = 200 amps**
200 amps x 200% = 400 amp circuit breaker

BRANCH CIRCUIT TAPS

The general Code rules 210-19 and 240-3 require the use of an overcurrent device at the point of which a conductor receives its supply. Exceptions to this rule are made in the case of taps.

GENERAL RULE

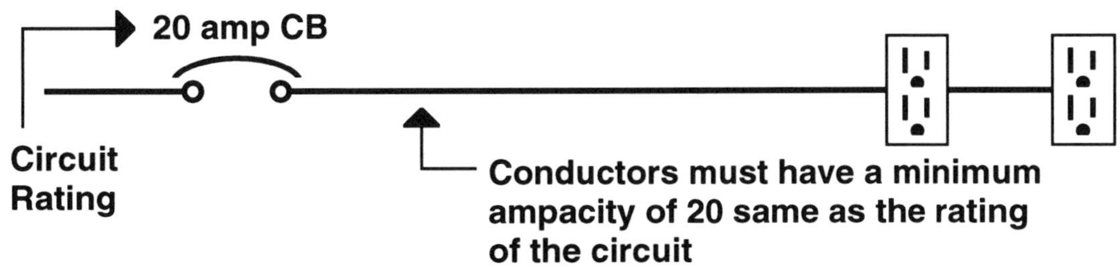

The general rule is when two different sizes of wire are connected together, the rating of the overcurrent device cannot be greater than what is required for the smallest wire as shown below.

GENERAL RULE

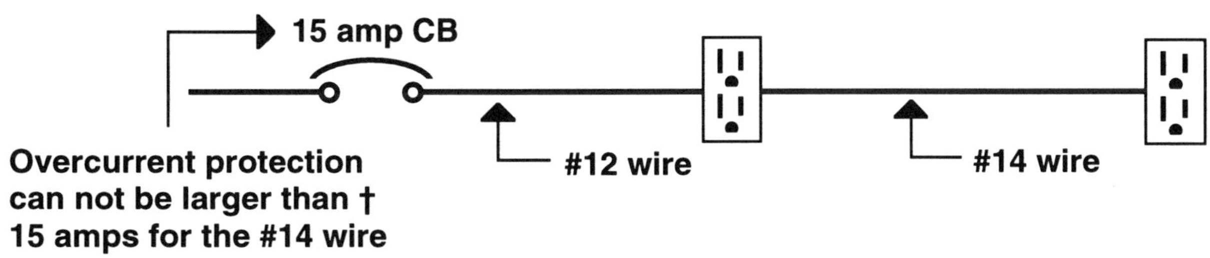

There are a number of exceptions to this general rule. Often it is necessary to connect a wire to another continuous wire. This is called a tap connection.

In some cases it's not necessary to run wires of the same size as the branch circuit wires to individual loads on the circuit. A good example of this exception to the general rule is the table lamp in the living room. The cord to the lamp may be a #18 wire plugged into a 20 amp branch circuit of #12 wires.

210-19b ex.1 permits branch circuit tap conductors supplying electric ranges, wall-mounted ovens, and counter-mounted electric cooking units supplied from a 50 amp branch circuit to have an ampacity of not less than 20.

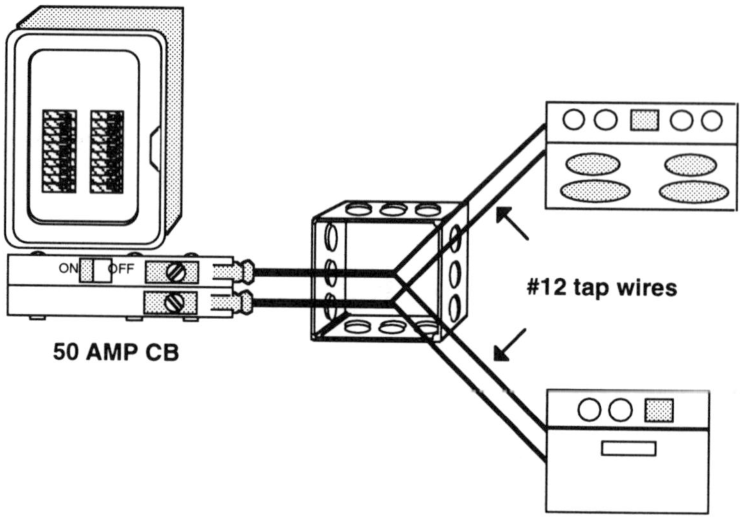

Section 210-19c ex.1 states tap conductors for branch circuits serving other than cooking appliances shall have an ampacity not less than 15 for circuits rated less than 40 amperes and not less than 20 for circuits rated 40 or 50 amperes and only where these tap conductors supply any of the following loads:

Fixtures with taps not longer than 18"

Fixtures having taps as provided in 410-67

Individual outlets with taps not over 18"

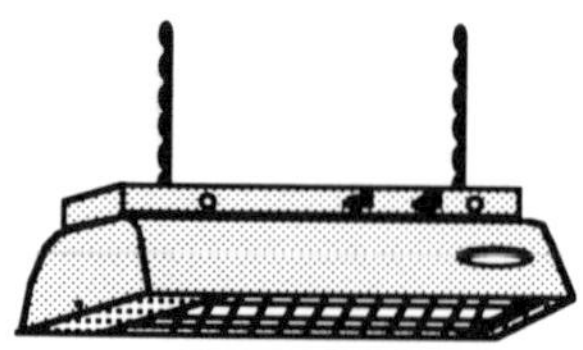

Infrared lamp industrial heating appliances

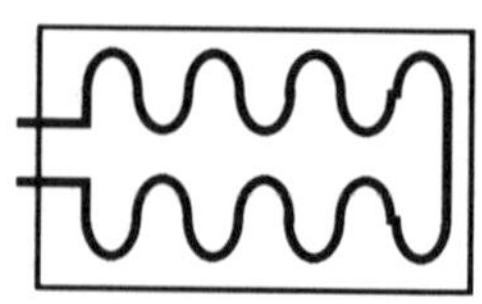

Nonheating leads of de-icing and snow-melting mats and cables

SHORT-CIRCUIT PROTECTION

SHORT-CIRCUIT PROTECTION

The selection of overcurrent protective devices must include careful designing of the ability of the devices to operate properly and safely on ground-faults and short-circuits.

Normally a circuit draws current in proportion to the voltage applied and the impedance of the load. When a short circuit occurs, the source voltage no longer encounters the opposition to the current flow that the normal load had presented. The load is *by-passed* and large currents are impressed upon the circuit. The *speed* in which the overcurrent device opens is critical for the protection of the equipment.

Whereas overload currents occur at rather modest levels, the short-circuit or fault current can be many hundreds of times larger than the normal operating current. A high level fault may be 50,000 amps or more. If not cut off within a few thousandths of a second, there can be severe damage, melting of conductors, arcing and fires.

The point to point method of calculating short-circuit currents is based on computation of the two main circuit impedance parameters, transformers and conductors. Of these two components, the transformer is generally the major short-circuit current factor for faults near the service entrance. The percent of impedance of the transformer can vary considerably. The transformer specification of the drawing should always be checked.

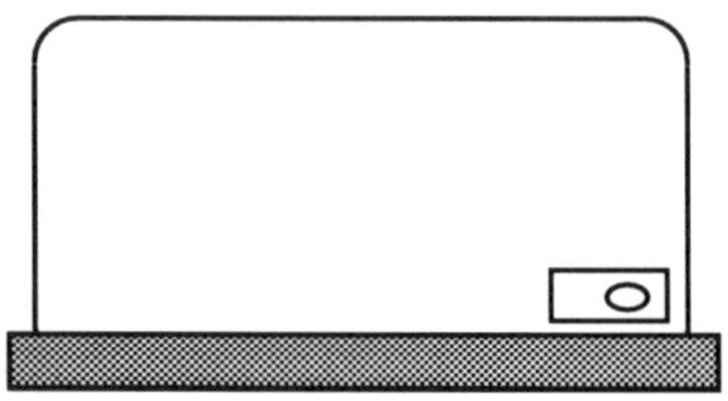

Transformer impedance (Z) helps determine what the short circuit current will be at the transformer secondary. Transformer impedance is determined as follows:

The transformer secondary is shorted. Voltage is applied to the primary which causes full load current to flow in the secondary. The applied voltage divided by the rated primary voltage is the impedance of the transformer.

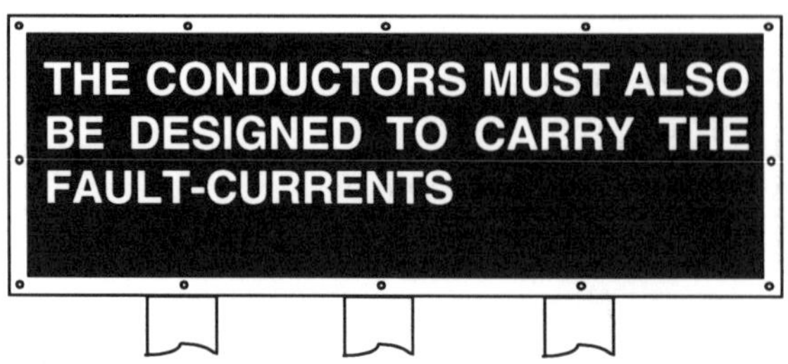

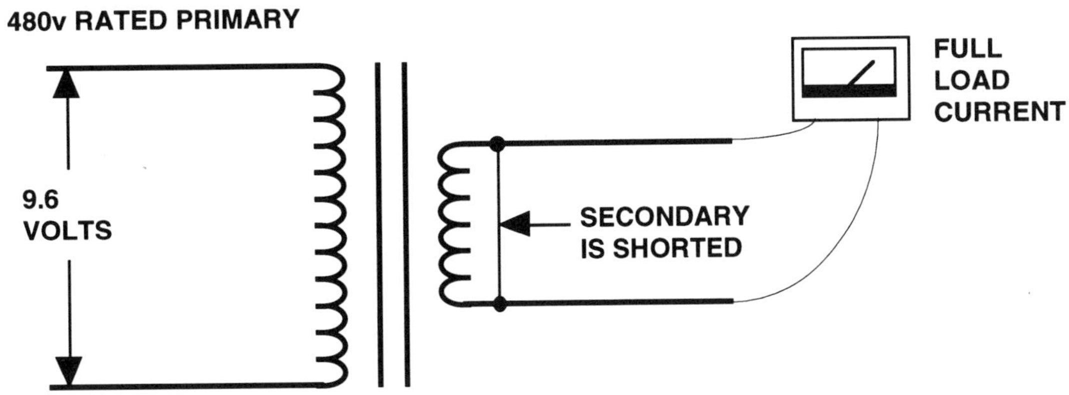

For a 480 volt rated primary, if 9.6 volts causes secondary full load current to flow through the shorted secondary, the transformer impedance is 9.6/480 = .02 = 2% impedance.

First step is to determine the transformer full load amperes from either the nameplate or by using the formula:

3ø Transformer $I = \dfrac{\text{kva} \times 1000}{E \times 1.732}$ 1ø Transformer $I = \dfrac{\text{kva} \times 1000}{E}$

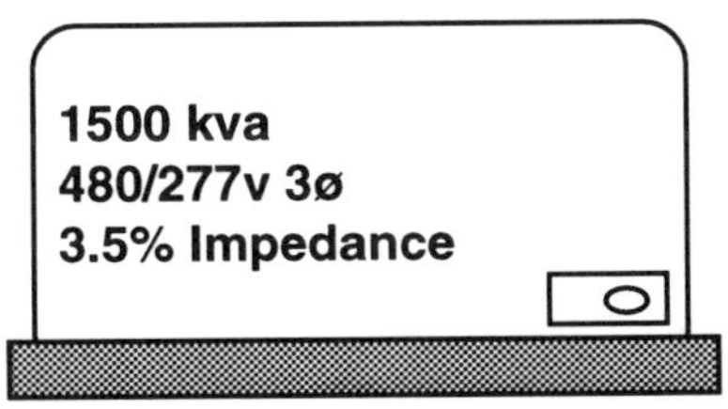

$\dfrac{1500 \times 1000}{480\text{v} \times 1.732} =$ 1804 amperes

Next step is to find the multiplier. Divide 100 by the % of impedance. 100/3.5 = 28.57

The available short-circuit current from this transformer would be 1804 amperes x 28.57 = 51,540 amperes.

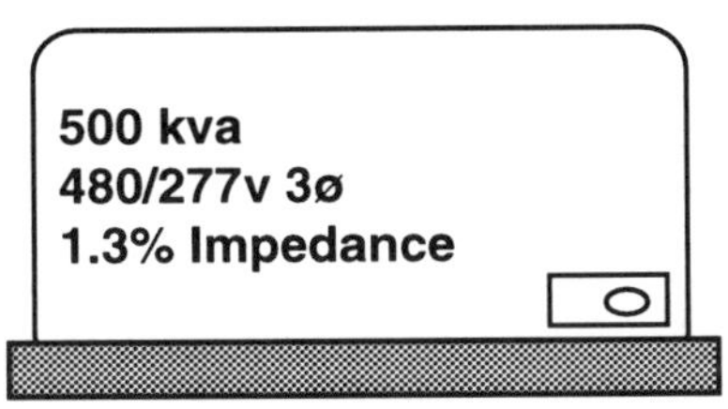

$\dfrac{500 \times 1000}{480\text{v} \times 1.732} =$ 601 amperes

The multiplier is 100/1.3 = 76.9 x 601 amperes = 46,217 amperes short-circuit current.

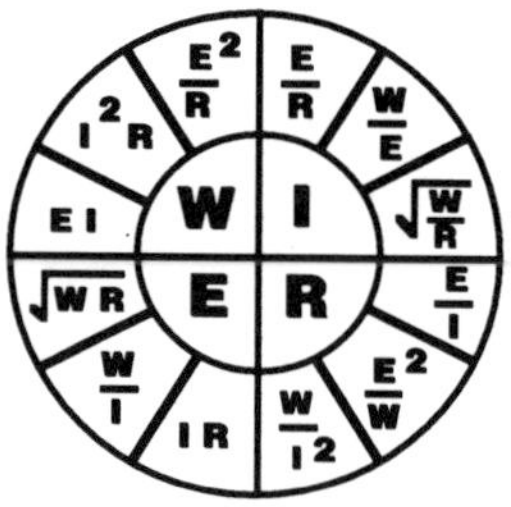

Neglecting conductor impedances simplifies short-circuit calculations in circuits.

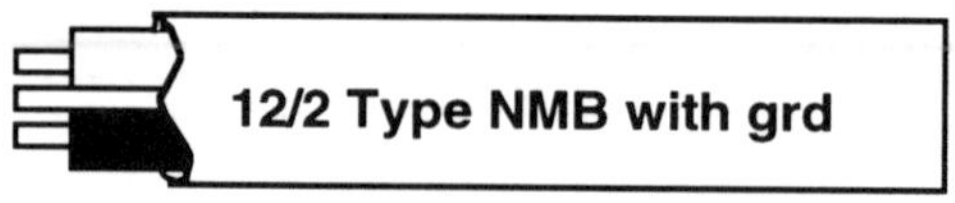

The 120 volt branch circuit shown below is 50 feet in distance from the source to the load. This circuit would contain 100 feet of conductor. A #12 solid conductor has a resistance of 1.93Ω per thousand feet. 1.93Ω x .050' = .0965Ω per 50 feet or .193Ω for 100 feet. The load has a resistance of 7.5Ω. The total resistance in the circuit is .0965 + .0965 + 7.5 = 7.693Ω.

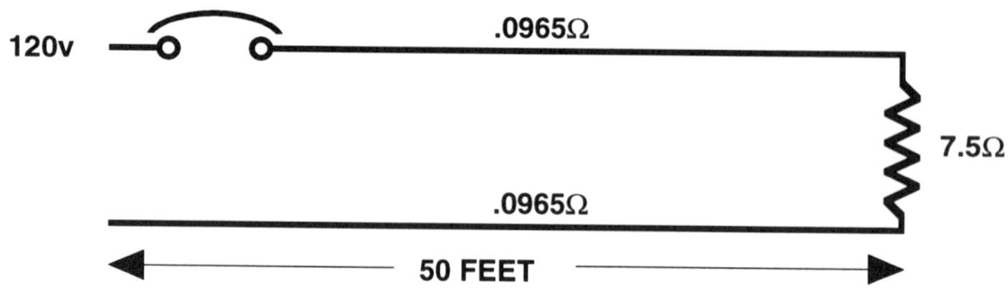

To find the current flowing in this circuit divide 120v by 7.693Ω = 15.6 amps.

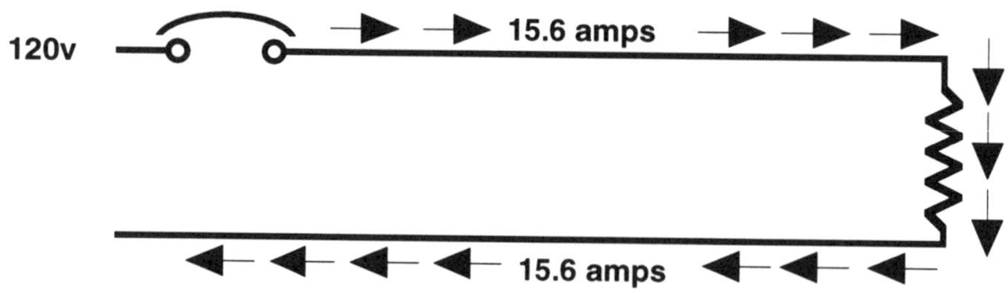

The 15.6 amps is the normal current. The 20 amp current rating of a breaker is the current level which the breaker is designed to carry continuously. The second type of rating given to a breaker is the interrupting capacity (AIC). It is the highest short-circuit current at a rated voltage that the breaker can safely interrupt. If a fault current exceeds the interrupting rating of the breaker, it can be damaged or destroyed. Severe equipment damage and personal injury can result.

Short-circuits are a result of an abnormal occurrence which causes the current to take a path outside the usual conducting path through the resistance of the load.

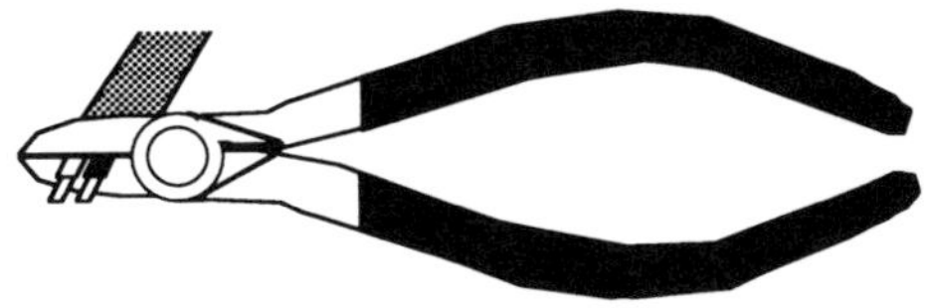

An example an electrician can relate to would be cutting into the black and white wire at the same time with the power on.

If the wire was accidentally cut at the load, the circuit would only have the resistance of the wire, as the load resistance is bypassed through the pliers which has less resistance.

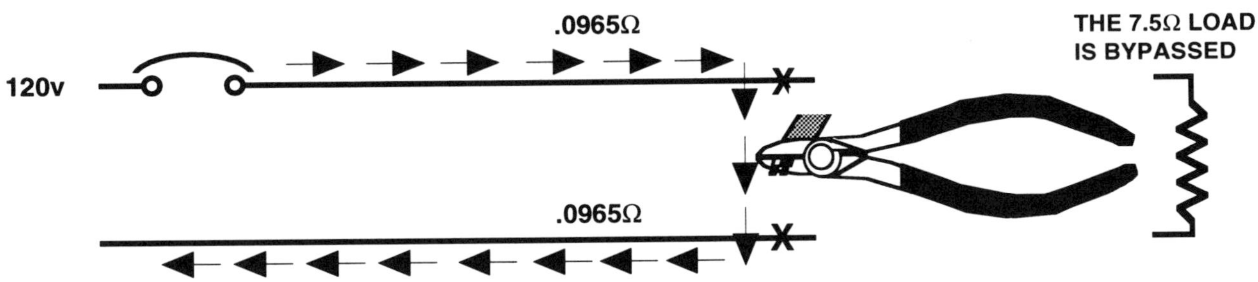

The resistance of the circuit is .0965Ω + .0965Ω = .193Ω. I = 120v/.193Ω = 622 amps.

If the wire was cut 4 feet from the breaker, the circuit would see over 7000 amperes!

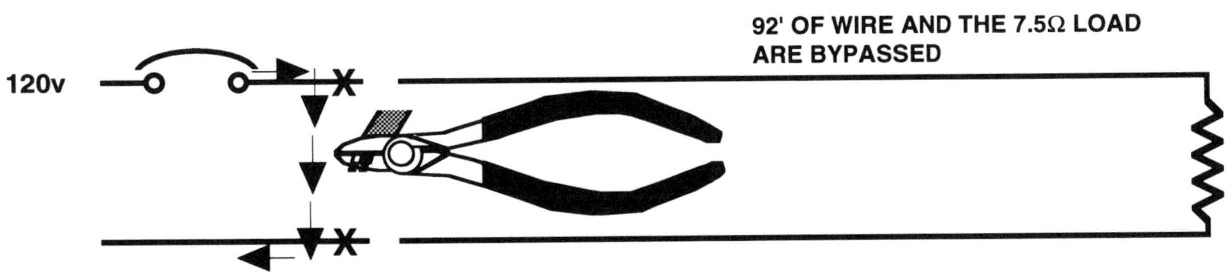

The resistance of the circuit is 1.93Ω per thousand feet x .008 feet of wire = .01544Ω.

I = 120v/.01544Ω = 7,772 amperes.

Because the load is bypassed, a very large magnitude of fault current can occur. In recent years many utility companies have increased their generating capacity and, consequently, available short-circuit currents have also increased. Newer transformers with lower impedance ratings have allowed for greater magnitudes of fault current. These factors made the proper selection of an overcurrent device with a proper interrupting rating extremely critical.

A ground fault is a phase (line) conductor contacting the metal raceway enclosure (conduit), motor frame, etc.

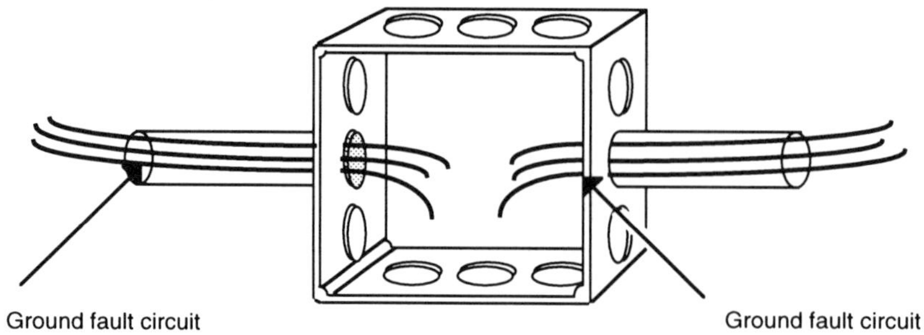

A fault may start as a ground fault and develop into a short circuit or vice versa. The first situation is the more common. It is very important to understand the difference between a ground fault and a short circuit as it is a more serious condition to protect against an insulation failure of a conductor to a metal enclosure.

When a fault occurs in a circuit, it is only during the time period while the fault exists that a potential hazard is present. That is why it is so important to clear the fault as quickly as possible by blowing the fuse or tripping the circuit breaker.

The fault current must return to the overcurrent protection (fuse or circuit breaker) in order to blow the fuse or trip the breaker. The grounding circuit is the return path to the overcurrent device.

Without this grounding circuit, a fault current will generate heat and eventually enough heat damage will cause the insulation to break down causing a short circuit which will cause the fuse to blow or the circuit breaker to trip.

A ground fault that is not allowed to clear itself is a very dangerous situation. A fault current may exist for minutes, hours, or days. During this time that the fault is uncleared, faults can cause a fire or even a fatal shock. Electrical accident statistics have proven that many personal injuries are caused by electric shock as a result of making contact with metal that is not normally energized. An effective grounding path would eliminate these injuries.

The interrupting rating of most branch-circuit, molded case, circuit breakers like a 20 amp used in residential services are 10,000 amperes. The rating is usually expressed as "10,000 AIC" (AIC is the abbreviation for "amperes interrupting capacity.") Larger circuit breakers may have ratings of 14,000 or higher. Current-limiting fuses have an interrupting capacity of 200,000 amperes.

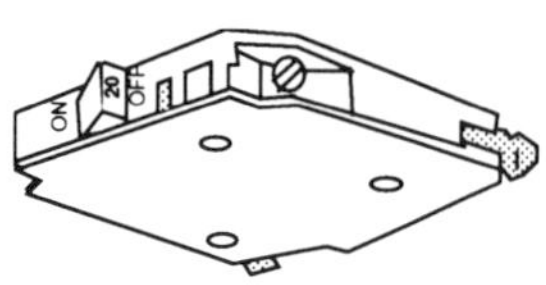

The maximum available fault current must be known in order to comply with Code Section 110-9. The transformer is generally the major short-circuit current factor for faults near the service entrance. Given the full-load transformer secondary amperage and % impedance of a transformer the available short-circuit current delivered at the secondary terminals can be calculated.

Approximate Short-circuit Currents Available From Transformers

	KVA	Full Load Amps	% Impedance	Short-circuit amps
240/120v 1ø	25	104	1.6	6,500
	50	209	1.7	12,289
	100	417	1.6	26,063
208/120v 3ø	150	417	2.0	20,850
	500	1388	2.0	69,400
	1000	2776	5.0	55,520
	2500	6950	5.0	139,000
480/277v 3ø	150	181	1.2	15,077
	500	601	1.3	46,217
	1000	1203	5.0	24,060
	2500	3007	5.0	60,140

Motors also add current to the short-circuit currents.

To adequately protect components, it is very important that the overcurrent protective device be current-limiting. Current-limiting overcurrent devices cut the short-circuit current off before the first full cycle can reach its maximum value.

If an overcurrent device cuts off a short-circuit current in less than one-half cycle, before it reaches its total available (and highly destructive) value, the overcurrent device is a "current-limiting" device.

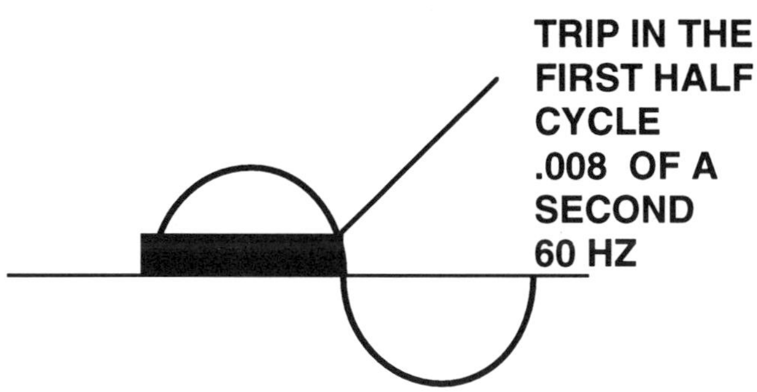

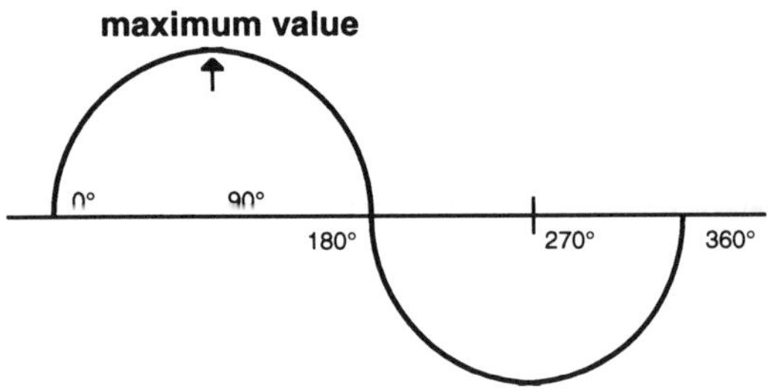

With a ground-fault condition, the higher the faulted current flows the quicker the breaker will trip. If the breaker does not trip quickly and the current remains for several cycles you are generating heat which will cause damage.

The chart below shows the approximate current that conductors can carry during a ground-fault or short-circuit condition without damage to the insulation.

Copper, 75° C Thermoplastic Insulated Cable Damage Table

Copper Wire Size 75° C Thermo-plastic	Maximum Short-Circuit Withstand Current In Amperes For					
	1/8 Cycle	1/4 Cycle	1/2 Cycle	1 Cycle	2 Cycles	3 Cycles
#14	4800	3400	2400	1700	1200	1000
#12	7600	5400	3800	2700	1900	1550
#10	12000	8500	6020	4300	3000	2450
#8	19200	13500	9600	6800	4800	3900
#6	30400	21500	15200	10800	7600	6200
#4	48400	34200	24200	17100	12100	9900

A #12 conductor with 75°C insulation can withstand 3800 amps, but only for 1/2 of a cycle and on a 60 cycle system that equals eight thousandths of one second, which is a very short time period.

Available short-circuit current and clearing time of the overcurrent device must be considered so the withstand rating of the conductor is not exceeded.

The 75°C conductor has three withstand classifications:

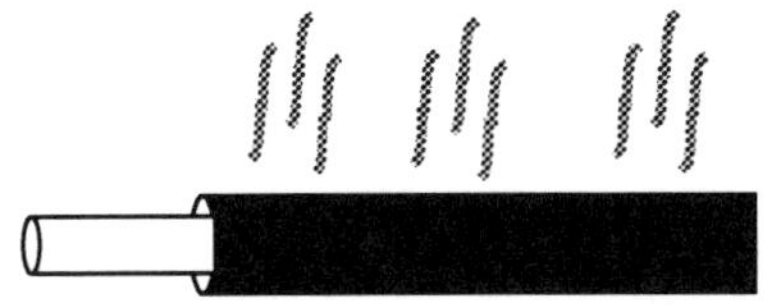

(1) INSULATION DAMAGE 150°C or 302°F
(2) TERMINAL DAMAGE 250°C or 482°F
(3) MELTING 1083°C or 1981°F

Shown below are examples using a #6 THW (75°C) conductor which has a normal ampacity of 65 amps.

INSULATION DAMAGE

#6 THW can withstand 15,218 amps for 1/2 cycle

TERMINAL DAMAGE (loosening of lugs)

#6 THW can withstand 22,090 amps for 1/2 cycle

MELTING

#6 THW can withstand 39,704 amps for 1/2 cycle

IF THE CURRENT CONTINUED FOR 5 SECONDS (300 cycles) THE #6 THW WOULD MELT AT 1620 AMPS

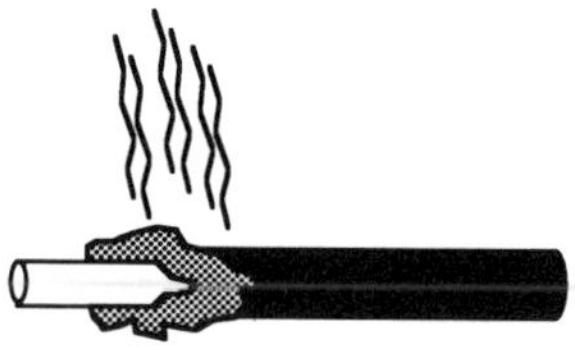

WITHSTAND RATINGS

INSULATION DAMAGE

THHN COPPER CONDUCTORS

WIRE SIZE	AREA CM	1/4 cycle amperes	1/2 cycle amperes	1 cycle amperes	3 cycle amperes	30 cycle amperes
#14	4110	4581	3239	2291	1322	418
#12	6530	7279	5147	3639	2101	664
#10	10380	11570	8181	5785	3340	1056
#8	16510	18403	13013	9201	5312	1680
#6	26240	29248	20681	14624	8443	2670
#4	41740	46525	32898	23262	13431	4247
#3	52620	58652	41473	29326	16931	5354
#2	66360	73967	52303	36984	21353	6752
#1	83690	93284	65962	46642	26929	8516
#1/0	105600	117706	83230	58853	33979	10745
#2/0	133100	148358	104905	74179	42827	13543
#3/0	167800	187036	132254	93518	53993	17074
#4/0	211600	235857	166776	117928	68086	21531
#250	250000	278659	197042	139329	80442	25438
#300	300000	334391	236450	167195	96530	30526
#400	400000	445854	315267	222927	128707	40701
#500	500000	557318	394083	278659	160884	50876
#600	600000	668781	472900	334391	193061	61051
#750	750000	835977	591125	417988	241326	76314
#1000	1000000	1114636	788167	557318	321768	101752

WITHSTAND RATINGS

INSULATION DAMAGE

THHN **ALUMINUM** CONDUCTORS

WIRE SIZE	AREA CM	1/4 cycle amperes	1/2 cycle amperes	1 cycle amperes	3 cycle amperes	30 cycle amperes
#14	4110	2995	2118	1497	865	273
#12	6530	4758	3365	2379	1374	434
#10	10380	7564	5348	3782	2184	690
#8	16510	12031	8507	6015	3473	1098
#6	26240	19121	13521	9561	5520	1746
#4	41740	30416	21507	15208	8780	2777
#3	52620	38344	27113	19172	11069	3500
#2	66360	48356	34193	24178	13959	4414
#1	83690	60985	43123	30492	17605	5567
#1/0	105600	76951	54412	38475	22214	7025
#2/0	133100	96990	68582	48495	27999	8854
#3/0	167800	122276	86462	61138	35298	11162
#4/0	211600	154193	109031	77096	44512	14076
#250	250000	182175	128817	91087	52589	16630
#300	300000	218610	154580	109305	63107	19956
#400	400000	291480	206107	145740	84143	26608
#500	500000	364350	257634	182175	105179	33260
#600	600000	437219	309161	218610	126214	39912
#750	750000	546524	386451	273262	157768	49891
#1000	100000	728699	515268	364350	210357	66521

WITHSTAND RATINGS

TERMINAL DAMAGE

BARE COPPER CONDUCTORS

WIRE SIZE	AREA CM	1/4 cycle amperes	1/2 cycle amperes	1 cycle amperes	3 cycle amperes	30 cycle amperes
#14	4110	4893	3460	2447	1413	447
#12	6530	7774	5497	3887	2244	710
#10	10380	12358	8738	6179	3567	1128
#8	16510	19656	13899	9828	5674	1794
#6	26240	31239	22090	15620	9018	2852
#4	41740	49693	35138	24846	14345	4536
#3	52620	62646	44297	31323	18084	5719
#2	66360	79004	55864	39502	22806	7212
#1	83690	99635	70453	49818	28762	9095
#1/0	105600	125720	88897	62860	36292	11477
#2/0	133100	158459	112048	79230	45743	14465
#3/0	167800	199771	141259	99885	57669	18236
#4/0	211600	251916	178131	125958	72722	22997
#250	250000	297632	210458	148816	85919	27170
#300	300000	357159	252549	178579	103103	32604
#400	400000	476212	336732	238106	137470	43472
#500	500000	595264	420916	297632	171838	54340
#600	600000	714317	505099	357159	206206	65208
#750	750000	892897	631373	446448	257757	81510
#1000	100000	1190529	841831	595264	343676	108680

WITHSTAND RATINGS

TERMINAL DAMAGE

BARE **ALUMINUM** CONDUCTORS

WIRE SIZE	AREA CM	1/4 cycle amperes	1/2 cycle amperes	1 cycle amperes	3 cycle amperes	30 cycle amperes
#14	4110	3734	2641	1867	1078	341
#12	6530	5933	4195	2967	1713	542
#10	10380	9431	6669	4716	2723	861
#8	16510	15001	10607	7501	4330	1369
#6	26240	23842	16859	11921	6883	2176
#4	41740	37926	26817	18963	10948	3462
#3	52620	47811	33808	23906	13802	4365
#2	66360	60296	42635	30148	17406	5504
#1	83690	76042	53770	38021	21951	6942
#1/0	105600	95950	67847	47975	27698	8759
#2/0	133100	120937	85515	60468	34911	11040
#3/0	167800	152466	107809	76233	44013	13918
#4/0	211600	192263	135950	96131	55502	17551
#250	250000	227154	160622	113577	65574	20736
#300	300000	272584	192746	136292	78688	24883
#400	400000	363446	256995	181723	104918	33178
#500	500000	454307	321244	227154	131147	41472
#600	600000	545169	385493	272584	157377	49767
#750	750000	681461	481866	340730	196721	62209
#1000	100000	908615	642488	454307	262294	82945

WITHSTAND RATINGS

MELTING OF CONDUCTORS *1083°C or 1981°F*

BARE COPPER CONDUCTORS

WIRE SIZE	AREA CM	1/4 cycle amperes	1/2 cycle amperes	1 cycle amperes	3 cycle amperes	30 cycle amperes
#14	4110	8795	6219	4397	2539	803
#12	6530	13973	9881	6987	4034	1276
#10	10380	22212	15706	11106	6412	2028
#8	16510	35330	24982	17665	10199	3225
#6	26240	56151	39704	28075	16209	5126
#4	41740	89319	63158	44659	25784	8154
#3	52620	112601	79621	56300	32505	10279
#2	66360	142003	100411	71001	40993	12963
#1	83690	179087	126634	89543	51698	16348
#1/0	105600	225972	159786	112986	65232	20628
#2/0	133100	284819	201397	142409	82220	26000
#3/0	167800	359073	253903	179536	103655	32779
#4/0	211600	452800	320178	226400	130712	41335
#250	250000	534971	378282	267486	154433	48836
#300	300000	641965	453938	320983	185319	58603
#400	400000	855954	605251	427977	247093	78138
#500	500000	1069942	756564	534971	308866	97672
#600	600000	1283931	907876	641965	370639	117206
#750	750000	1604914	1134845	802457	463299	146508
#1000	100000	2139885	1513127	1069942	617732	195344

WITHSTAND RATINGS

MELTING OF CONDUCTORS *660°C or 1220°F*

BARE ALUMINUM CONDUCTORS

WIRE SIZE	AREA CM	1/4 cycle amperes	1/2 cycle amperes	1 cycle amperes	3 cycle amperes	30 cycle amperes
#14	4110	4865	3440	2432	1404	444
#12	6530	7729	5465	3864	2231	706
#10	10380	12286	8687	6143	3547	1122
#8	16510	19541	13818	9771	5641	1784
#6	26240	31058	21961	15529	8966	2835
#4	41740	49404	34934	24702	14262	4510
#3	52620	62281	44039	31141	17979	5685
#2	66360	78544	55539	39272	22674	7170
#1	83690	99056	70043	49528	28595	9042
#1/0	105600	124988	88380	62494	36081	11410
#2/0	133100	157537	111396	78769	45477	14381
#3/0	167800	198608	140437	99304	57333	18130
#4/0	211600	250450	177095	125225	72299	22863
#250	250000	295900	209233	147950	85419	27012
#300	300000	355080	251080	177540	102503	32414
#400	400000	473440	334773	236720	136670	43219
#500	500000	591800	418466	295900	170838	54024
#600	600000	710161	502159	355080	205006	64828
#750	750000	887701	627699	443850	256257	81036
#1000	100000	1183601	836932	591800	341676	108047

THE FEEDER

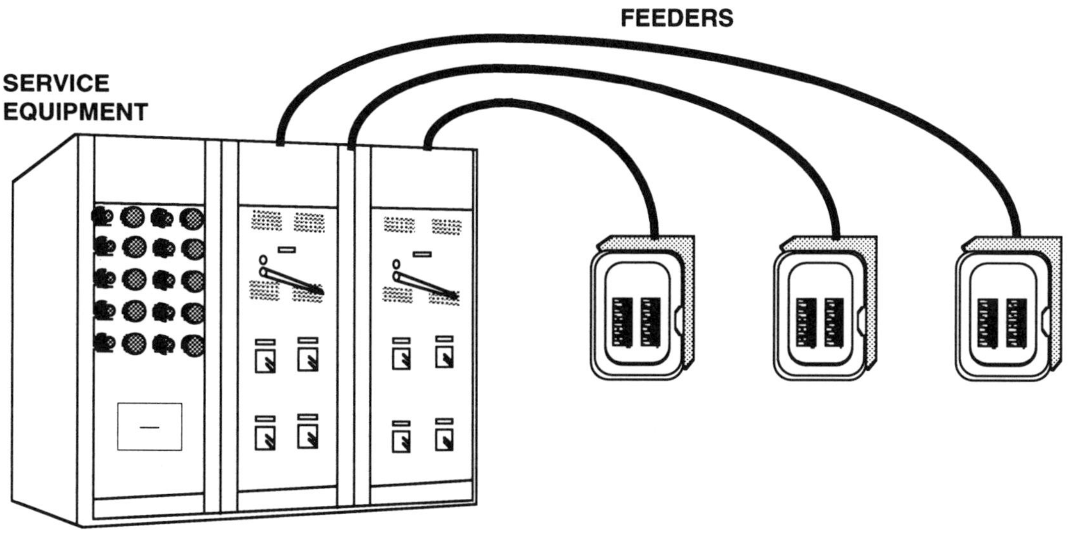

DISTRIBUTION VOLTAGE LEVELS

SINGLE PHASE 240/120v

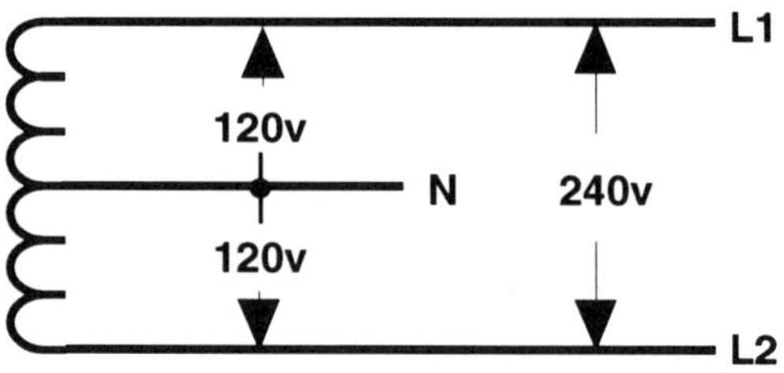

This single-phase distribution is used where the total load is small, the most common application is residential.

THREE-PHASE 208/120v

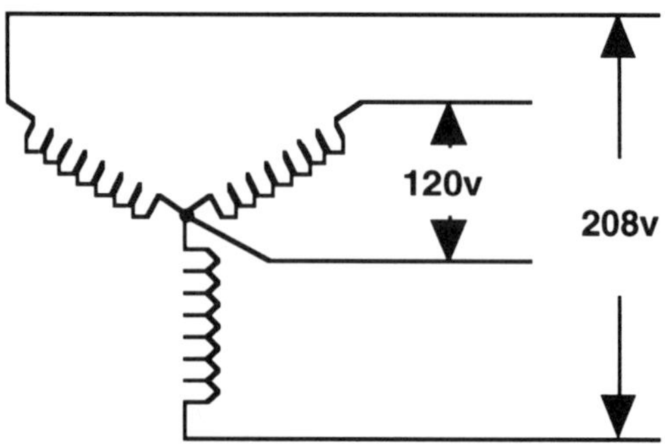

The three-phase, 208/120v, 4-wire is the most common type of system for commercial and industrial buildings with limited electrical loads.

THREE-PHASE 240v

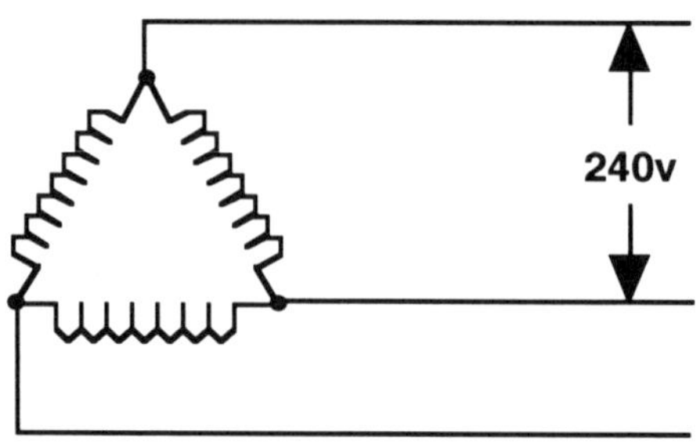

This is a common system for power loads in commercial and industrial buildings. It may operate ungrounded or corner grounded.

THREE-PHASE 480/277v

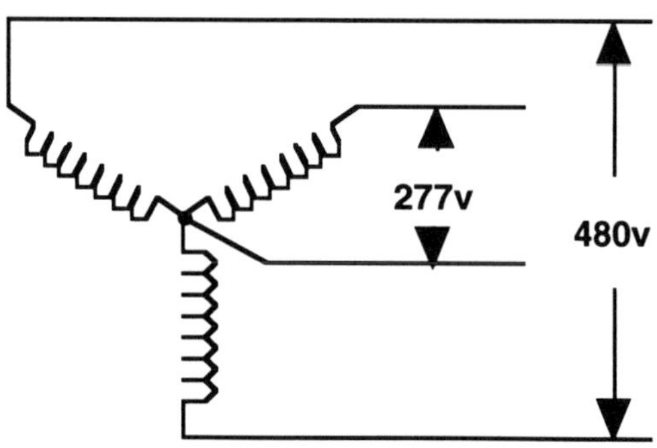

This is a very popular system for commercial and industrial buildings. Required 208/120v 4-wire circuits are derived from step-down transformers for lighting and receptacle circuits.

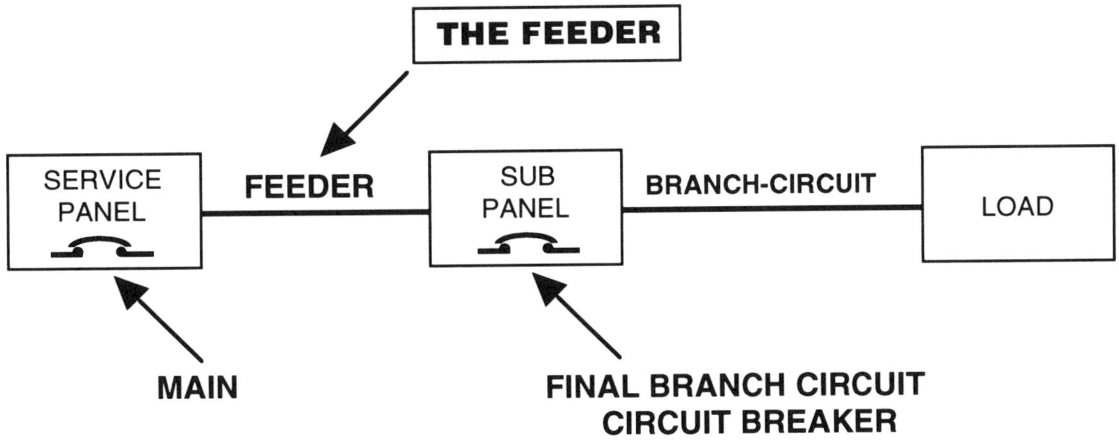

Definition of feeder: All circuit conductors between the service equipment and the final branch circuit overcurrent device.

The feeder is the distribution of power from the service to the subpanels throughout the building to provide the right amount of current and voltage to the utilization equipment being fed from each panelboard.

Since feeders require a large quantity of the material and labor in most buildings, they should rank high in priority of proper design.

The cost and efficiency of the system depend heavily on the selection of the proper size of feeders.

Many variables must be considered when designing feeder circuits.

The feeder must be sized to carry a computed load that is not less than the sum of the branch circuit loads it supplies. The feeder must also have a capacity to meet the minimum lighting requirements from Table 220-3b which are based on a volt-amps per square foot. If the actual connected load is larger than the Table 220-3b requirements, the actual connected load must be used as Table 220-3b is the *minimum* requirement.

Where a feeder supplies continuous loads or any combination of continuous and noncontinuous loads, the *rating of the overcurrent device* shall not be less than the noncontinuous load plus 125% of the continuous load.

The Code permits demand factors to be applied in certain conditions for a feeder. In certain applications not all the connected loads are being used at the same time, so the Code permits the size of the feeder to be *reduced*.

CONNECTED LOAD

DEMAND FACTOR

In an office building a panelboard has 8 - 20 amp 120 volt circuits feeding a total of 100 receptacles.

For buildings other than a house, the Code requires a *minimum* of 180va per yoke to be calculated for receptacles.

The *maximum* number of receptacles on a 20 amp branch circuit would calculate:

$$20 \text{ amp} \times 120\text{v} = \frac{2400\text{va}}{180\text{va}} = 13 \text{ receptacles}$$

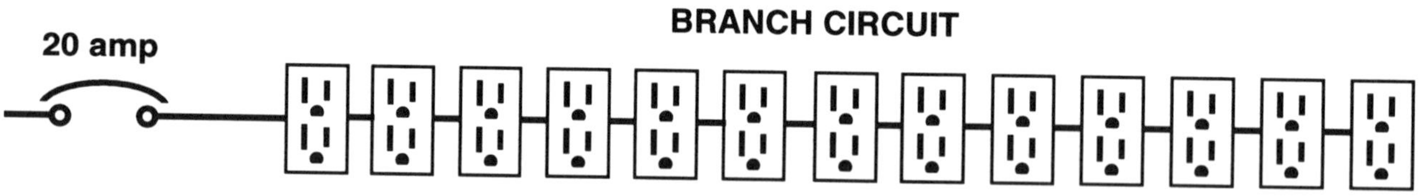

Code Table 220-13 permits a *reduction* in the size of the feeder to the panelboard supplying these 8-20 amp branch circuits.

The minimum calculated load would be 100 receptacles x 180va = 18,000 va or 18 kva.

Table 220-13 states the first 10 kva cannot be reduced, but any load over 10 kva can be reduced 50%. 10 kva + 4 kva (8kva x 50%) = 14 kva.

The feeder size can be reduced from 18 kva to 14 kva per the *demand factor* from Table 220-13.

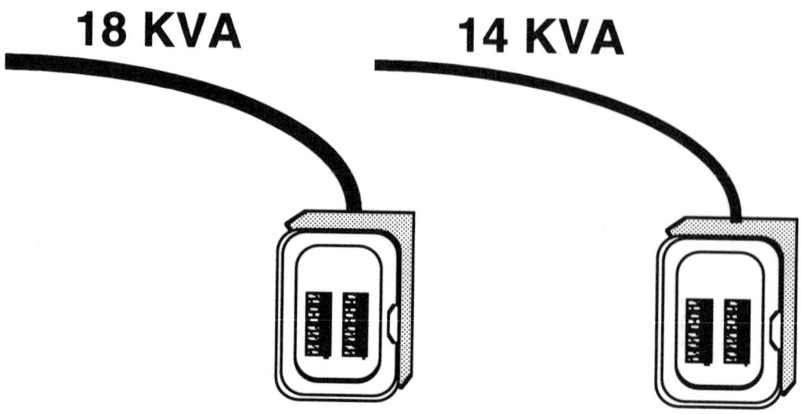

The demand factors the Code allows *reduces* the size of the feeder. This will provide the *minimum* size feeder required by the Code.

Proper design would require the feeder to be designed with allowance for future growth. Minimum sizing of feeder conductors or conduits has proved to be a major shortcoming of past electrical design work.

Based on experience, modern design practice dictates sizing to allow an increase of 25 to 50% in the load on the feeder, to provide for additional branch circuit loads in the future.

Extra capacity for load growth should automatically be included in the feeder conductors when they are initially installed. Selecting conduits larger than initially required will allow larger conductors to be installed in the future. Designing spare conduits in which conductors can be pulled at a later date is a must.

There are advantages to paralleling feeder conductors in the design of high-current-capacity feeders. Smaller conductors can carry more amps per circular mil than large conductors. The designer must remember that by reducing the total circular mil area when paralleling smaller conductors will produce a higher resistance in the conductors and a greater voltage drop.

Example: To supply a 500 amp load, a #1000 kcmil THW has an ampacity of 545 amps and a resistance of .0129Ω per k/ft., a #250 kcmil THW has an ampacity of 255 amps and a resistance of .0515Ω per k/ft.

By paralleling two #250 kcmil THW conductors the ampacity would be 510 amps. The resistance in parallel would be .0515Ω/2 = .02575Ω. The resistance for two paralleled #250 conductors is almost double the .0129Ω for a #1000 kcmil.

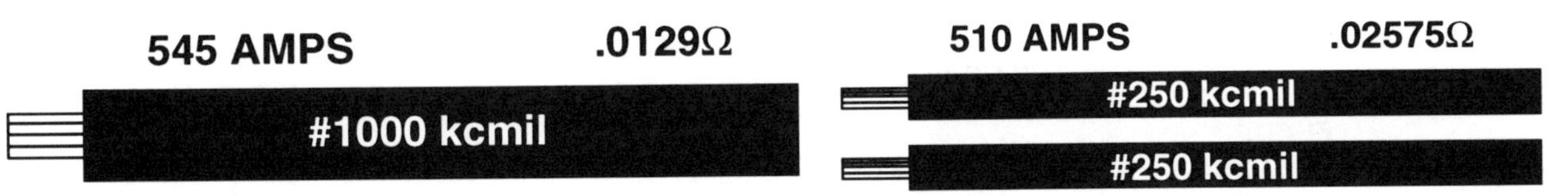

VD = I x R   500 amp load x .0129Ω = 6.45 vd   500 amp load x .02575Ω = 12.88 vd

Also note that when conductors are paralleled, more space may be required at equipment terminals to bend and install the conductors.

MOTOR FEEDER

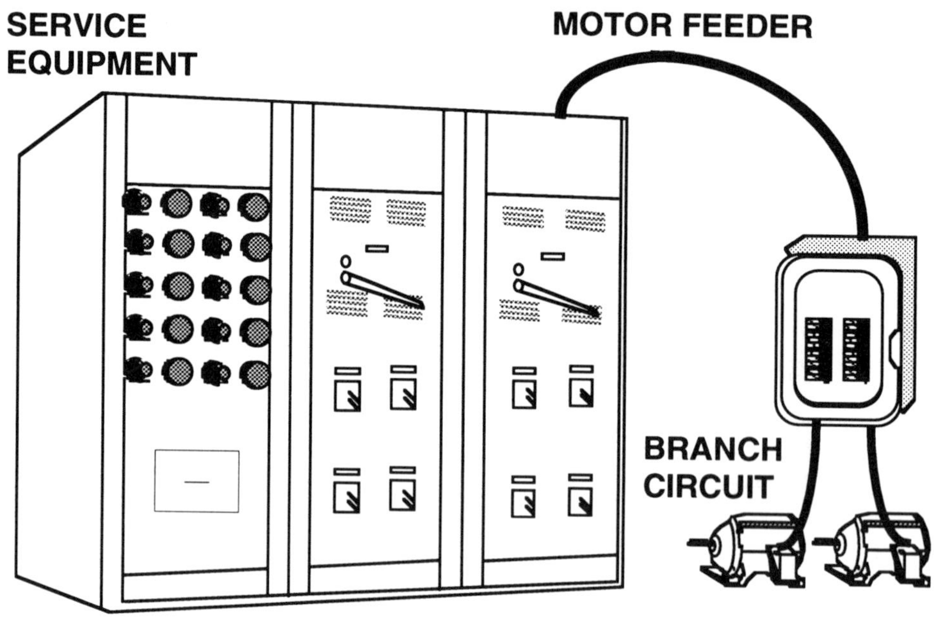

To size the feeder conductors, section 430-24 requires the largest rated motor in F.L.C. to be increased 25% and add the F.L.C. of all the other motors connected to the feeder for the total ampacity required.

To size the feeder overcurrent protection, 430-62 requires adding the largest branch circuit *overcurrent device* to all the other motor F.L.C. connected to the feeder to find the amperage required. Section 240-6 lists standard sizes of overcurrent devices.

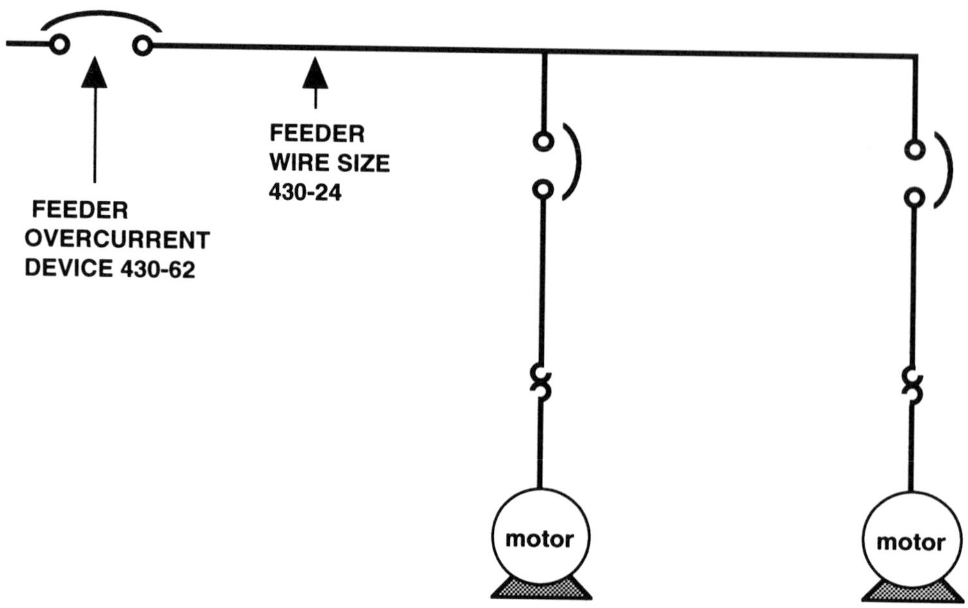

Example: To size the feeder wire, take the F.L.C. of the largest motor 14a x 125% = 17.5a + 11a + 7.6a = 36.1 required ampacity. Table 310-16 requires a 60°C #8 copper.

To size the nontime delay fuse, take the largest branch circuit fuse which is a 45 amp + 11a + 7.6a = 63.6 amps. •You *cannot* go up to the next higher standard size for feeder protection. You must go *down* to a 60 amp nontime delay fuse.

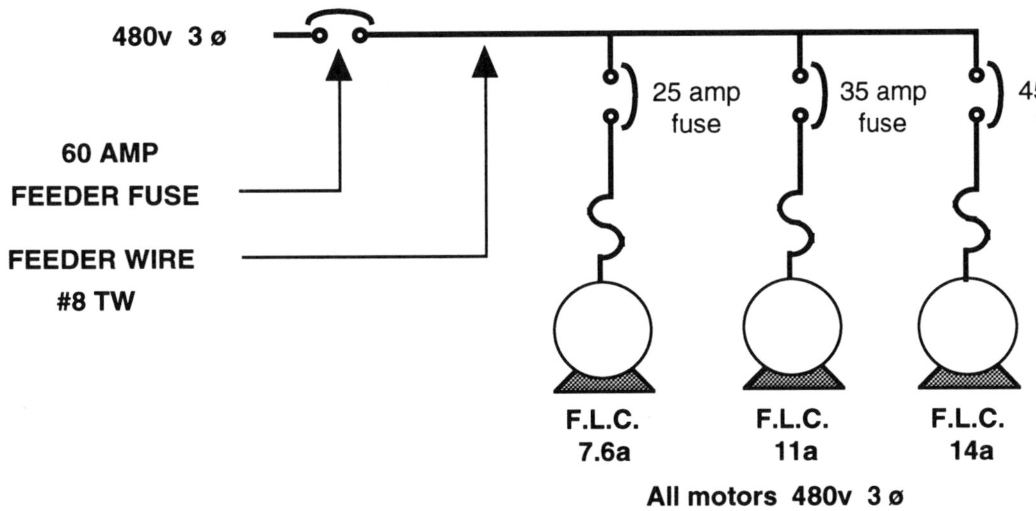

Example: The one-line sketch below shows a three-phase, 4-wire feeder, supplying one three-phase motor and three single-phase 120 volt motors.

The feeder wire and breaker are sized to only *two* motors, not four motors. By drawing a sketch showing how the motors are connected is very helpful in feeder sizing. (see next page for sketch).

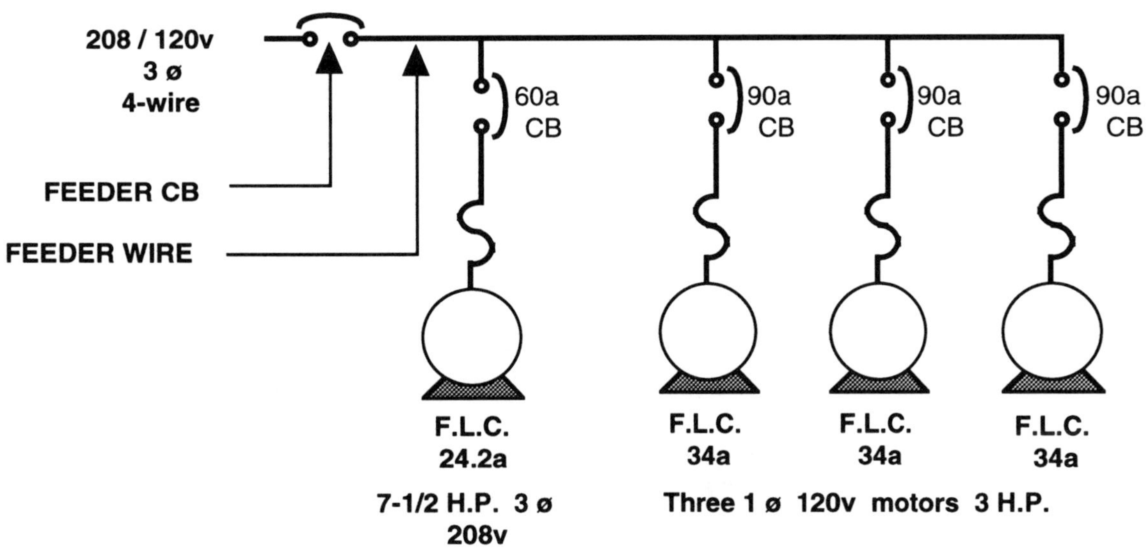

•Draw a sketch:

As you can see, by connecting the motors, Phase "A" has two connections, Phase "B" has two connections, Phase "C" has two connections. The 120 volt motors have connections to the neutral.

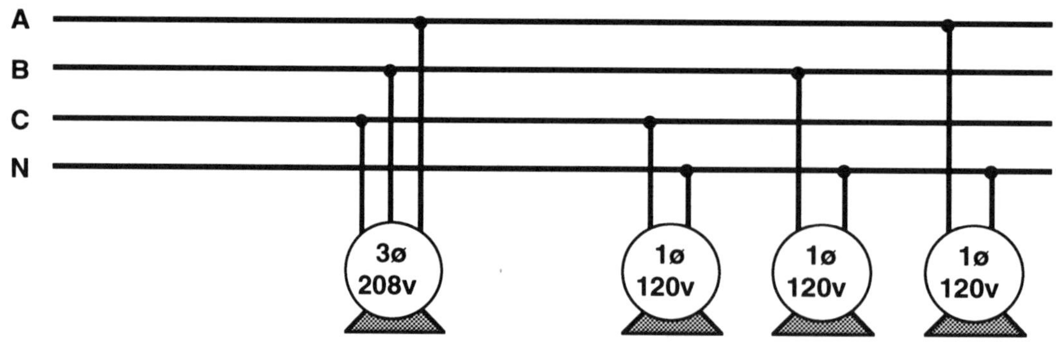

To size the feeder wire, take the largest motor 34a x 125% = 42.5a + 24.2a = 66.7 required ampacity. Table 310-16 requires a #4 TW minimum.

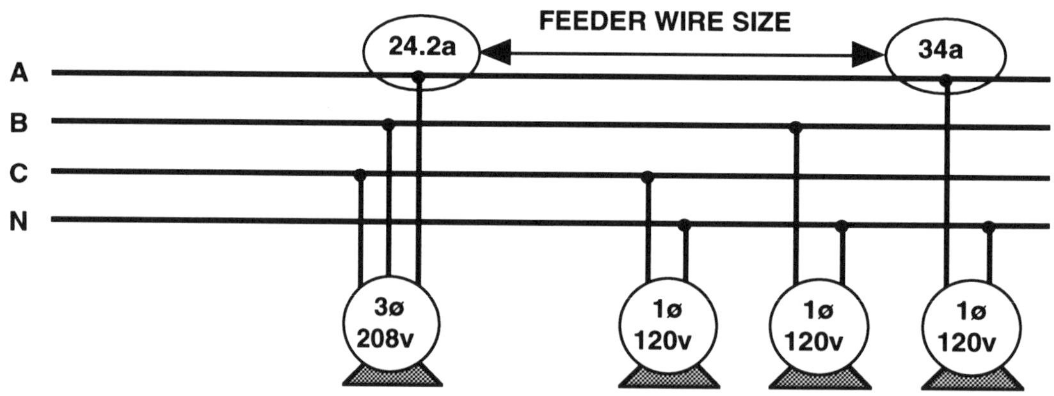

To size the inverse time circuit breaker take the largest branch circuit breaker 90a + 24.2a = 114.2 amps. Section 240-6 lists a 110 amp as the next *lower standard size.*

FEEDER TAPS

To meet the practical demands of field application feeder taps are permitted by the Code.

Feeder conductors shall be permitted to be tapped, without overcurrent protection at the tap where all of the following conditions are met:

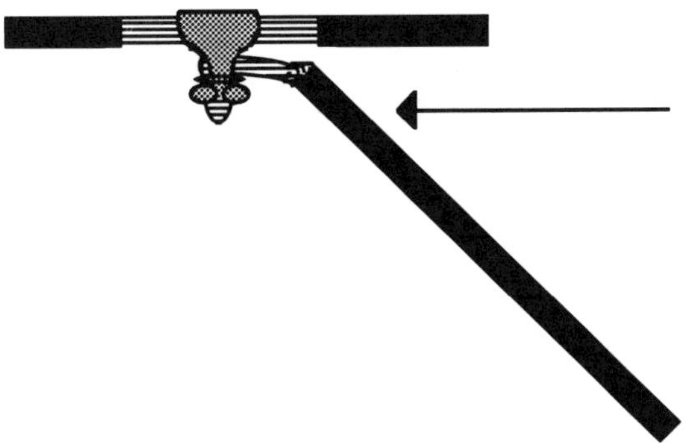

FEEDER TAPS NOT OVER 10' LONG

(1) Length of tap conductors cannot exceed 10'.

(2) The ampacity of the tap conductors can not be less than the total combined loads supplied by the tap conductors or less than the rating of the device supplied by the taps, or not less than the rating of the O.C.P. device at the termination of the taps.

(3) The taps can not extend beyond the switchboard, panelboard, or control device they supply.

(4) Except at the point of connection to the feeder, the tap conductors shall be enclosed in a raceway.

(5) For field installations where the tap conductors leave the enclosure in which the tap is made, the rating of the O.C.P. device on the line side of the taps shall not exceed 1000% of the tap conductors ampacity.

Short circuit protection for feeder taps is obtained by limiting the maximum length of each tap conductor (not raceway) to 10 feet. This assures a minimum circuit impedance corresponding to the practical circuit conditions to be encountered. Any short circuit in this short length of tap conductor would open the feeder overcurrent device.

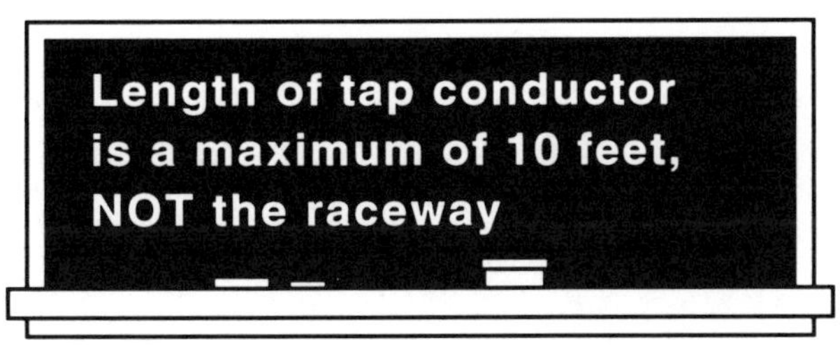

Feeder taps *not over 25'* long, conductors shall be permitted to be tapped, without overcurrent protection at the tap, to a feeder where all of the following conditions are met:

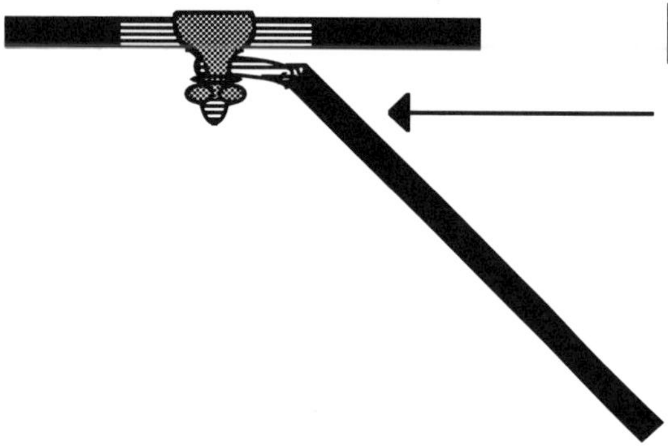

FEEDER TAPS NOT OVER 25' LONG

(1) Length of tap conductors cannot exceed 25'.

(2) The ampacity of the tap conductors cannot be less than 1/3 of the rating of the overcurrent device protecting the feeder conductors.

(3) The tap conductors terminate in a single CB or single set of fuses that will limit the load to the ampacity of the tap conductors. This device shall be permitted to supply any number of additional overcurrent devices on its load side.

(4) The tap conductors shall be suitably protected from physical damage or are enclosed in a raceway.

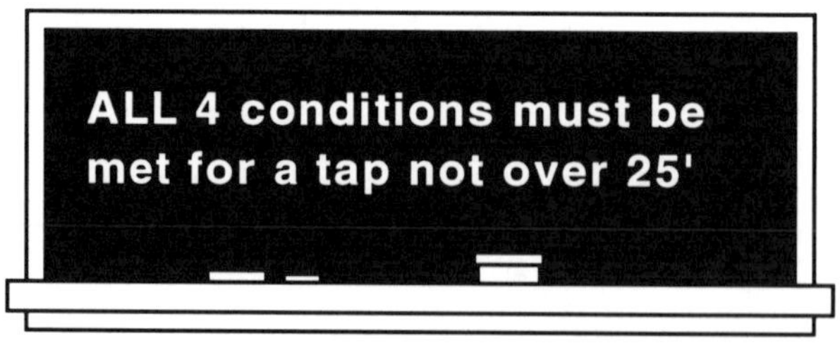

Transformer feeder taps with primary plus secondary *not over 25'* long, conductors shall be permitted to be tapped, without overcurrent protection at the tap, to a feeder where all of the following conditions are met:

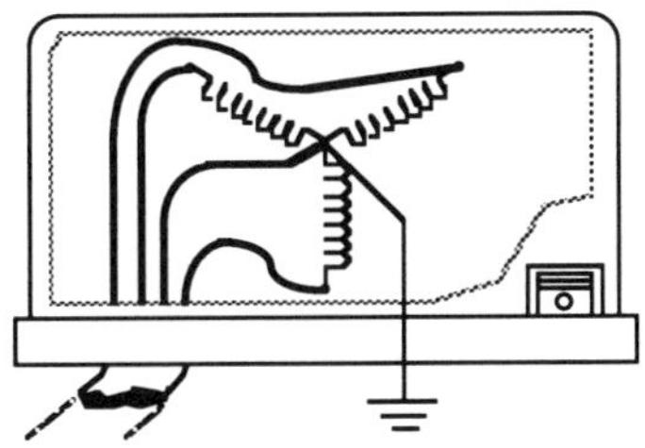

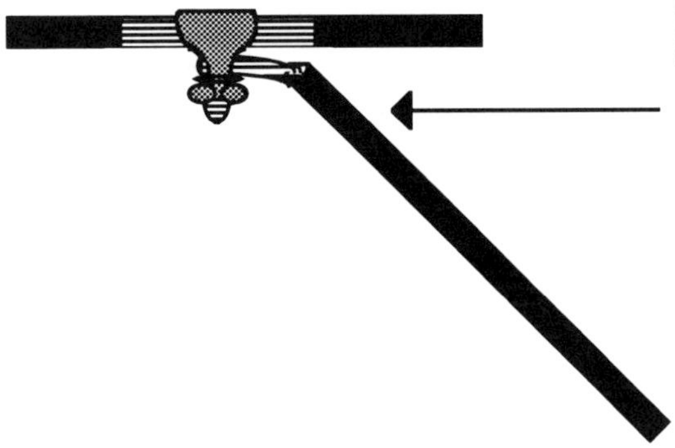

FEEDER TAPS NOT OVER 25' LONG

(1) The conductors supplying the primary of a transformer shall have an ampacity at least 1/3 of the rating of the overcurrent device protecting the feeder conductors.

(2) The conductors supplied by the secondary of the transformer shall have an ampacity that, when multiplied by the ratio of the secondary-to-primary voltage, is at least 1/3 of the rating of the overcurrent device protecting the feeder conductors.

(3) The total length of one primary plus one secondary conductor, excluding any portion of the primary conductor that is protected at its ampacity, is not over 25'.

(4) The primary and secondary conductors are suitably protected from physical damage.

(5) The secondary conductors terminate in a single CB or set of fuses that will limit the load to that allowed in section 310-15.

ALL 5 conditions must be met for a tap not over 25' for transformers

Feeder taps *over 25'* long, conductors over 25 feet shall be permitted to be tapped from feeders in *high bay manufacturing buildings* over 35 feet high at walls, where conditions of maintenance and supervision ensure that only qualified persons will service the systems. Conductors tapped, without overcurrent protection at the tap, to a feeder shall be permitted to be not over 25 feet long *horizontally* and *not over 100 feet total length* where all of the following conditions are met:

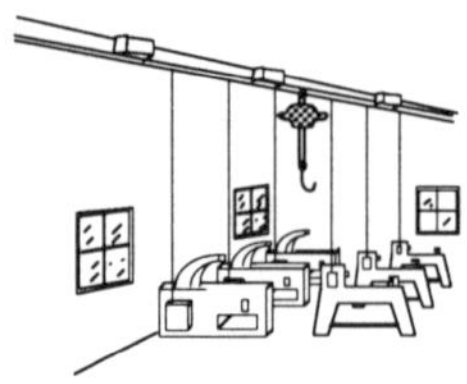

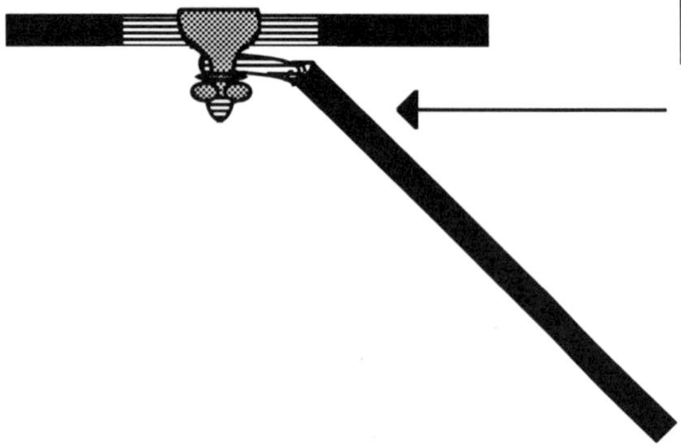

FEEDER TAPS OVER 25' LONG

(1) The ampacity of the tap conductors is not less than 1/3 of the rating of the overcurrent device protecting the feeder conductors.

(2) The tap conductors terminate at a single CB or a single set of fuses that will limit the load to the ampacity of the tap conductors. This single overcurrent device shall be permitted to supply any number of additional overcurrent devices on its load side.

(3) The tap conductors are suitably protected from physical damage or are enclosed in a raceway.

(4) The tap conductors are continuous from end-to-end and contain no splices.

(5) The tap conductors are sized #6 copper or #4 aluminum or larger.

(6) The tap conductors do not penetrate walls, floors, or ceilings.

(7) The tap is made no less than 30' from the floor.

MOTOR FEEDER TAPS

Feeder tap conductors shall have an ampacity not less than required by Part B (430-22 & 430-24), shall terminate in a branch circuit protective device and in addition, shall meet *one* of the following requirements:

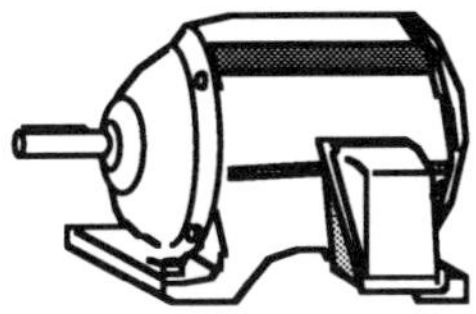

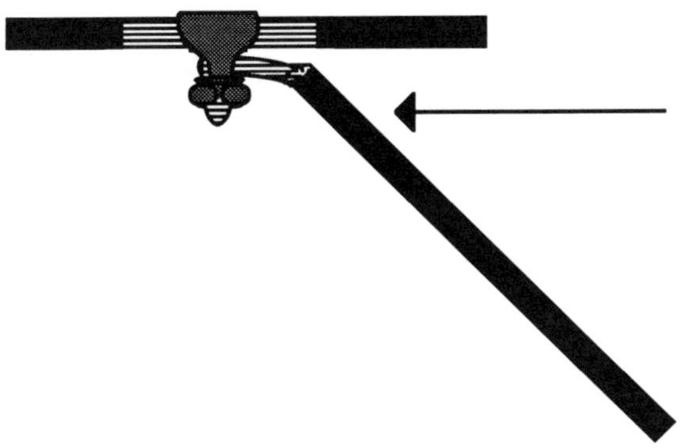

MOTOR FEEDER TAPS

(1) The tap conductors shall be enclosed by either an enclosed controller or by a raceway. Be not more than 10 feet in length. For field installation be protected by an overcurrent device on the line side of the tap conductor, the rating or setting of which shall no exceed 1000% of the tap conductor ampacity.

OR

(2) Have an ampacity of at least 1/3 that of the feeder conductors, be suitably protected from physical damage or enclosed in a raceway and be not more than 25' in length.

OR

(3) Have the same ampacity as the feeder conductors.

- **There is an exception for feeder taps OVER 25' in high-bay manufacturing buildings.**

Without the 1000% limitation of the overcurrent device (10:1 ratio) it would allow a tap of any size. A #10 wire could be tapped from a bus having overcurrent protection set at 1200 amps. This is a 40:1 ratio. Under a fault condition, the #10 wire would burn off, creating other shock and fire hazards. The 1000% limitation is a much safer design.

(1) The 10' or less tap conductors shall be enclosed by either an enclosed controller or by a raceway. When installed in a raceway or enclosed controller the taps can be the same size as the branch circuit conductors.

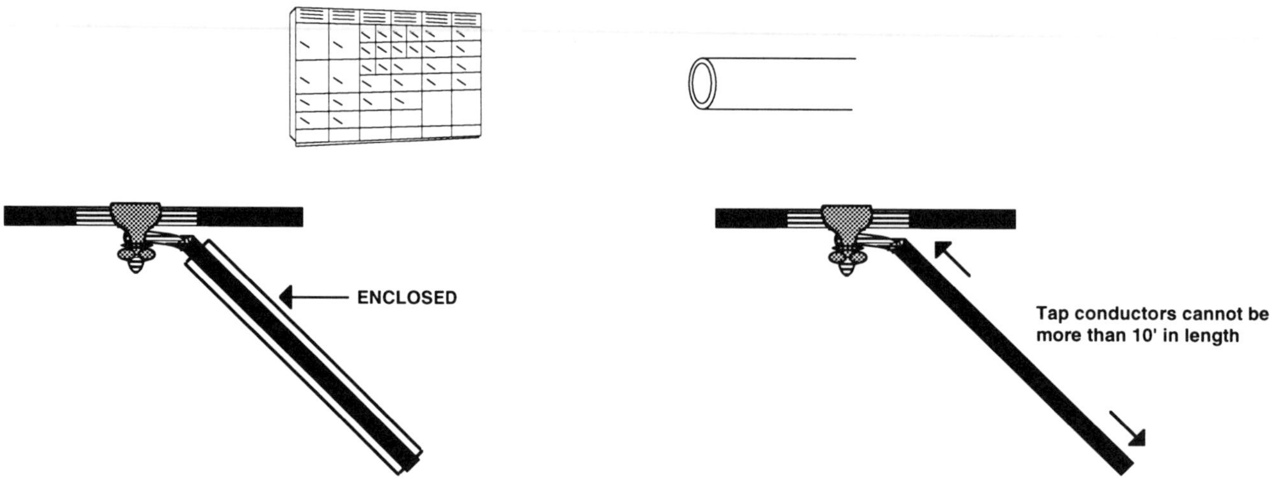

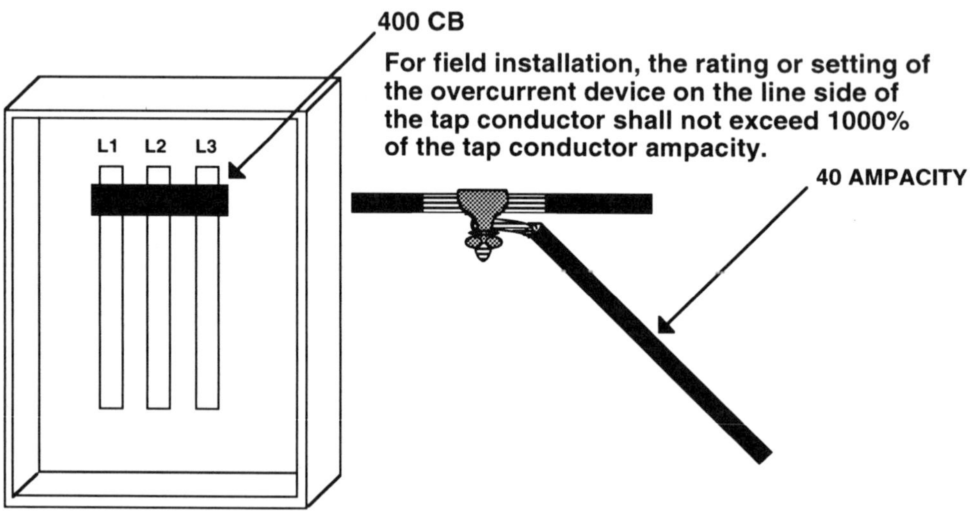

OR (2) Have an ampacity at least 1/3 that of the feeder conductors, be protected from physical damage and be not more than 25' in length. **OR** (3) Have the same ampacity as the feeder conductors.

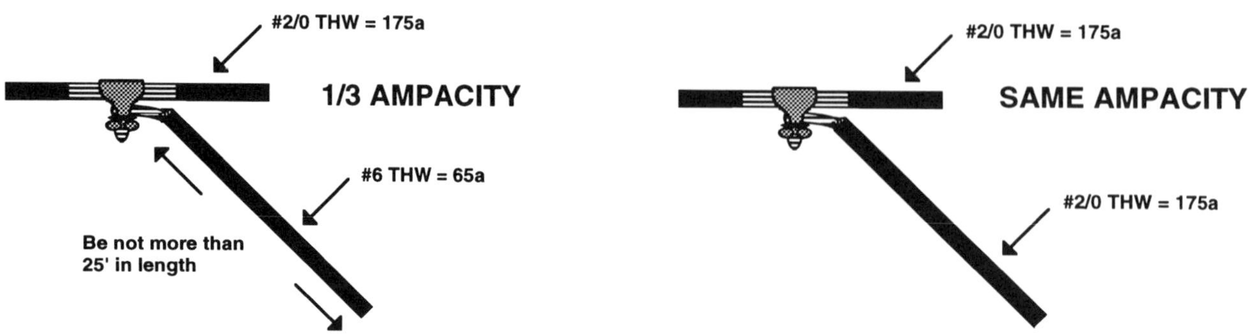

THE SERVICE

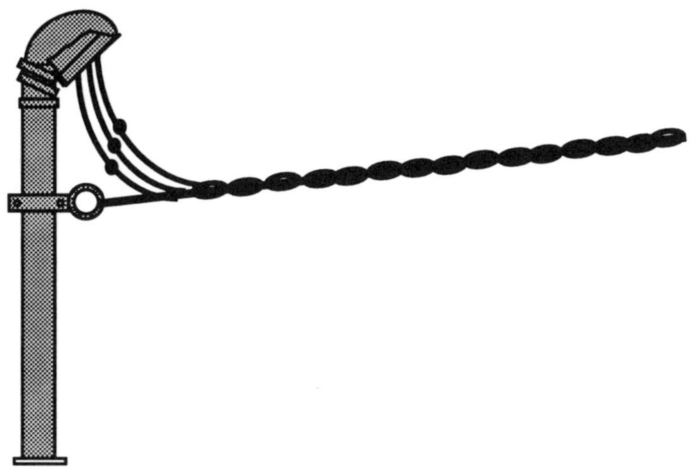

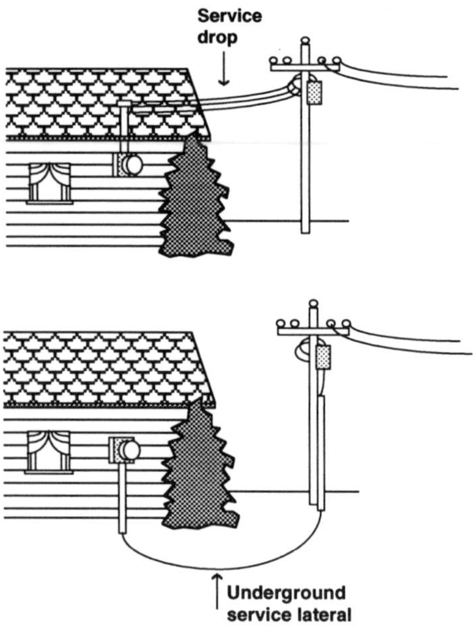

THE SERVICE

The general rule is that a building shall be supplied by only *one* service. The definition of "service" is the conductors and equipment for delivering energy from the electricity supply system to the wiring system of the premises served. The service to a building may be either overhead or underground. Overhead is referred to as "service drop" and underground as "service lateral".

There are seven exceptions to the general rule that covers two or more service drops or laterals to a single building. If more than one service is permitted to the building by one of the exceptions then a permanent plaque or directory is required at each service drop or lateral explaining what section of the building the service supplys.

SERVICE #1 of two services. Service #1 will disconnect the 1st floor. SERVICE #2 is located in the northeast corner of this building.

← **The directory must be made of material to withstand the environment so that it remains legible.**

The service conductors are required to adequate capacity for the total load current that might be drawn *simultaneously*. The sizing of the service conductors involves a calculated step-by-step procedure. When a total load has been established, the required current-carrying capacity is easily determined by dividing the total load in kva by the service voltage.

I have written a book on these step-by-step procedures required by the Code for the *minimum* service size. I suggest reading this book "Calculations For The Electrical Exam".

Article 220 is shown below as the different procedures are shown in a format.

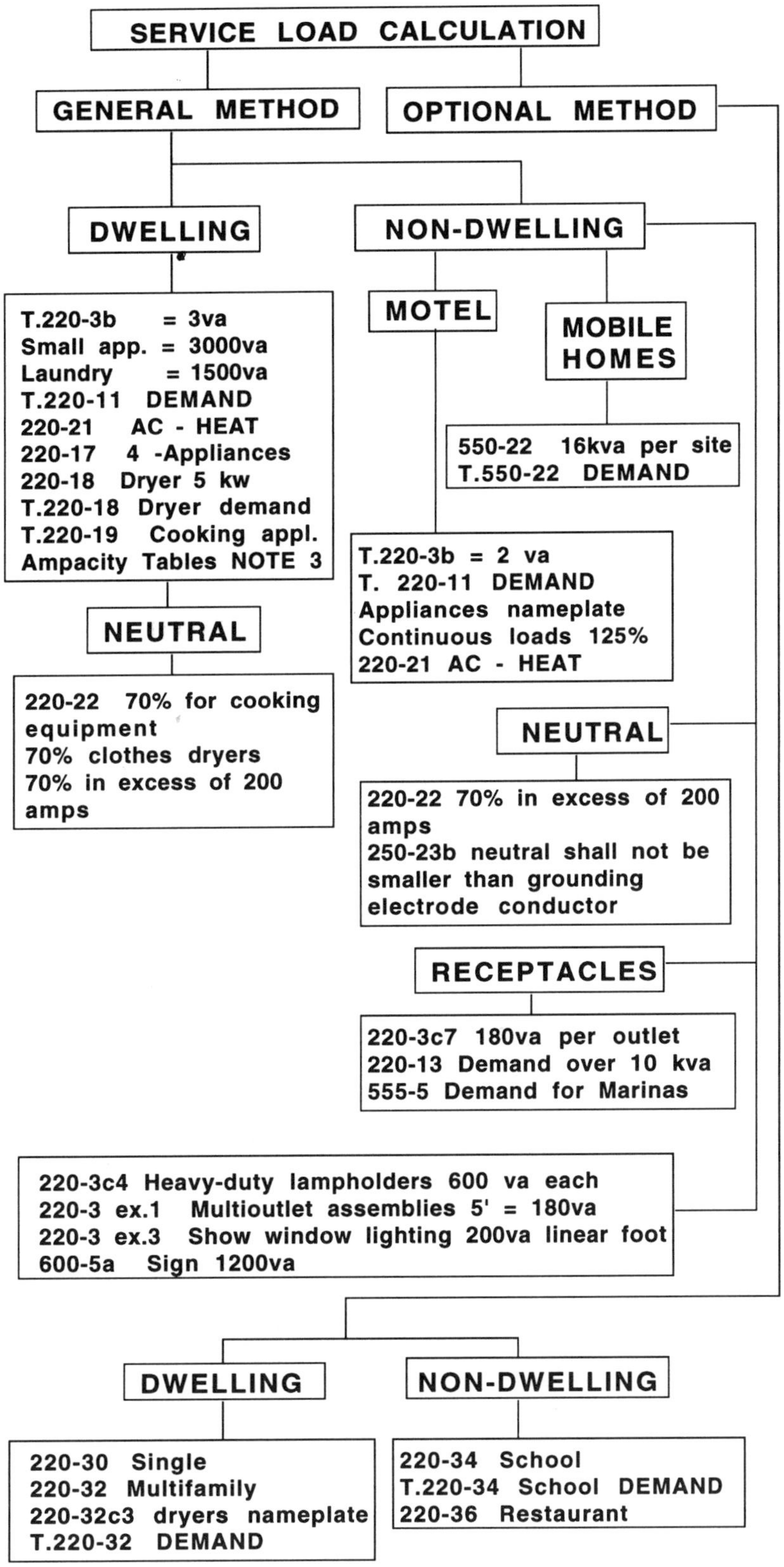

The temperatures recorded below are from an August afternoon with a temperature of 96°F.

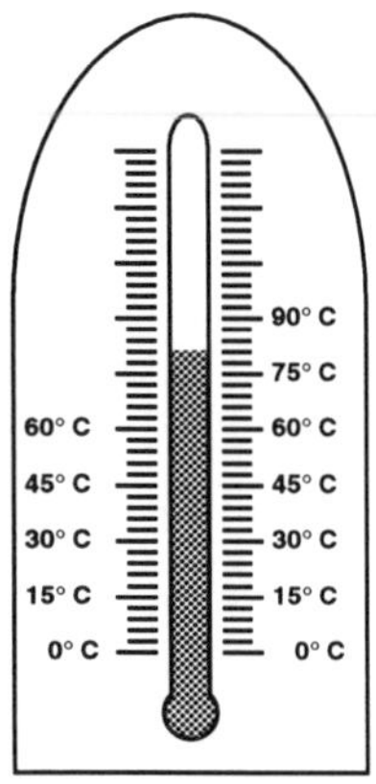

Metal raceways and enclosures exposed to direct sunlight can result in high ambient temperatures which must be considered when designing the service.

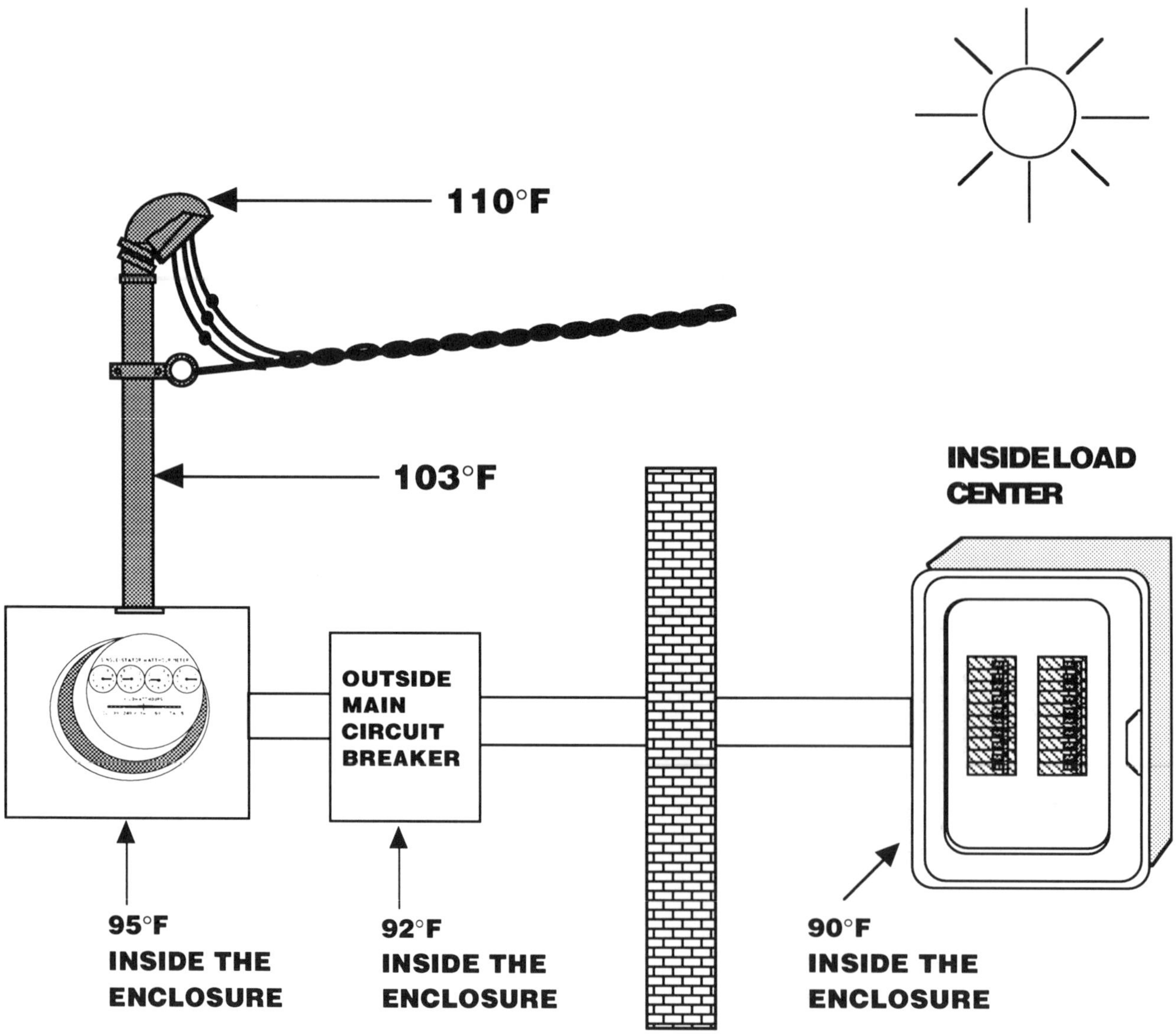

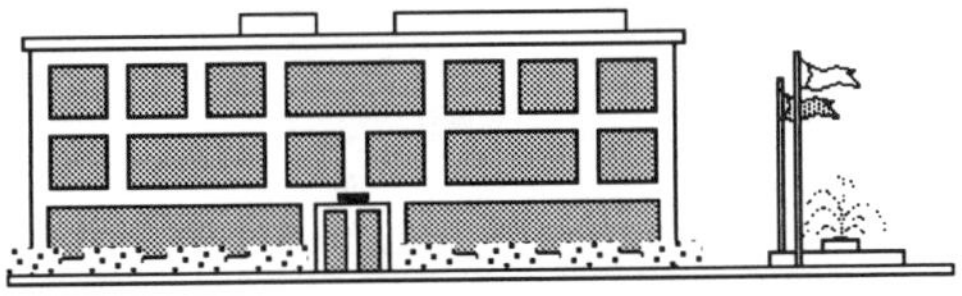

OFFICE BUILDING

A 50,000 sq.ft. office building is required by Table 220-3b to have a *minimum* lighting load of 3.5 volt-amps per square foot. Assume 208/120v, 3ø service.

607 AMPS

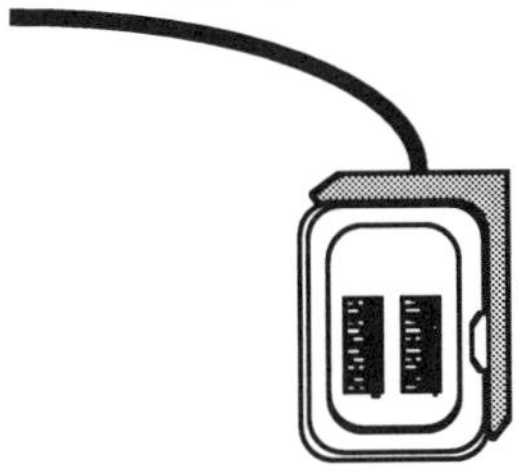

607 AMPS

50,000 sq.ft. x 3.5va x 125% = 218,750va/208x1.732 = 607 amps. The service conductors and panelboard busbars must have their load-current rating based on the total "computed load" of 607 amps.

Instead of determining the number of lighting branch circuits to the computed load the Code states in Section 220-4d that branch circuits need only be installed to serve the *connected* load.

20a CB

20a x 80% CONTINUOUS LIGHTING = 16 AMPS PER CIRCUIT

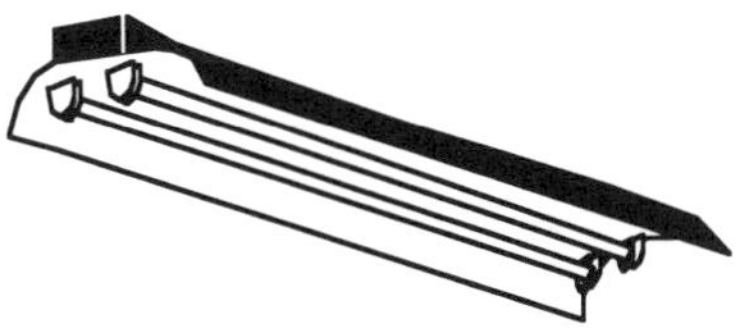

ACTUAL CONNECTED LOAD
90,000va

The office has an *actual connected* lighting load of 90,000va. To determine the number of branch circuits required for the lighting load the 90,000va would be divided by 1920va for each circuit. •A 20a circuit x 120 volts = 2400va per circuit x 80% continuous load = 1920va per circuit. 90,000va/1920va = 46.8 circuits/3 = 15.6 circuits per phase.

If the computed load of 218,750va was used: 218,750/2400va = 91.1 circuits/3 = 30.3 circuits per phase.

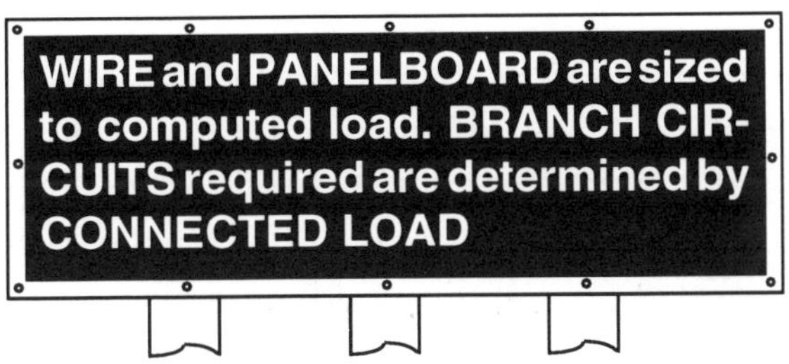

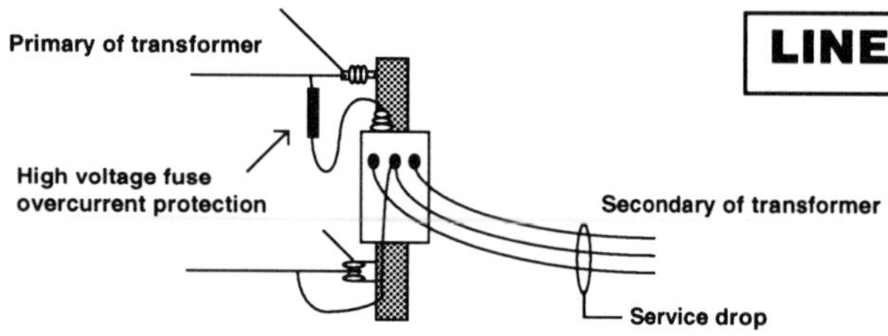

LINE - LOAD

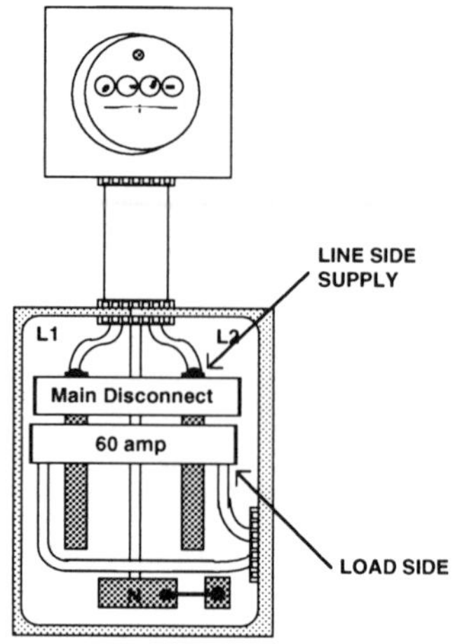

The *line* side consists of those conductors that are tapped or spliced to the service drop and run to the service equipment.

The *load* side consists of the conductors run from the over-current devices in the service equipment.

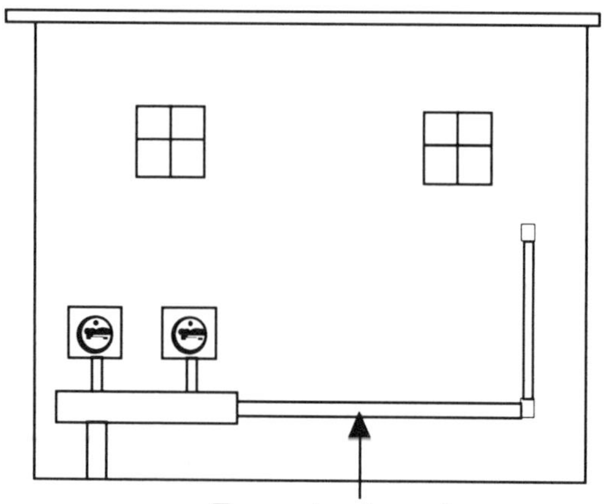

Two sets of service entrance conductors

The sketch shows two sets of service-entrance conductors supplying two apartments, in the same raceway. One set provides service to a downstairs apartment and the other to an upstairs apartment. This installation complies with Code section 230-7 which states, " Conductors *other* than service conductors shall not be installed in the same service raceway or cable."

The bottom sketch on the left is a Code violation as the service and fused feeder conductors are fed through the same nipples back to the gutter. The sketch on the right meets the requirements of the Code which states "Only *service* conductors are permitted in the same raceway." Once the service conductors are connected to a fuse or breaker, they become *feeder conductors* and are not permitted to be run with service conductors.

VIOLATION

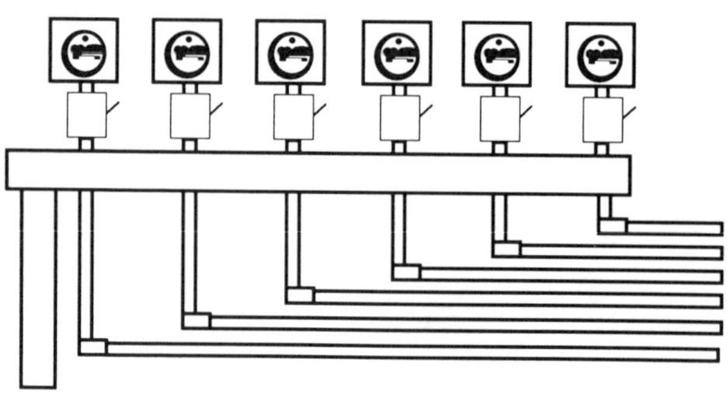

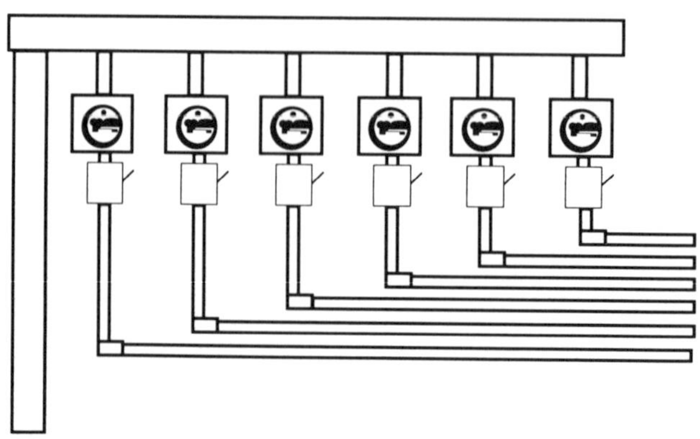

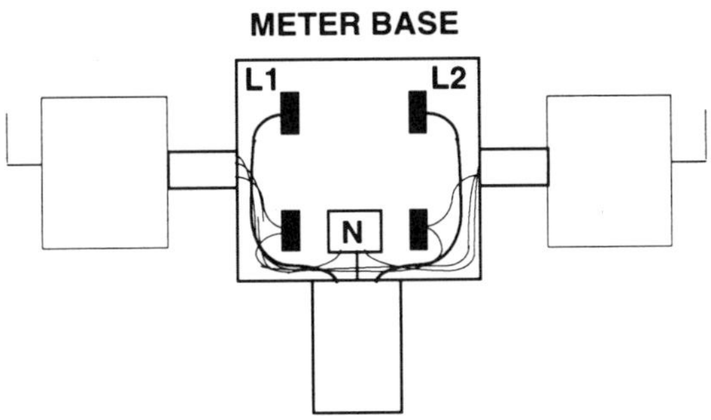

An underground service lateral feeds the bottom of the meter base. From the meter two service disconnects are fed. The service lateral conductors from the utility are crossing service conductors from the meter to the service disconnects. The service lateral conductors to the meter are unmetered, and the service conductors to the disconnects are metered. This is not prohibited by the Code as all the conductors are *service* conductors.

The service shown below has three #250 THHN conductors with an ampacity of 290 amps each. The four fused disconnects total 400 amps which exceeds the 290 ampacity of the conductor. This is permitted per Section 230-90 ex.3 which states "Not more than 6 circuit breakers or 6 sets of fuses shall be permitted as the overcurrent device to provide the overload protection. The *sum* of the ratings of the circuit breakers or fuses shall be permitted to *exceed* the ampacity of the service conductors, provided the *calculated load* does not exceed the ampacity of the conductors."

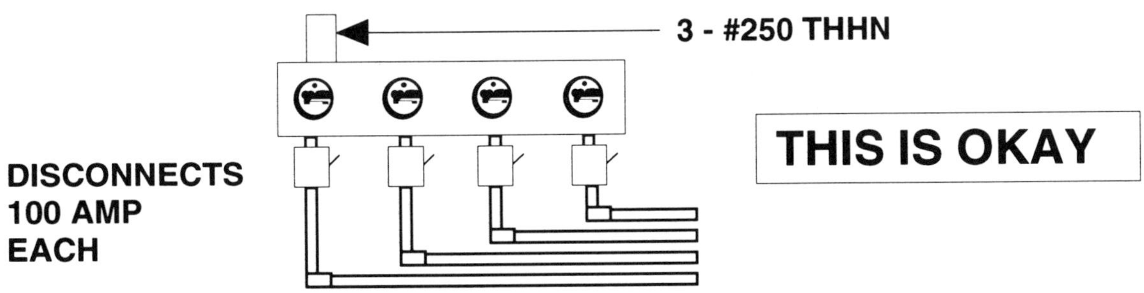

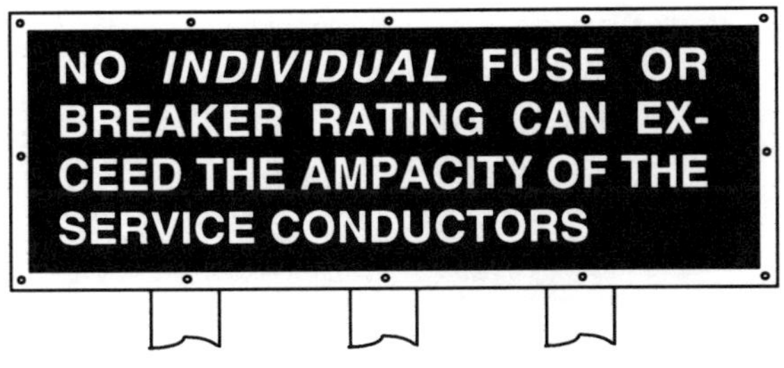

Shown below is a connected load of 275 amps. The example shows two different situations. The panelboard on the left has a main breaker of 400 amps the panelboard on the right has NO main breaker. The four breakers serve as the required disconnecting means.

The sizing of the conductors to the panelboard are different depending on which type of panelboard you connect the 275 amp load.

If the panelboard has a 400 amp main breaker the conductors can be protected to the next higher standard size per section 240-3b. Any conductor with an ampacity of over 350 amps would be protected by the next higher standard size of 400 amps. A #400 kcmil THHN has an ampacity of 380 and can be loaded to the 75°C ampacity of 335.

The main lugs only panelboard without a main overcurrent device would require the conductors to be sized to carry the load of 275 amps. A #300 kcmil THHN conductor has an ampacity of 320 and can be loaded to the 75°C ampacity of 285.

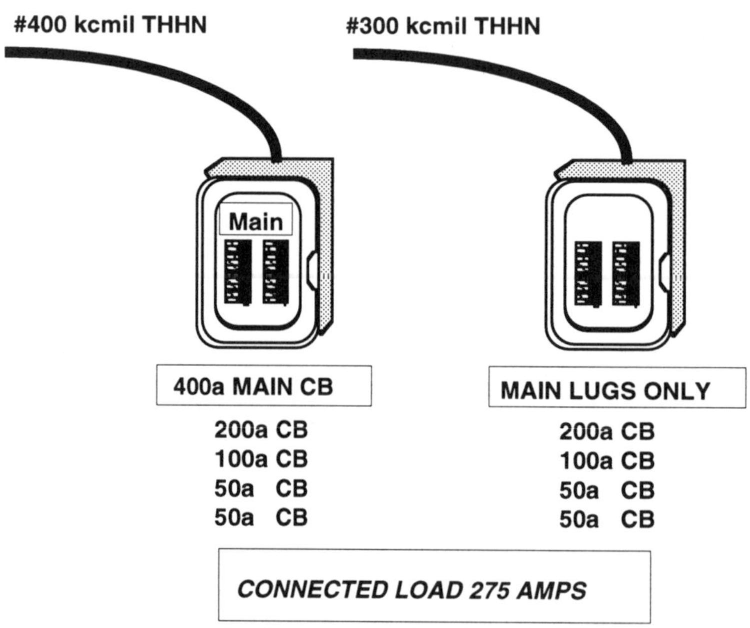

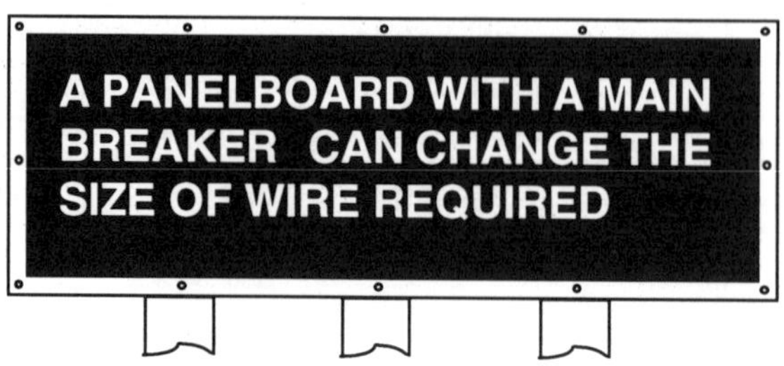

The Code requires the conductor to be sized to the *computed* load. The example below shows a Delta high-leg which would permit the high-leg conductor to be sized smaller than the other two phase conductors as the high-leg has a smaller computed load.

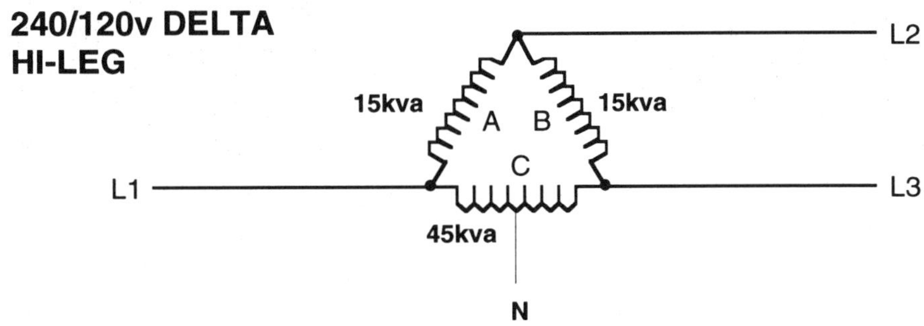

Phase A current = 15000va/240v = 62.5 phase amps
Phase B current = 15000va/240v = 62.5 phase amps
Phase C current = 45000va/240v = 187.5 phase amps

L1 current = 62.5a + 187.5a = 250a/2 = 125a x 1.732 = **216.5 amperes in L1**
L2 current = 62.5a + 62.5a = 125a/2 = 62.5a x 1.732 = **108.25 amperes in L2**
L3 current = 62.5a + 187.5a = 250a/2 = 125a x 1.732 = **216.5 amperes in L3**

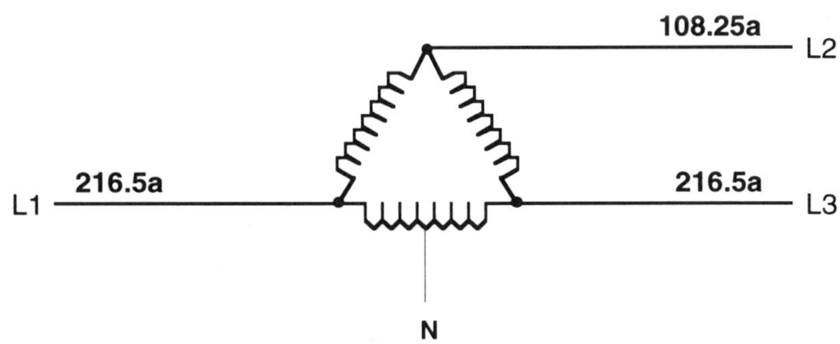

SERVICE PANELBOARDS

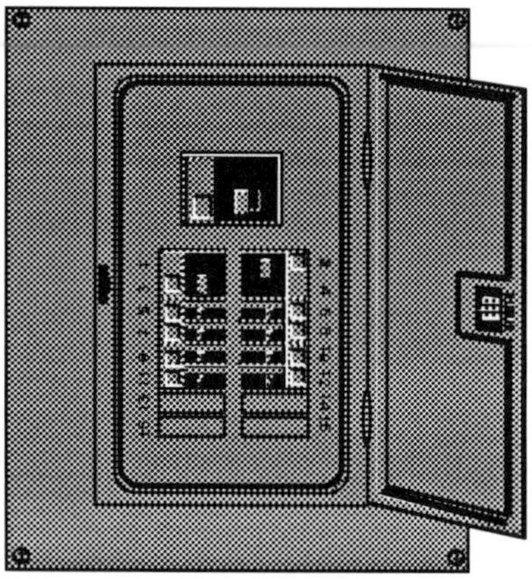

"SUITABLE FOR USE AS SERVICE EQUIPMENT"

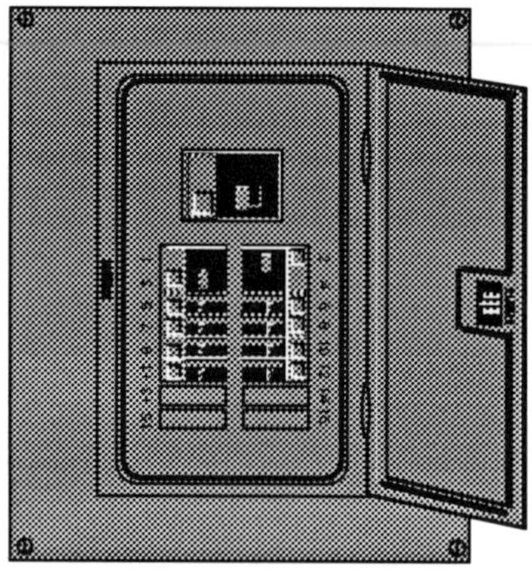

"SUITABLE *ONLY* FOR USE AS SERVICE EQUIPMENT"

The panelboard marked "SUITABLE FOR USE AS SERVICE EQUIPMENT" includes an isolated neutral bus that may be bonded to the metal enclosure through a bonding screw or strap. Where used for *service* equipment the neutral *must* be bonded. If the panelboard is *not* used for service equipment and is used as a subpanel. The neutal must be *isolated* from the enclosure.

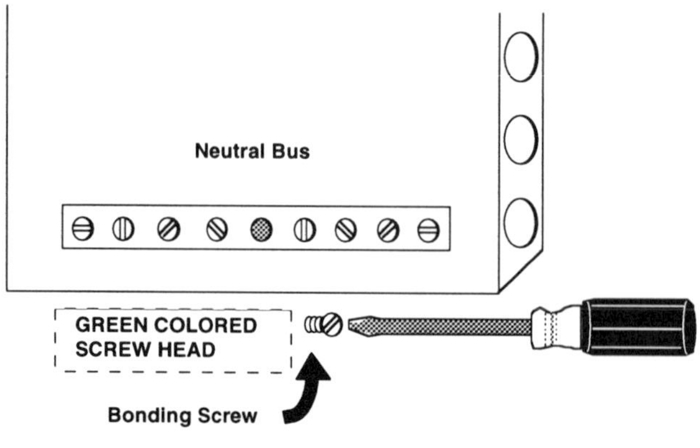

A panelboard that is marked "SUITABLE ***ONLY*** FOR USE AS SERVICE EQUIPMENT" has the neutral bus riveted or *permanently bonded* to the metal enclosure at the factory and cannot be used as a subpanel because there is no way to isolate the neutral bus from the enclosure.

THE PANELBOARD

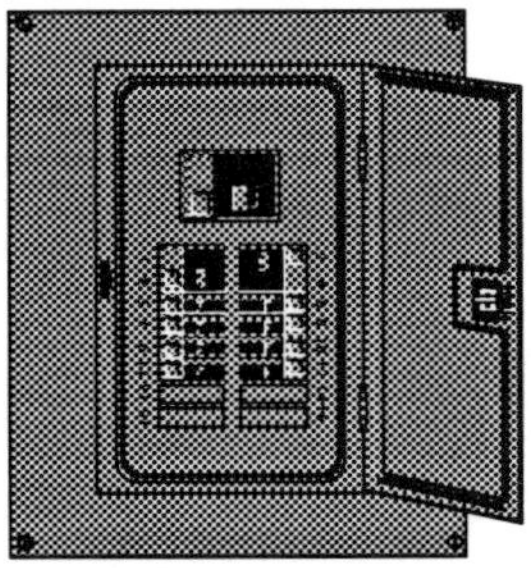

Definition: A single panel or group of panel units designed for assembly in the form of a single panel; including buses, automatic overcurrent devices, and equipped with or without switches for the control of light, heat, or power circuits; designed to be placed in a cabinet or cutout box placed in or against a wall or partition and accessible only from the front.

The term *loadcenter* is used to describe a light-duty type of panelboard used for residential and smaller commercial and industrial applications.

A panelboard provides overcurrent protection for the circuits contained within. A panelboard generally contains a means for switching the circuit on or off. Some of the older panelboards have a snap switch and a fuse. Most panelboards today contain circuit breakers which provides both overcurrent protection and a means for switching the circuit.

The selection of a panelboard is determined first by the number of circuits it must supply. Then the panelboard size is determined by the calculated load that it will supply. The busbars in the panelboard must have sufficient current rating for the calculated load of all the circuits.

Example: A new home has a total load computed load by Article 220 of 87 amps. The panelboard is a 30 circuit with 150 amp main lugs. The feeder to the panelboard is #1/0 aluminum which is protected by a 125 amp main circuit breaker located on the exterior of the house.

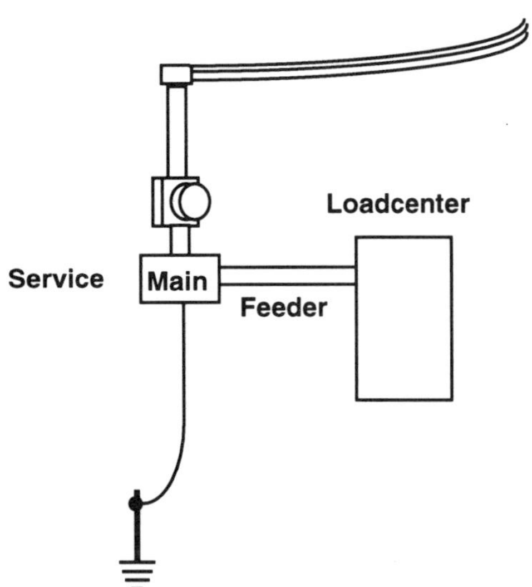

Section 384-13 requires that panelboards have a rating not less than the minimum feeder or service *capacity* required for the load computed by Article 220. The key word is *capacity* not ampacity. The calculated feeder capacity is 87 amps, the busbars must have a rating of at least 87 amps.

ALL PANELBOARDS ARE REQUIRED TO HAVE BUSBARS WITH A CURRENT CAPACITY NOT LESS THAN THE REQUIRED LOAD COMPUTED FROM ARTICLE 220

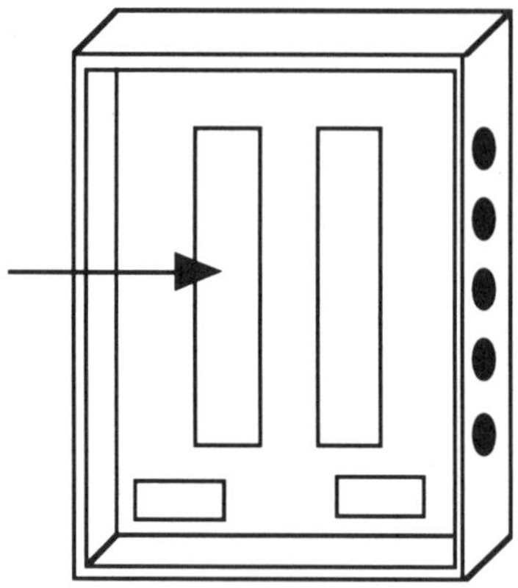

A panelboard may have a busbar current rating *larger* than the calculated feeder capacity but never less than the calculated load. The calculated load generally is smaller than the actual connected load.

The calculated load is 87 amps, Note 3 to Table 310-16 permits an ampacity of 125 amps for #1/0 AL, the overcurrent protection for the feeder conductors is 125 amps.

THE OVERCURRENT PROTECTION CANNOT BE LARGER THAN THE RATING OF THE PANELBOARD

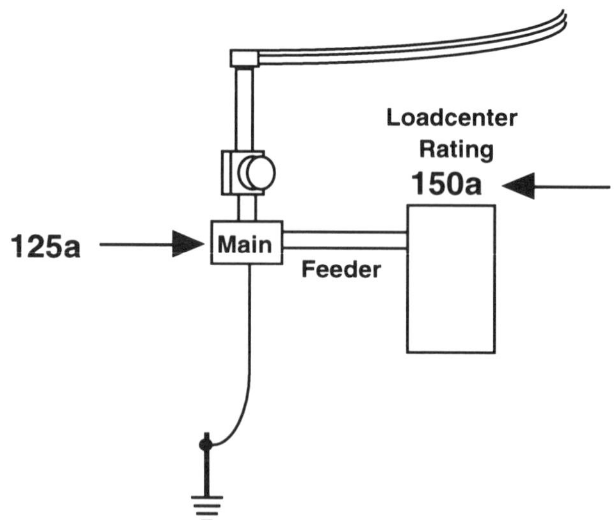

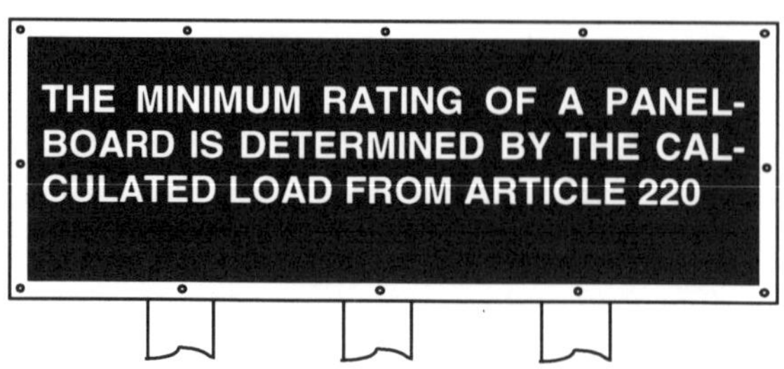

Often the question arises, can you install 100 amp service cable to a 200 amp, 30 circuit main lug panel, with a 100 amp main breaker? The answer is *YES*!

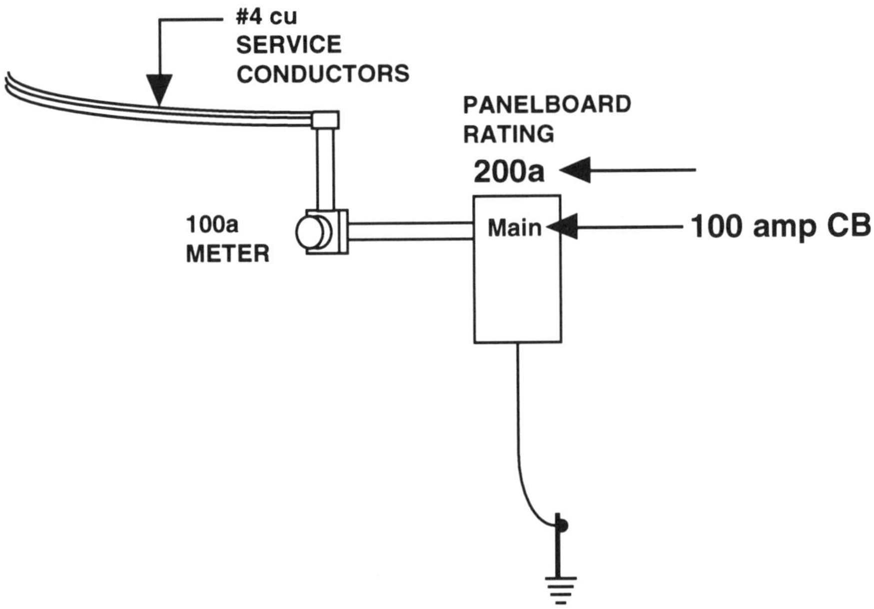

Sometimes you find there is a need for a panelboard with a main breaker that has a current rating LESS than the normally required rating of the panelboard to have the number of branch circuits required.

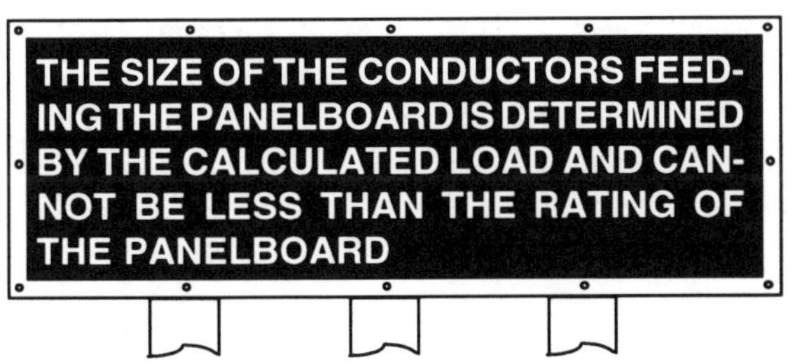

Section 384-16 states each lighting and appliance panelboard shall be individually protected on the supply side by *not more than two* main circuit breakers or two sets of fuses having a combined rating not greater than that of the panelboard. The exception to this rule is, individual protection for a lighting and appliance panelboard shall *not* be required if the panelboard *feeder* has overcurrent protection not greater than the rating of the panelboard.

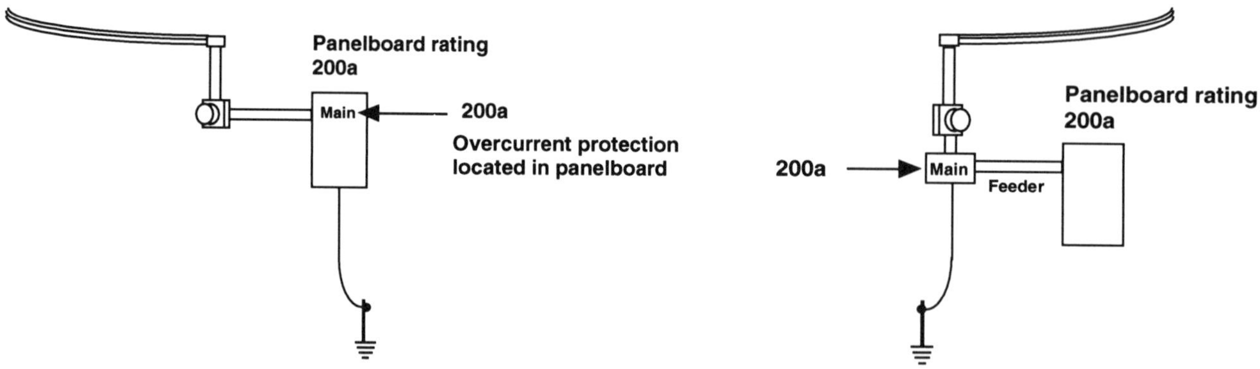

A panelboard might be a service equipment panelboard, distribution panelboard, or a branch-circuit panelboard. A single panelboard could combine two of these functions or all three within a single cabinet.

If the panelboard is used as a service equipment panelboard, it must be marked *suitable for service equipment.*

The most common panelboard is the lighting and appliance panelboard. No more than 42 overcurrent devices (in addition to the main) may be installed.

It is very important to understand the definition of a *lighting and appliance* branch circuit panelboard.

Section 384-14 defines a lighting and appliance panelboard as one having *more* than 10% of its overcurrent devices rated *30 amps or less*, for which *neutral connections* are provided.

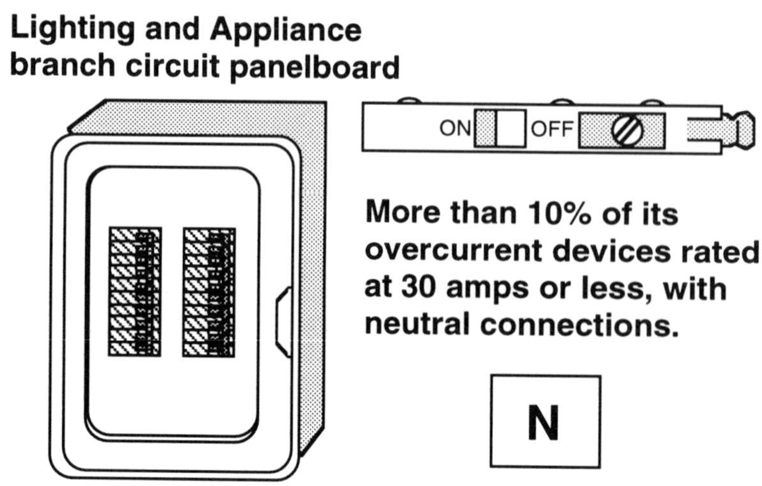

SINGLE-PHASE 3-WIRE

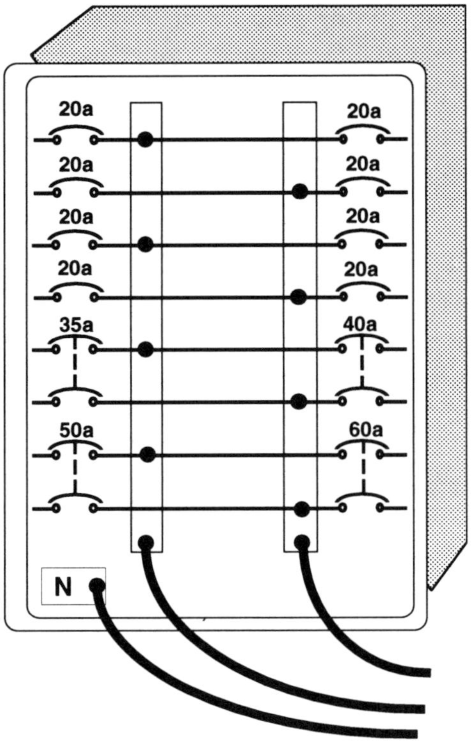

Some panelboards may appear to be power or distribution panelboards, but they are technically lighting and appliance panelboards. This is important because it indicates whether or not *main overcurrent protection* is required for a panelboard.

The single-phase panelboard has 16 overcurrent devices (each pole of a circuit breaker is counted as a separate device, 2-pole breaker counts two, 3-pole breaker counts three). 10% of 16 = 1.6 or 2. This panelboard easily qualifies as a lighting and appliance panelboard and would require overcurrent protection.

THREE-PHASE 3-WIRE

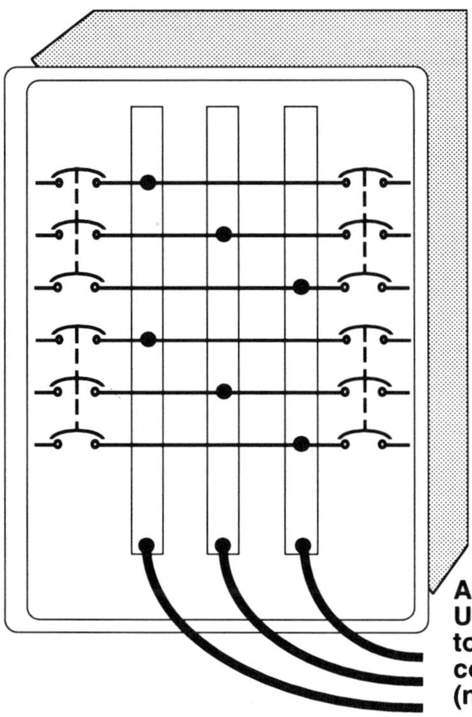

The three-phase panel feeds lights but is not considered a lighting panel as there are no neutral connections. Main protection is not required.

THREE-PHASE 4-WIRE

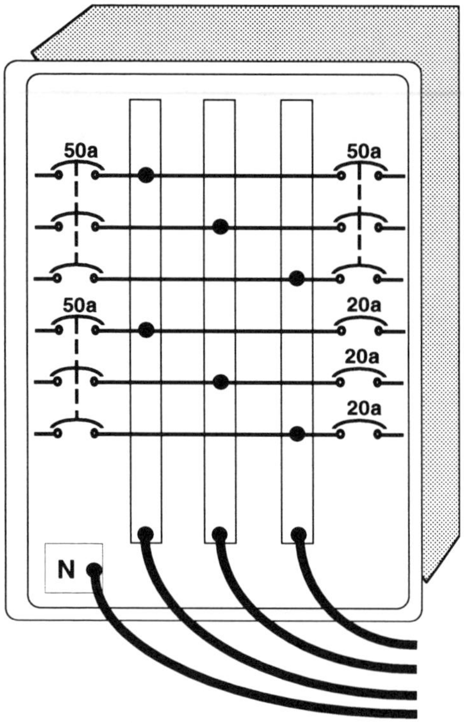

This 3ø, 208/120v, 4-wire panelboard contains 3-50 amp three-phase motor circuits and 3-20 amp single-phase 120v receptacle circuits. Even though it contains no lighting circuits it is classified as a lighting and appliance panelboard because more than 10% of the overcurrent devices are 30 amps or less, and neutral connections are provided. Main overcurrent protection must be provided for the panel busbars.

THREE-PHASE 4-WIRE

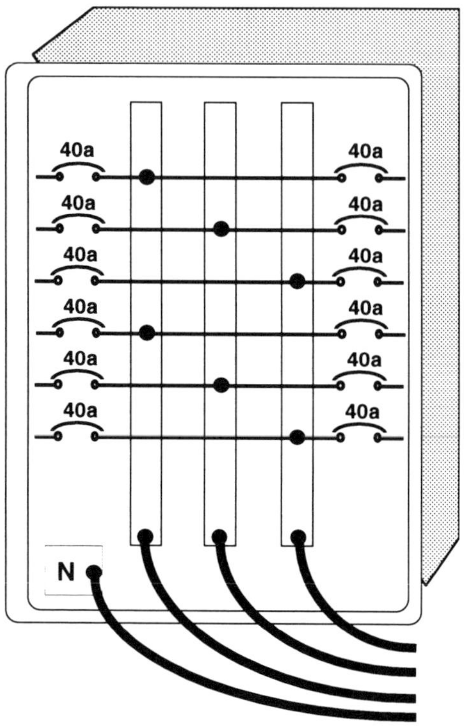

This 3ø, 208/120v, 4-wire panelboard contains 12-40 amp single-phase lighting circuits. Even though it contains all lighting circuits it is *not* classified as a lighting and appliance panelboard because all of the overcurrent devices are over 30 amps, even though neutral connections are provided. Main overcurrent protection is not required.

It is important to understand the difference between a lighting and appliance panelboard, a power panel, and a switchboard.

SWITCHBOARD

A large single panel, or assembly of panels on which are mounted, on the face or back, or both, switches, overcurrent and other protective devices, buses, and usually instruments. Switchboards are generally accessible from the rear as well as the front.

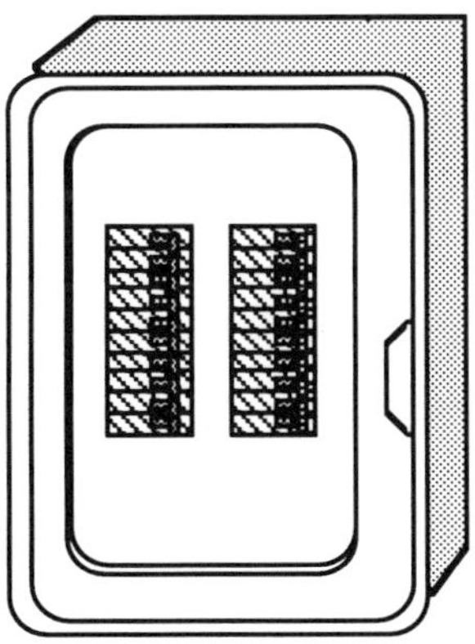

A lighting and appliance panelboard shall not contain more than 42 overcurrent devices (other than those provided for in the mains).

A lighting and appliance panbelboard shall be provided with a physical means to prevent the installation of more overcurrent devices than that number for which the panelboard was designed.

A 1-pole breaker is counted as one, a 2-pole breaker as two, and a 3-pole breaker as three.

•A *tandem* breaker takes one space and it has two circuits, it counts as *two* overcurrent devices.

ONE POLE

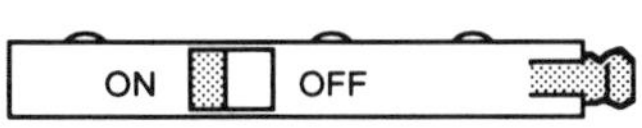

ONE CIRCUIT

ONE POLE TANDEM

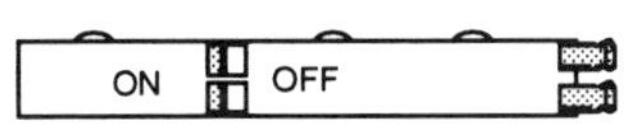

TWO CIRCUITS

MAINS RATING AMPS	SPACES	MAXIMUM NUMBER 1-POLE CIRCUITS	MAXIMUM NUMBER TANDEM BREAKERS
150	20	30	10
	24	24	0
	30	30	0
200	20	40	20
	24	24	0
	30	30	0
	30	40	10
	40	40	0

Loadcenters are designed, rated, and approved with a main ampere rating and number of spaces.

A 200 amp rated loadcenter has 20 spaces which could have a maximum of 40 single-pole circuits by using 20 tandem breakers. Each tandem breaker is two circuits, but takes only one space.

A 200 amp loadcenter with 30 spaces is rated for a maximum of 40 single-pole circuits. By using 20 single-pole breakers = 20 circuits, and 10 tandem breakers = 20 circuits, a total of 40 circuits, using 40 overcurrent devices in 30 spaces.

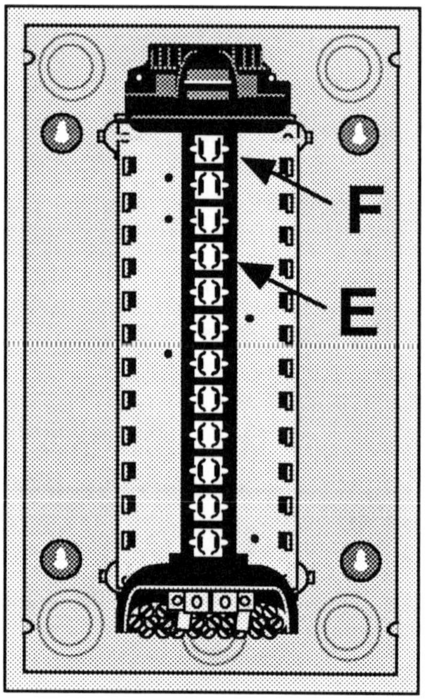

Class CTL is a UL designation for *circuit limiting* which the Code requires not more than 42 overcurrent devices.

The slots in the panelboard have different configurations to limit the total number of spaces to no more than 42.

Each "F" slot configuration will accept only one pole . Each "E" slot configuration will accept two poles.

TANDEM BREAKER

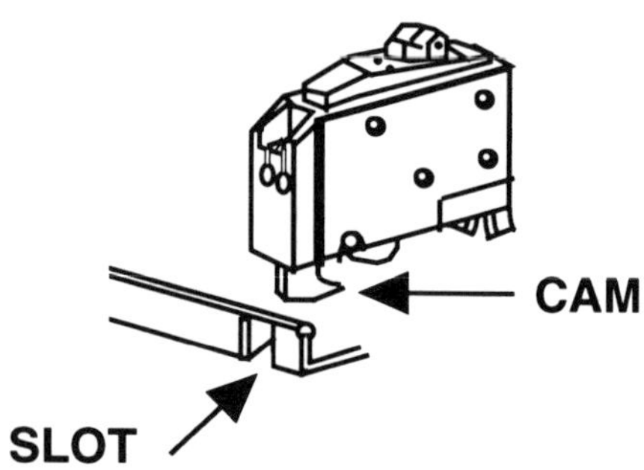

Not all panelboards are designed or listed for tandem breakers.

Panelboards which are designed to accept tandem breakers are designed so the mounting cam on the tandem breaker can only be installed in a position that has a mounting pan rail slot which will accept the cam. This prevents the panelboard from having over 42 overcurrent devices.

A *violation* of the UL listing and the Code occurs when the cam is broken off so the tandem breaker can be installed into a space that was not designed to receive it.

SERVICE DISCONNECT REQUIREMENTS

MAIN

Code Section 230-70 requires a means of disconnecting all conductors from the service entrance conductors. The location of the disconnecting means can be either inside or on the outside of a building. At either location it shall be readily accessible.

Each service is permitted *up to six* disconnecting means.

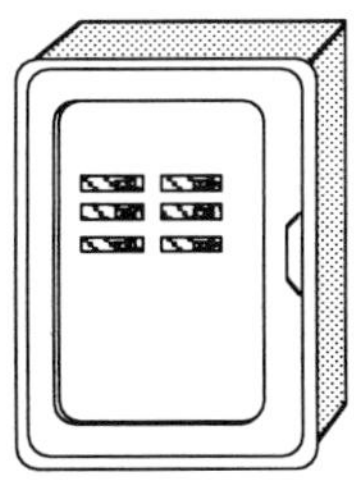

6 Breakers single enclosure

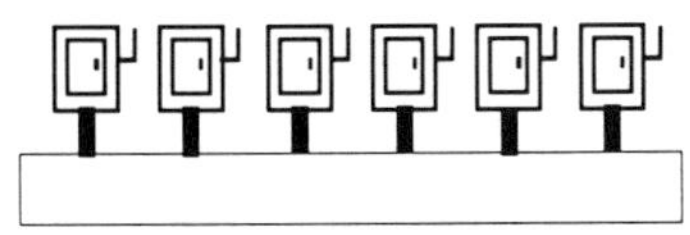

6 switches in a group of separate enclosures

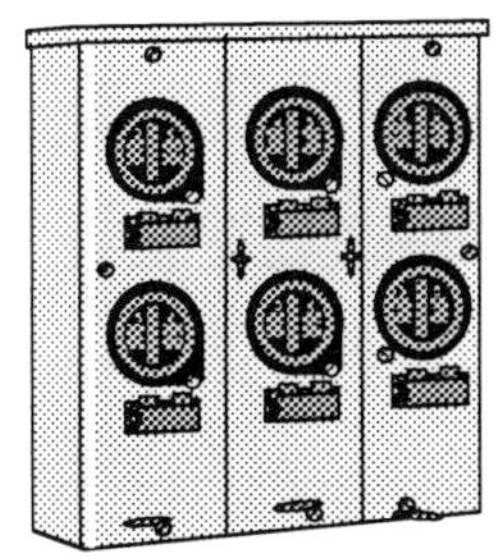

No more than 6 disconnects per service in any one location

Single-pole switches or circuit breakers with handle ties may be grouped as a single disconnect.

The Code allows six movements of the hand to disconnect the service.

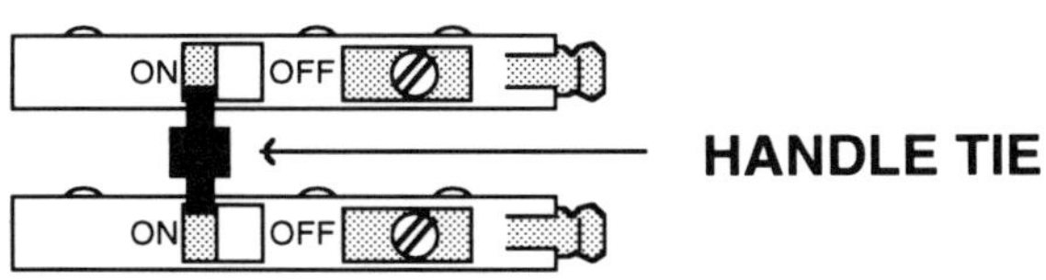

ONE MOVEMENT OF THE HAND

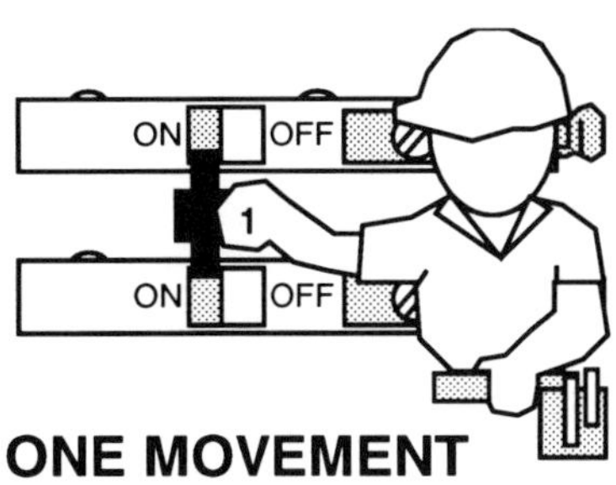

ONE MOVEMENT OF THE HAND

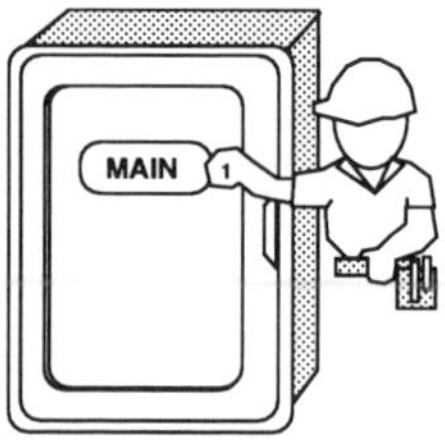

ONE MOVEMENT OF THE HAND

MAIN

The best situation is a main disconnect which disconnects the service with one switch or breaker.

LIGHTING AND APPLIANCE PANELBOARD

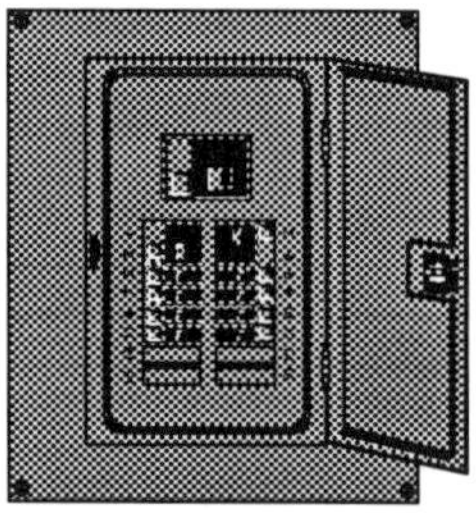

Section 384-16a requires each lighting and appliance branch-circuit panelboard to be protected on the supply side by *not more than two* main circuit breakers or two sets of fuses having a combined rating not greater than that of the panelboard. The definition of a lighting and appliance panelboard is one that has more than 10% of its overcurrent devices rated 30 amps or less with neutral connections.

The service disconnects are required to be grouped. Each disconnect shall be marked to indicate the load that it will shut off. If one of the disconnects is for a fire pump it can be located at a remote location separate from the other service disconnects.

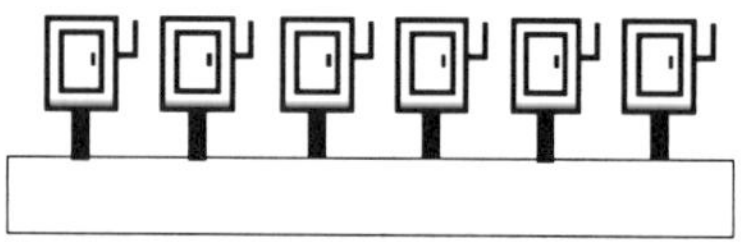

Service conductors must be readily disconnected from all loads while standing at one location

EACH DISCONNECT IS TO BE MARKED TO INDICATE THE LOAD SERVED

Section 230-72b requires that a disconnect used for emergency lighting, fire pumps, standby systems, etc. be located at a different location than that of the service disconnecting means. The reason is that a fire near the service equipment could knock out the emergency systems. There is no minimum distance stated in the Code, but proper design would suggest at least 10'.

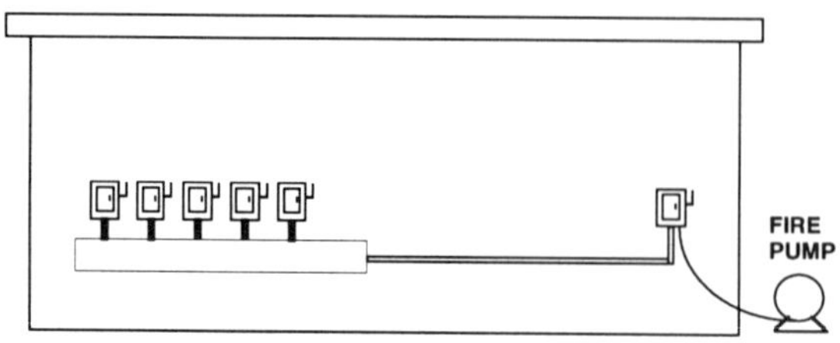

Sections 700-12d and 700-12e require emergency services to be separated from other services to prevent failure of both systems due to one fault.

SERVICE ENTRANCE CONDUCTORS

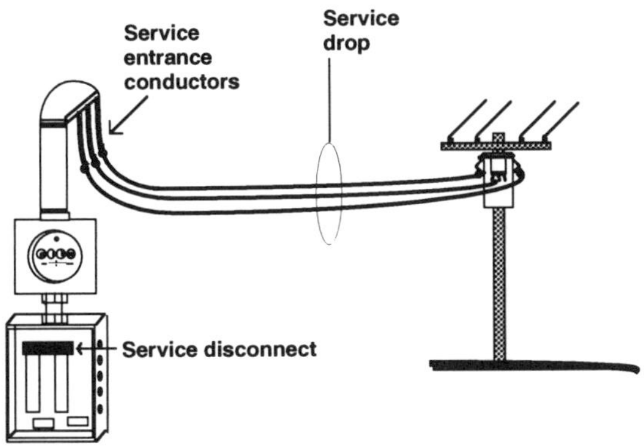

The general rule is the service drop or service lateral shall supply only one set of service entrance conductors.

An exception to the general rule is that multi-occupancy buildings shall be permitted to have one set of service entrance conductors run to each occupancy or to a group of occupancies.

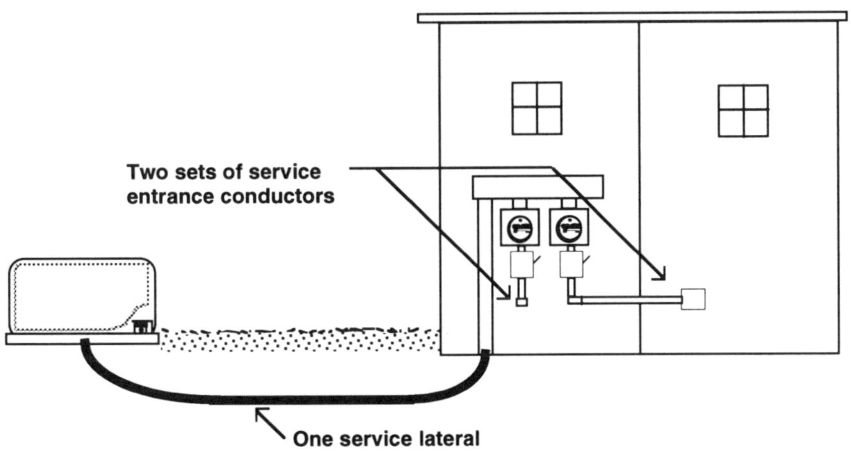

The Code permits more than one set of service entrance conductors to be tapped from a *single* service drop or service lateral.

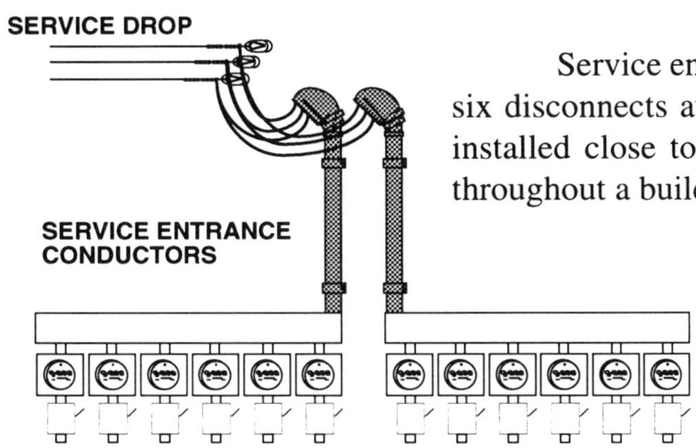

Service entrance conductors can be tapped to supply up to six disconnects at a common location. The disconnects must be installed close together at one location. They cannot be spread throughout a building.

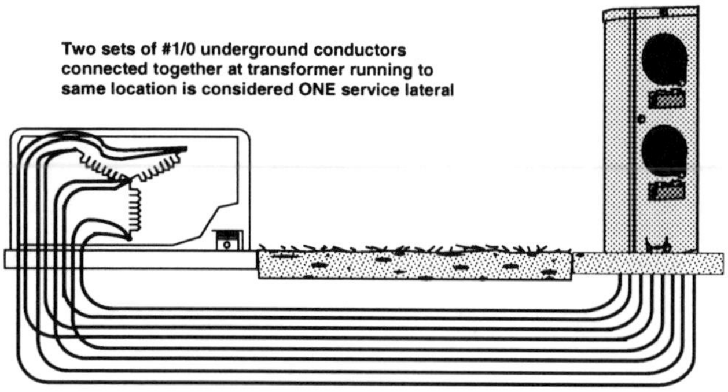

An exception allows two to six sets of service lateral conductors, #1/0 or larger, connected together at their supply end but not connected together at their load end to be considered as *one* service lateral.

The general rule is that a building shall be supplied by only *one* service. The definition of "service" is the conductors and equipment for delivering energy from the electricity supply system to the wiring system of the premises served. The service to a building may be either overhead or underground. Overhead is referred to as "service drop" and underground as "service lateral".

TWO BUILDINGS UNDER SAME ROOF

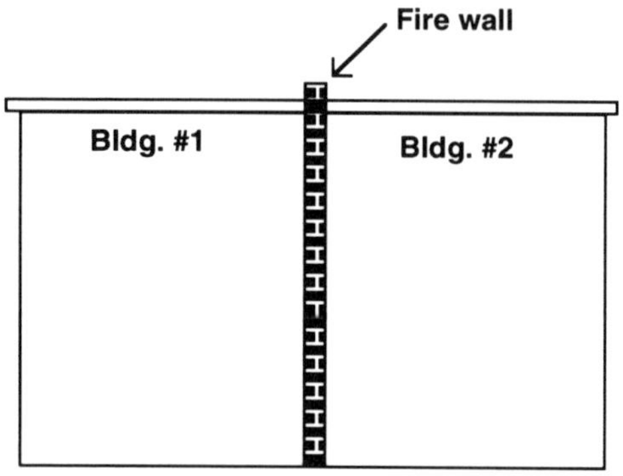

The definition of a building is a structure that stands alone or that is cut off from adjoining structures by fire walls with all openings therein protected by approved fire doors.

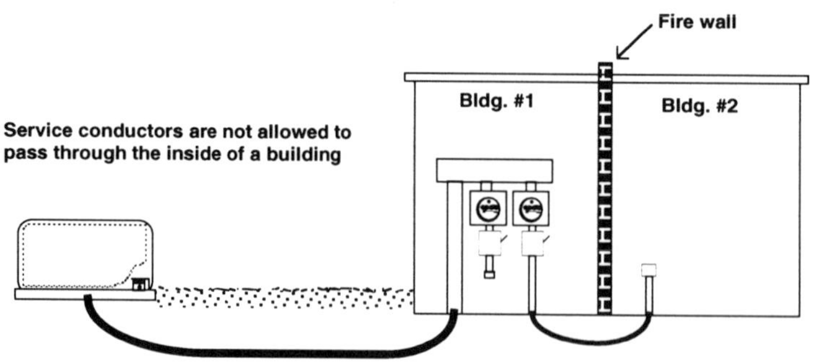

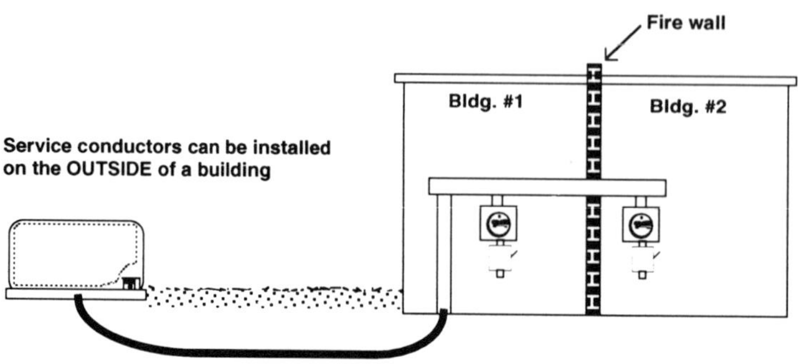

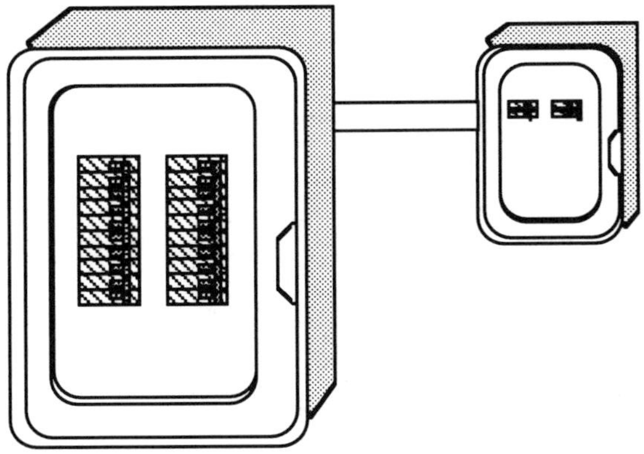

More than 10% of its overcurrent devices rated at 30 amps or less, with neutral connections.

A tap is made from the service panel to a four circuit "MAIN LUGS ONLY" panel using the four breakers at the disconnecting means. If the panel qualifies as a *lighting and appliance* panel, the maximum number of breakers that can be used as a disconnecting means is *two*. This panel would require a *main* breaker for individual protection.

DETACHED BUILDINGS

Section 225-8 requires that where *more than one* building or other structure is on the same property and under single management, each building or other structure served shall be provided with a means for disconnecting *all* ungrounded conductors. The disconnecting means shall be suitable for use as *service* equipment. The disconnect is to follow the same rules as a service disconnect in section 230-71 and 230-72.

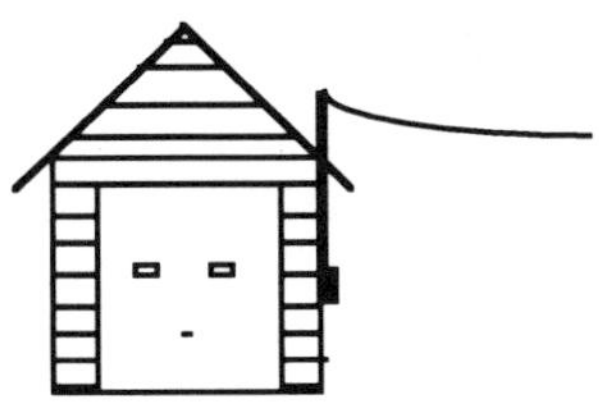

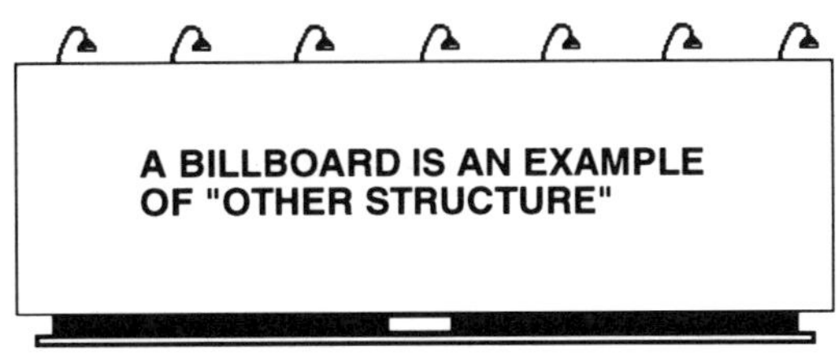

A SEPARATE DISCONNECT IS REQUIRED AT THE GARAGE LOCATION

If an outbuilding, such as a garage, has a panel installed, the disconnect would have to be installed either inside or outside at the garage. The garage *cannot* be disconnected from a panel at the house.

An exception allows for outbuildings on residential property, a snap switch or a set of 3-way or 4-way snap switches suitable for use on *branch circuits* shall be permitted as the disconnecting means. The switch would have to be located at the garage and readily accessible.

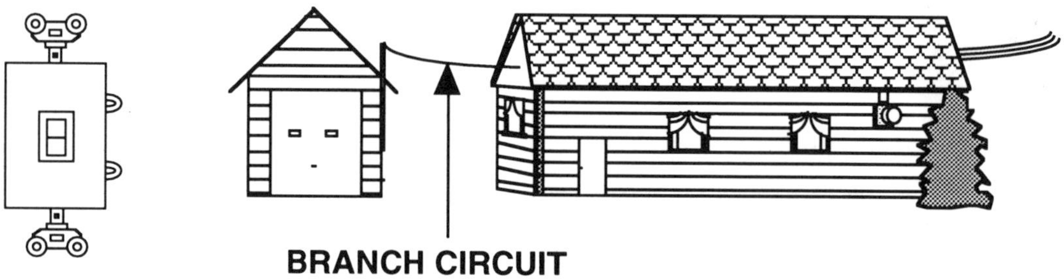

GROUNDING THE SERVICE

The sketch below shows the neutral bus in the service panel properly *bonded* to the panel and electrode. The service equipment enclosure is called the "bullseye" of grounding. The service equipment is the only location where all the grounding is connected together.

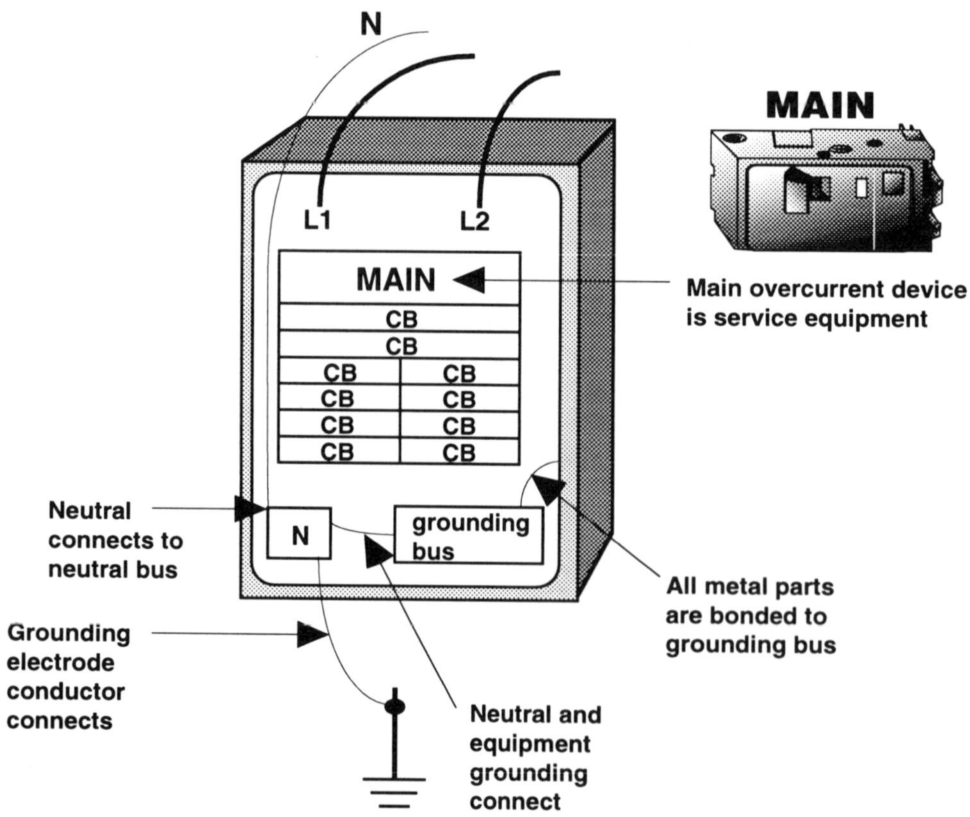

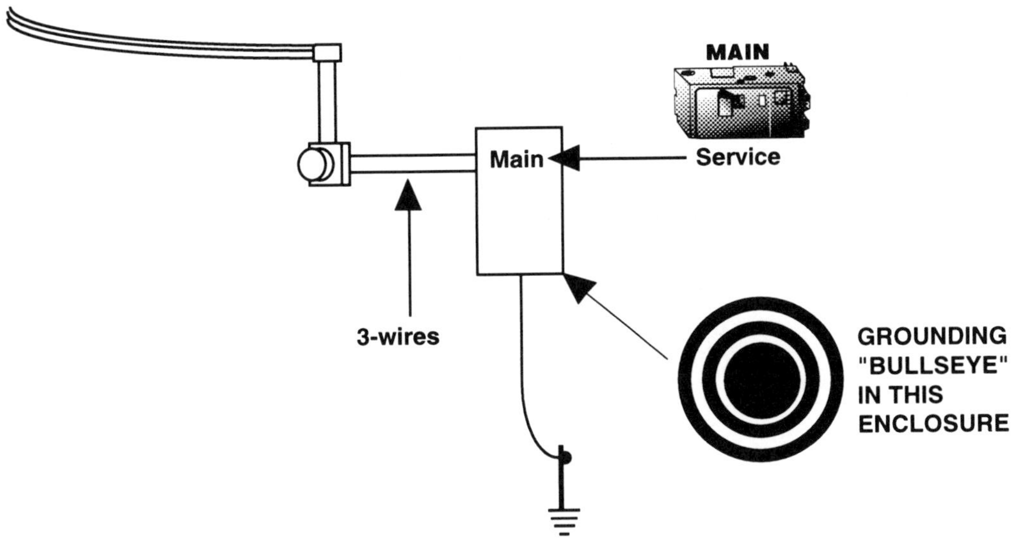

Section 250-81 lists the electrodes in parts a,b,c and d.

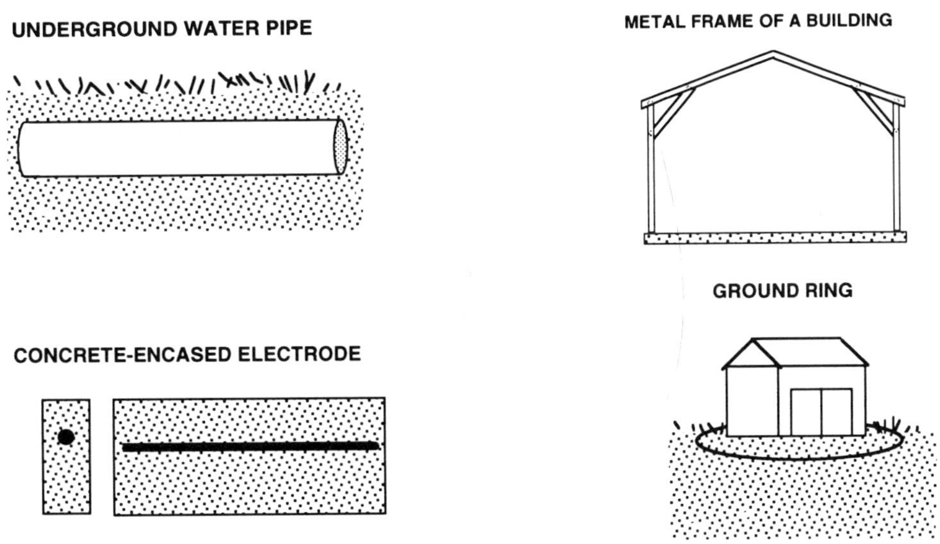

Section 250-83 b,c and d lists *made* electrodes.

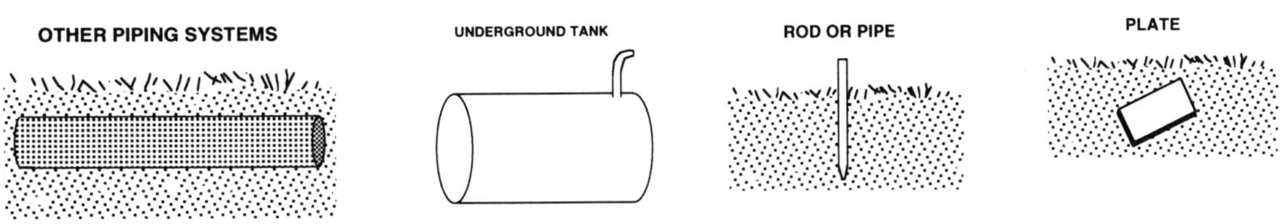

Section 250-84 states the requirements for resistance to earth of a *made* electrode. A *single* made electrode, either a ground rod, pipe, or plate that does *not* have a resistance to ground of *25 ohms or less* shall have one additional electrode in parallel spaced at least 6 feet apart. No ground resistance reading is required when a second electrode is added.

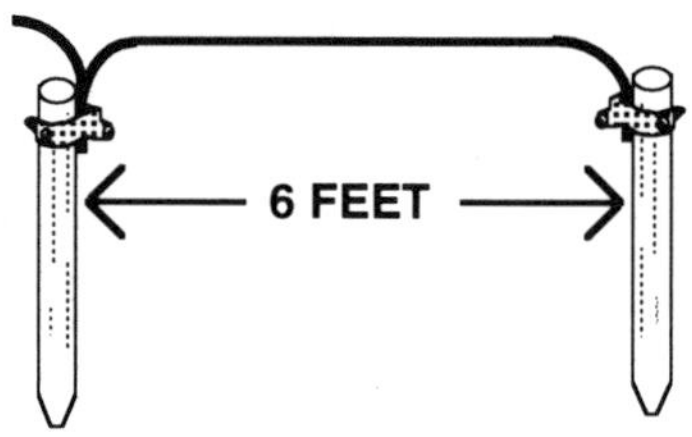

Two well-spaced ground rods driven into the earth provide parallel paths of resistance. The rule for two resistances in parallel does not apply in this situation. The total resistance is not one-half of the individual rod resistance to earth. Actually, the reduction for two equal-resistance ground rods is about 40%. If three rods are used, 60%, and if four rods are used the reduction is 66%.

When using multiple rods, they must be spaced further apart than the depth they are driven. If spaced too close together as shown below, the earth shell resistance area will be overlapped causing a higher resistance.

If you had two ground rods connected in parallel spaced 10 feet apart, the resistance is lowered approximately 40%. If the spacing is increased to 20 feet apart, the resistance is lowered to approximately 50%.

Ground rods are in parallel with each other as they make contact with earth. It's not like a circuit, when connected in series, the removal of one would open a series circuit. As shown below, with either connection the removal of a ground rod would not open the circuit to earth. The grounding electrode *conductor* shall be installed in one continuous length without a splice or joint.

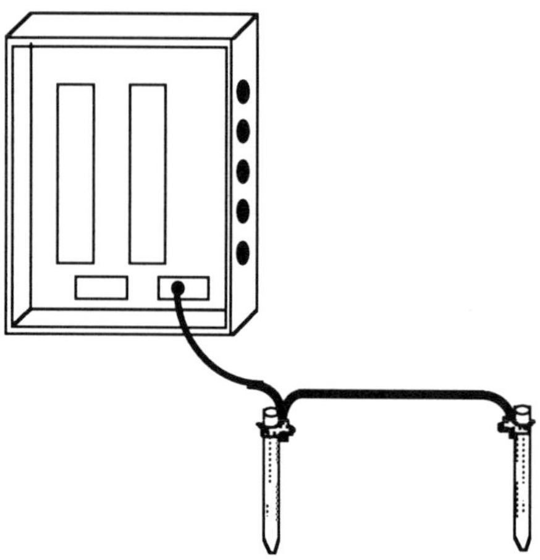

GROUND RODS ARE IN PARALLEL

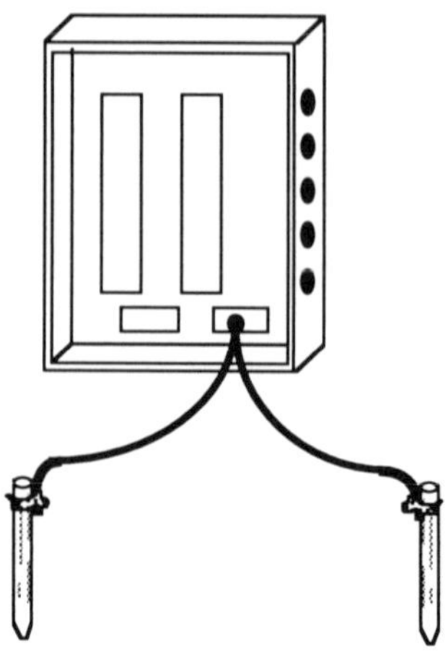

GROUND RODS ARE IN PARALLEL

Some areas have a local Code which requires the main disconnect to be on the outside of a building. In case of a fire, the fireman can disconnect the electrical service without going into the building.

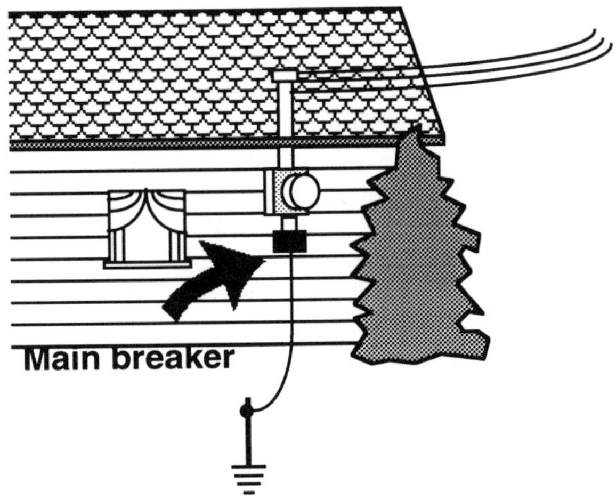

With the main breaker outside of the house, the wiring between the main breaker and the panelboard inside of the house is called a *feeder*. This requires 4-wires; two ungrounded hot wires, one neutral, and one equipment grounding conductor.

The neutral conductor, equipment grounding conductor, and grounding electrode conductor would all be bonded together in the main breaker panel on the *outside* of the house. The panelboard inside the house would have the neutral and equipment grounding conductor *isolated* from each other. The neutral would *not* be bonded to the panelboard. *Only at the service equipment is the neutral bus bonded to the metal cabinet.*

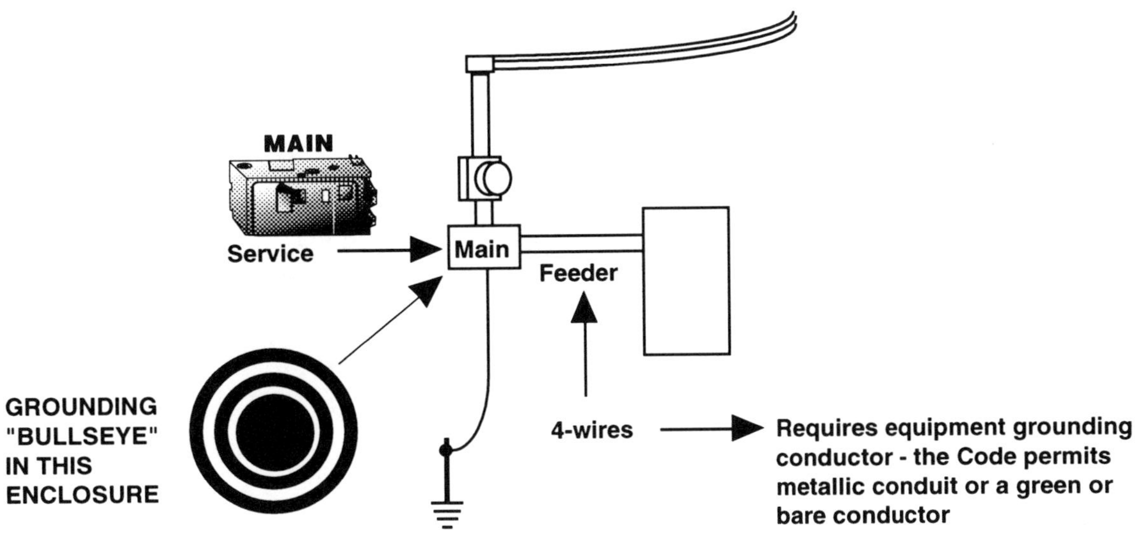

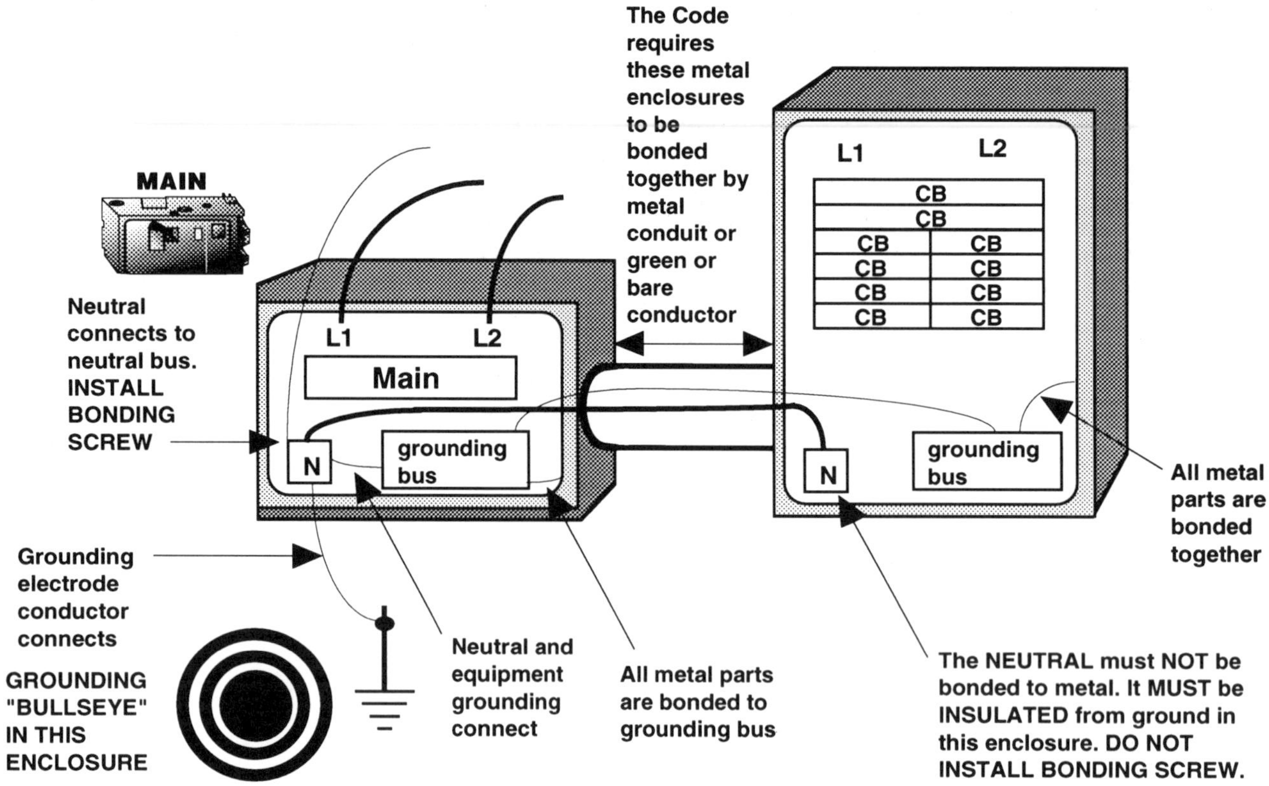

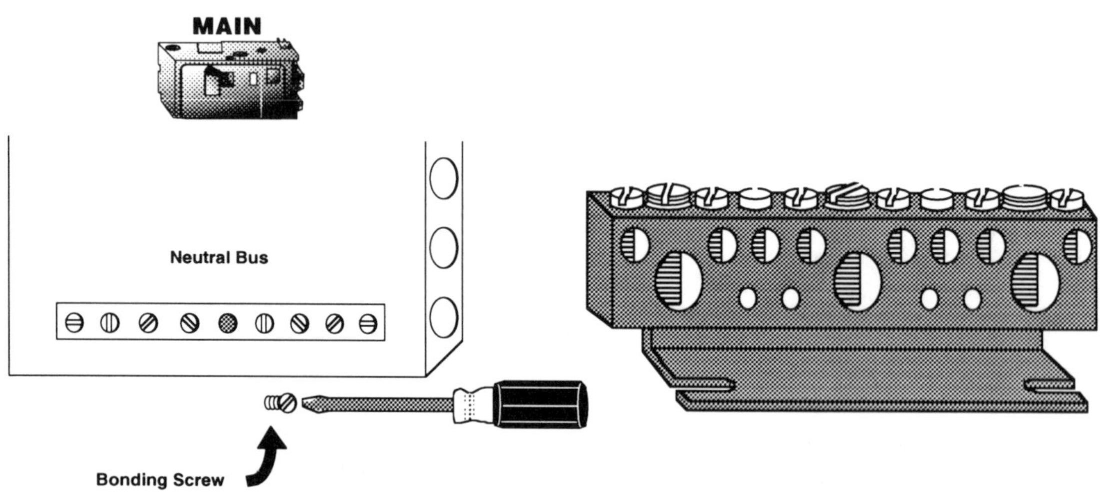

The neutral is only bonded to metal *at the service*; NOT at a sub-panel.This is one of the most violated areas in grounding, mainly because some people simply don't understand the functions of grounding. When incorrectly connecting the neutral to the sub-panel you would have currents flowing on all the metallic equipment connected to the panelboard.

GROUNDING THE SECOND BUILDING

The service is installed at the first building with a feeder to a panelboard in the second building. It is a feeder when it leaves the first building but must, in many respects, be treated as a service as it arrives at the second building.

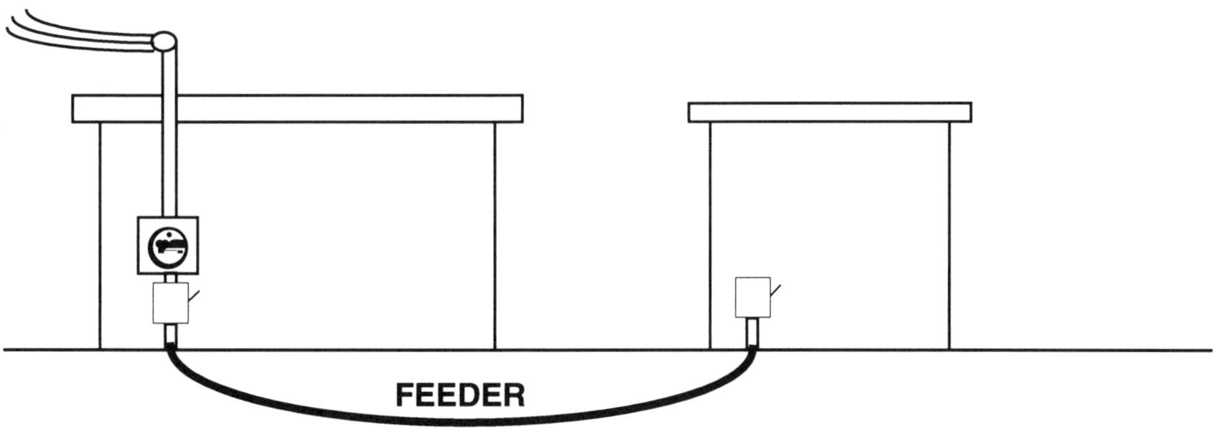

The second building requires a disconnecting means located inside or outside the building at a readily accessible location nearest the point of entrance. Up to six switches or breakers can be used unless the feeder terminates in a *lighting and appliance* panelboard, then the disconnects would be limited to *two*.

Section 215-6 requires the feeder to have a grounding means wherever it supplies branch circuits when equipment grounding conductors are required.

That equipment grounding means could be the metal raceway enclosing the feeder conductors, or in some cases it could be the *neutral* conductor.

Section 250-24 allows a choice for grounding of the *second* building. Either (1) bond the grounded (neutral) conductor to the disconnect enclosure and to the grounding electrode; or, (2) run an equipment grounding conductor with the feeder and bond it to the disconnect enclosure, leaving the grounded (neutral) conductor insulated at the second building.

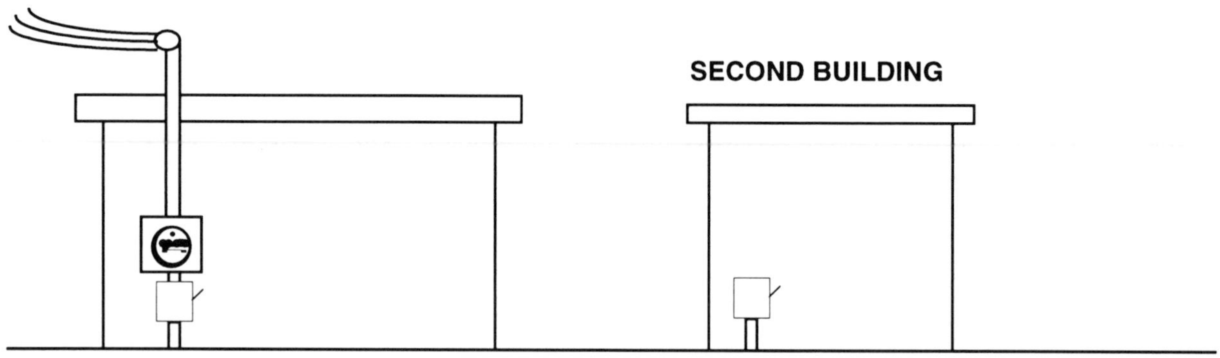

(1) Bond the grounded (neutral) conductor to the disconnect enclosure and to the grounding electrode.

OR

(2) Run an equipment grounding conductor with the feeder and bond it to the disconnect enclosure, leaving the grounded (neutral) conductor insulated at the second building.

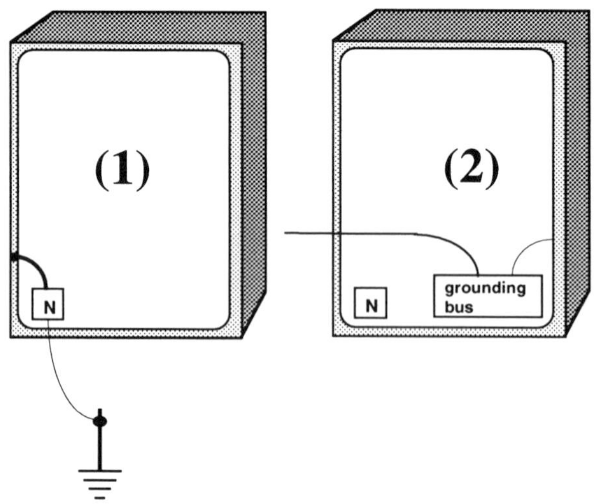

Where *livestock* is housed, that portion of the *equipment grounding* conductor run underground to the disconnecting means shall be insulated or covered copper. Metallic conduit is not permitted as the equipment grounding conductor when livestock is being housed.

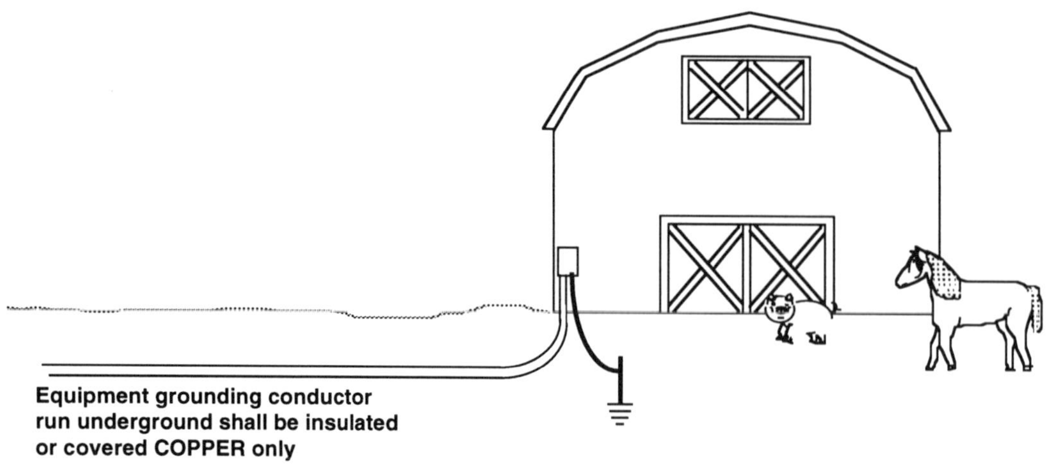

Section 547-8a ex. has special grounding requirements for agricultural buildings. Livestock have a greater sensitivity toward stray voltages, requiring special grounding in some cases.

At a second building, any existing grounding electrodes, such as an underground cold water piping system, must be bonded to the disconnect enclosure and to either the grounded (neutral) or to the equipment grounding conductor which is run with the feeder, depending on which method is used or required. If no grounding electrode exists at the second building, then one complying with Sections 250-81 or 250-83 shall be installed.

The grounded (neutral) conductor at the second building shall be bonded to the disconnect enclosure and grounding electrode, *EXCEPT:* (1) where ground fault protection for equipment is provided at the service at the first building; or (2) when it is required that for the protection of livestock in the second building any voltage appearing on the neutral (due to unbalanced current flowing in it) *not* be imposed on non-current-carrying conductive materials near the animals. In either case a separate equipment grounding conductor shall be run with the feeder and the neutral kept insulated at the second building.

The rule states that the grounding electrode conductor must be connected to the service conductor within the service disconnect or on the supply side of the service disconnect.

The sketch below shows an electrode connected at the meter on the pole. This does *not* satisfy the Code intent. The service disconnect is located in the barn and not on the pole. The Code states the connection of the grounding electrode conductor, grounding conductor, and grounded neutral conductor must be made on the supply side of the *service equipment* in the barn.

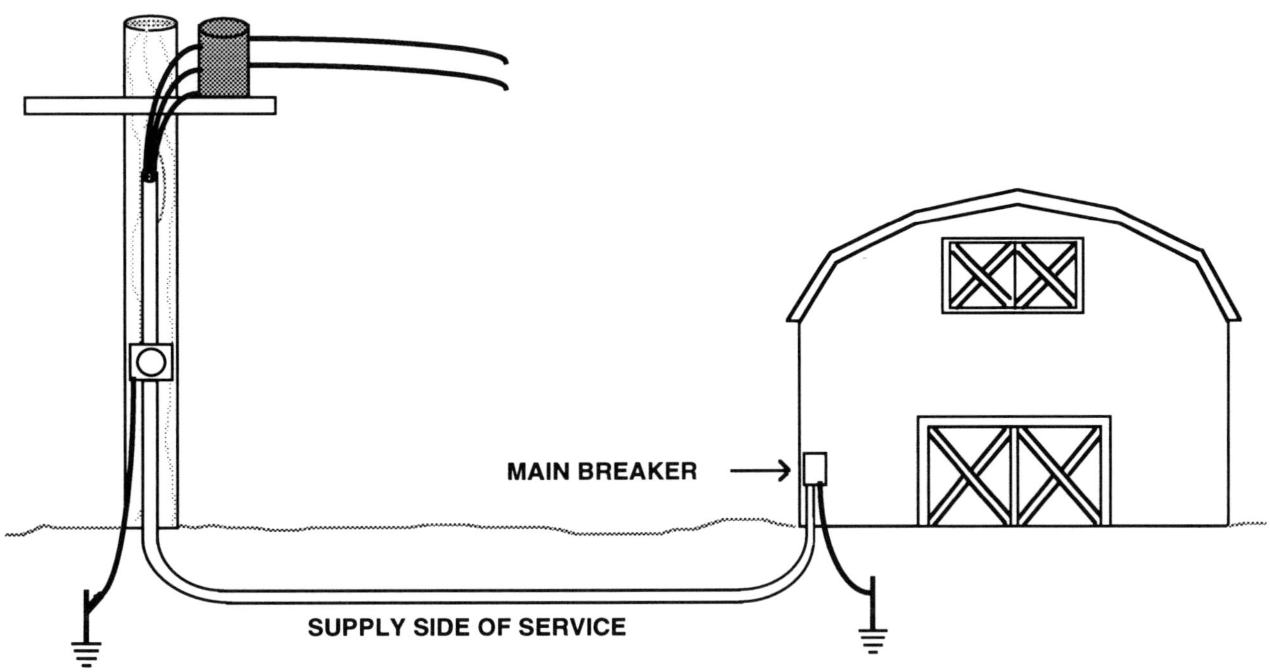

THE NEUTRAL

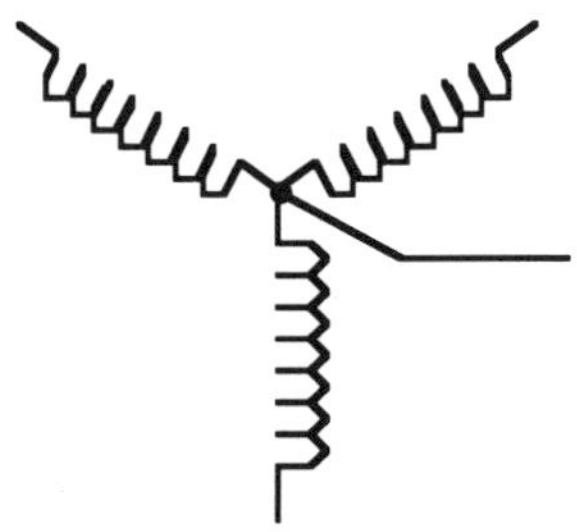

THE NEUTRAL

The definition of a neutral conductor is: It carries only the *unbalanced* current of the circuit.

It is very important in the designing of circuits to understand the definition of a neutral. You cannot have a neutral in a two-wire circuit; you must have at least three wires to have a neutral.

In a 2-wire, 120v circuit, the grounded conductor (identified white in color) is *not* a neutral conductor, it is the grounded conductor. In a 2-wire circuit the grounded conductor carries the same amount of current as the ungrounded (hot) conductor carries.

The "white" wire is *NOT* a neutral it is the grounded conductor.

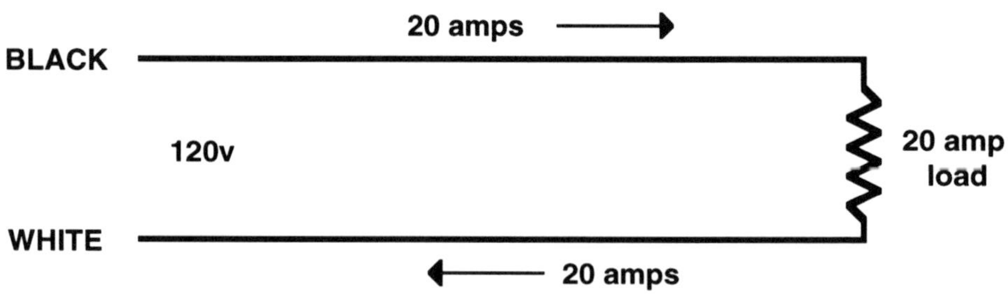

To have a neutral, you must must at least three wires.

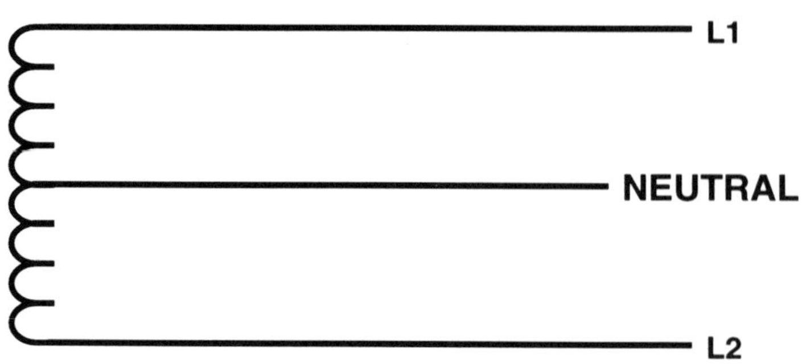

The sketch below shows two 20 amp loads using 3 wires, utilizing the neutral conductor.

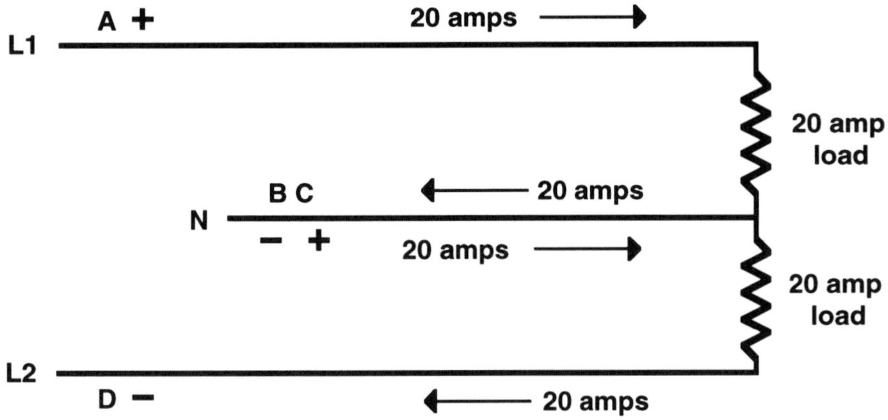

Each of the conductors, B and C, carries 20 amps, but note the direction of the arrows. The flow of current in conductor B at any *instant* is in an *opposite* direction of conductor C. So at any given instant, the single neutral conductor is said to be carrying 20 amps in one direction and 20 amps in the *opposite direction*. The two cancel each other and the flow of current in the neutral conductor is zero.

The system is actually 240 volts with the two 20 amp 120 volt loads in *series*.

Since the loads have equal resistance, the voltage of 240 would divide, 120 volts to each load. This is a *balanced* system. In a balanced system the neutral conductor carries zero current.

The reason for balancing loads on the neutral is to ensure that the neutral conductor never has to carry a large amount of current. If the loads are unbalanced, considerable current will flow in the neutral conductor. The neutral conductor is sized by the designer by balancing loads. If the loads are *not* connected as shown on the drawing schedule, overheating can occur causing insulation damage to the neutral conductor.

Shown below is how the AC sine wave looks on a 240 volt secondary winding.

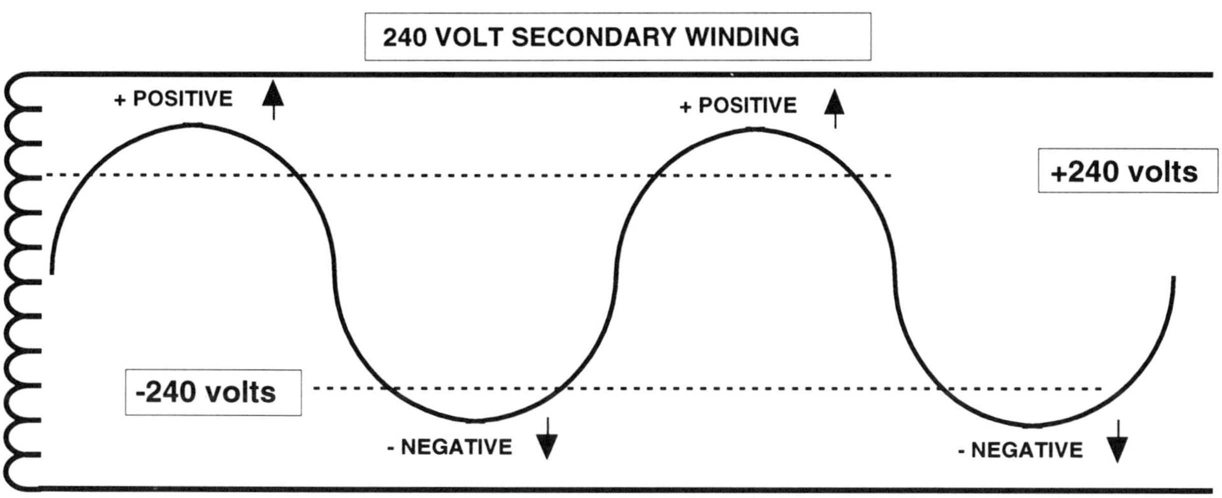

Shown below is the voltage relationship between L1 - L2 and the neutral. When the upper half of the winding (L1-N) is in the positive alternation, the lower half of the winding (L2-N) is in the negative alternation. These two sine waves (positive and negative) are said to be 180° out of phase with each other. When L1 is positive, L2 is negative.

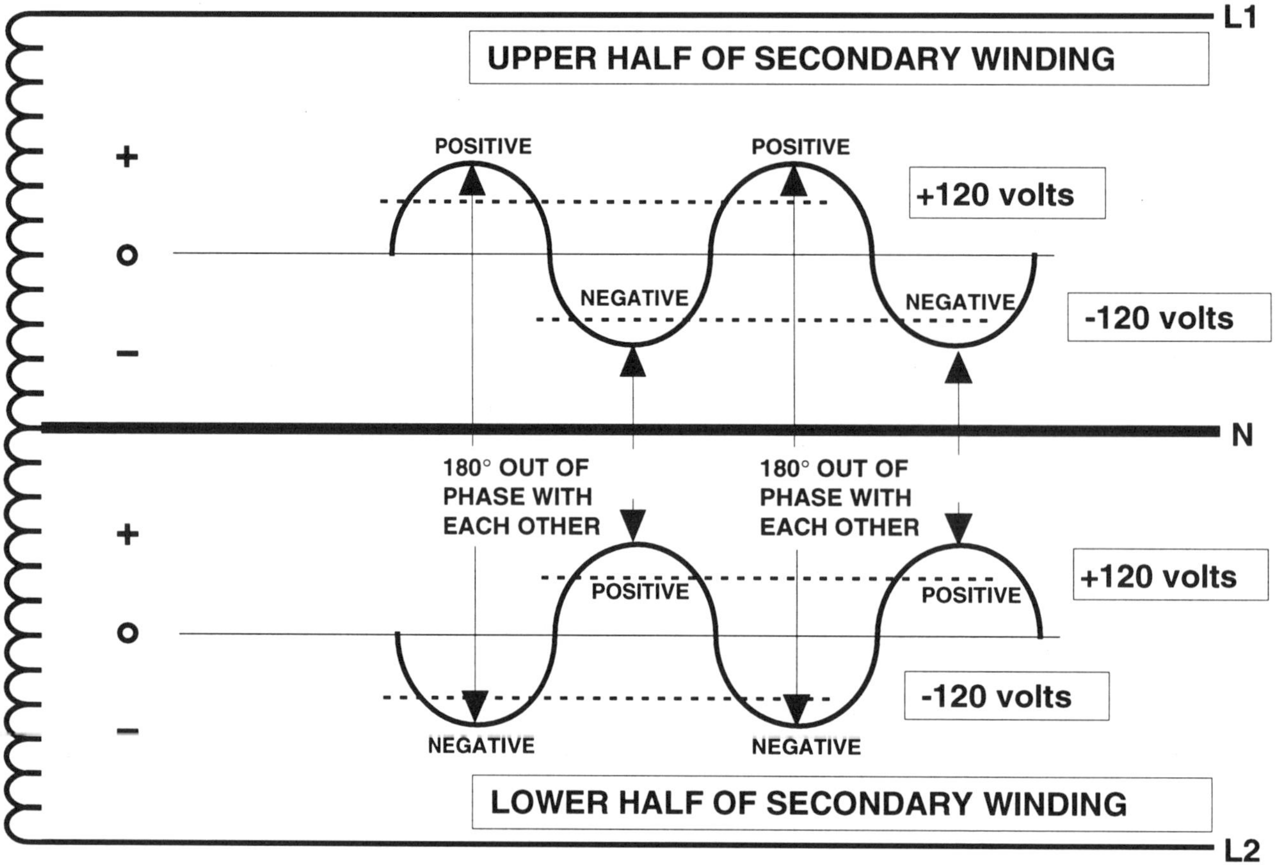

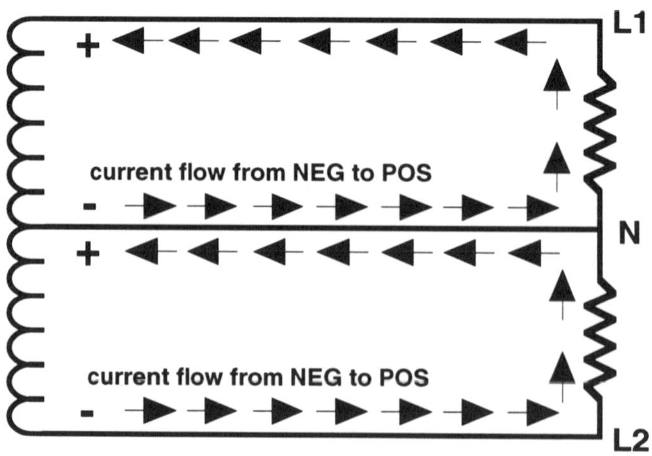

As you can see, current flow in the upper half of the winding (L1-N) is in one direction through the neutral conductor. In the lower winding (L2-N), the current flow is in the opposite direction through the neutral conductor. If the loads are the same the neutral would carry zero current as the upper circuit current (L1-N) cancels the lower circuit current (L2-N) in the neutral conductor.

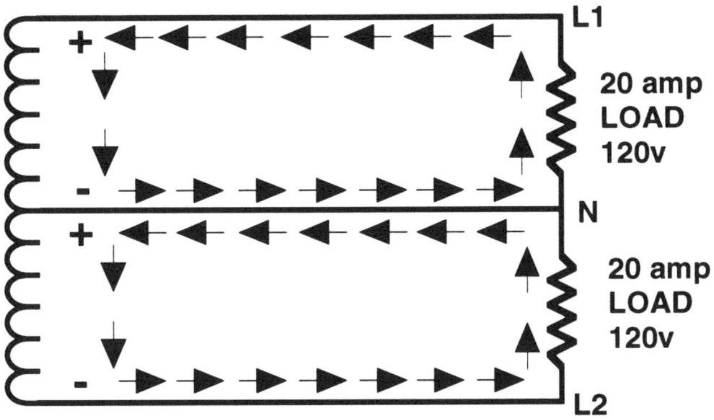

BALANCED LOADS

With both loads balanced at 20 amps each, the upper winding current of 20 amps is flowing in one direction in the neutral conductor while at the same time the lower winding current of 20 amps is flowing in the opposite direction in the neutral conductor. The currents cancel each other, so the neutral conductor is carrying zero current.

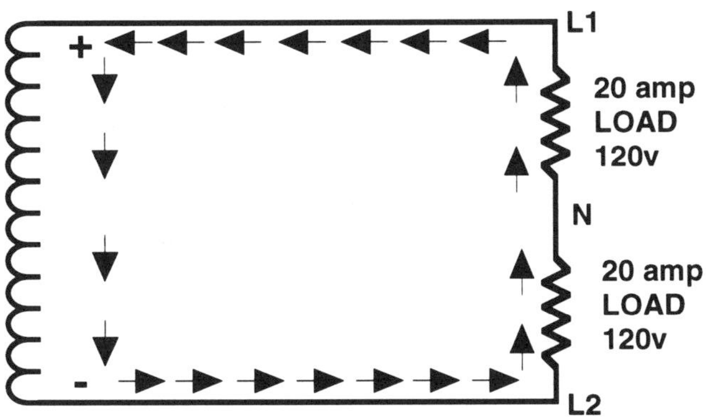

In fact, if the loads were **always** equal, you could remove the neutral conductor. Of course, this is never the case.

UNBALANCED LOADS

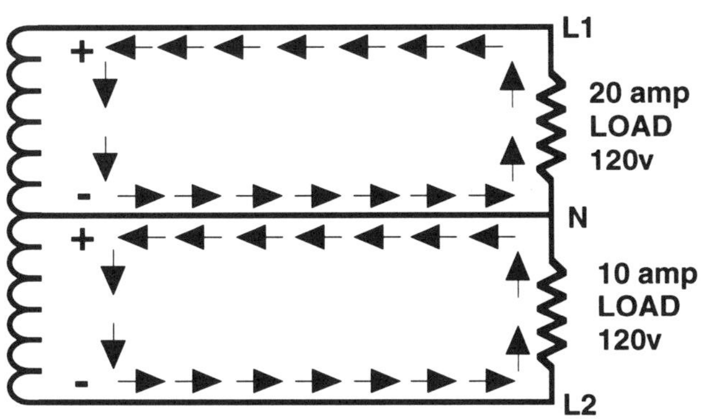

With an unbalanced condition as shown with one 20 amp load and one 10 amp load the neutral would carry the unbalance of 10 amps.

From L1 to N the neutral conductor has 20 amps flowing in one direction, from L2 to N the neutral conductor has 10 amps flowing in the opposite direction. The 10 amps from L2-N will cancel 10 of the 20 amps from L1-N, thus leaving the unbalance of 10 amps flowing in the neutral conductor.

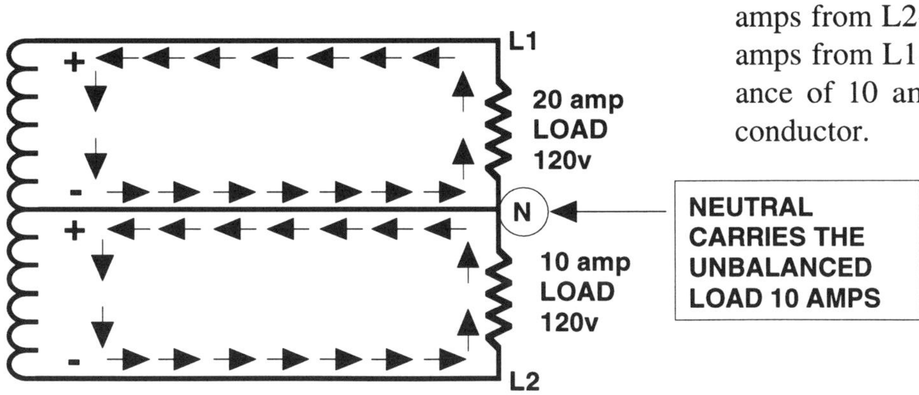

NEUTRAL SIZING

The Code permits, in some cases, a reduction in the size of the neutral conductor.

Section 210-19b ex.2 states on a *branch circuit* for cooking equipment the neutral shall have an ampacity not less than 70% of the branch circuit rating and not smaller than a #10 conductor.

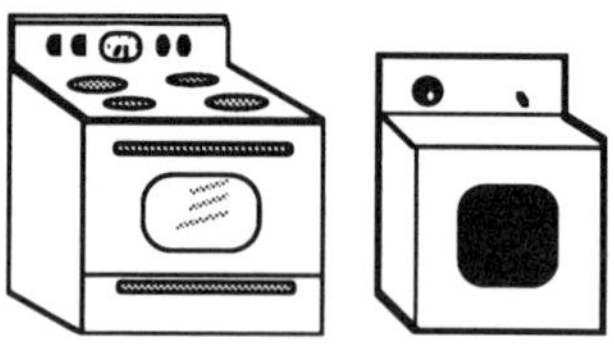

Section 220-22 permits a demand factor of 70% for *household* cooking equipment and *household* clothes dryers on feeder and service conductors.

Note 3 of the Tables of Ampacity permits the grounded (neutral) conductor to be smaller than the ungrounded (hot) conductors when sizing the service to a dwelling.

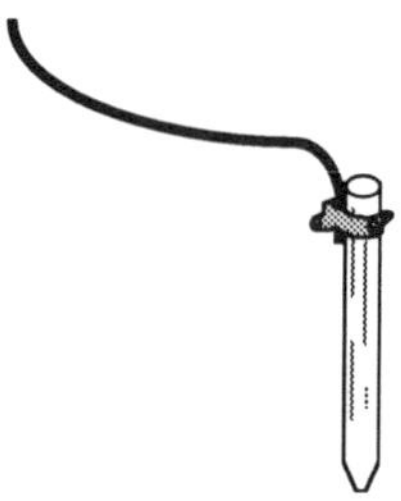

Section 250-23b states the grounded service conductor (neutral) cannot be smaller than the required grounding electrode conductor sized from Table 250-94.

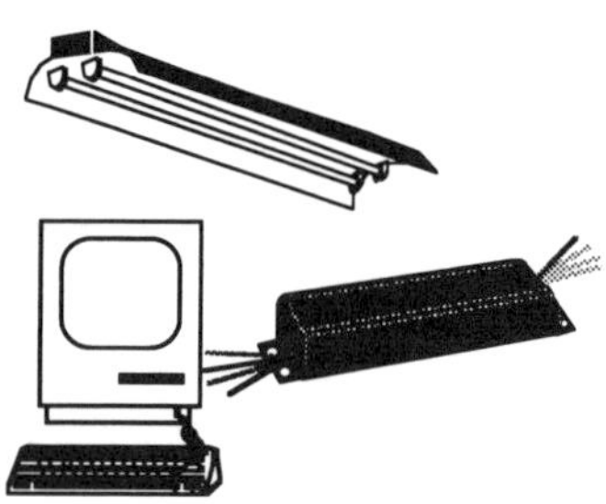

Section 220-22 permits a demand factor of 70% to be applied to the neutral in *excess* of 200 amps. *No* reduction of the neutral is permitted for harmonic loads such as fluorescent lighting, ballasts, computers, etc.

COUNTING THE NEUTRAL

In the designing of circuits the neutral conductor is counted in some instances and sometimes it is not *counted.*

Note 10a of the Table of Ampacity states: A neutral conductor that carries only the unbalanced current from other conductors of the same circuit need *not be counted* when applying the provisions of Note 8.

Note 11 states the grounding or bonding (green or bare) conductors shall not be counted when applying the provisions of Note 8.

Note 8 is an ampacity adjustment for current-carrying conductors. Notes 10a and 11 state these conductors are *not heat factors* and do not have to be included when applying Note 8.

The green or bare wire is designed only to carry fault current to open the overcurrent device within .008 of a second. This is not considered a *heat* problem when applying Note 8.

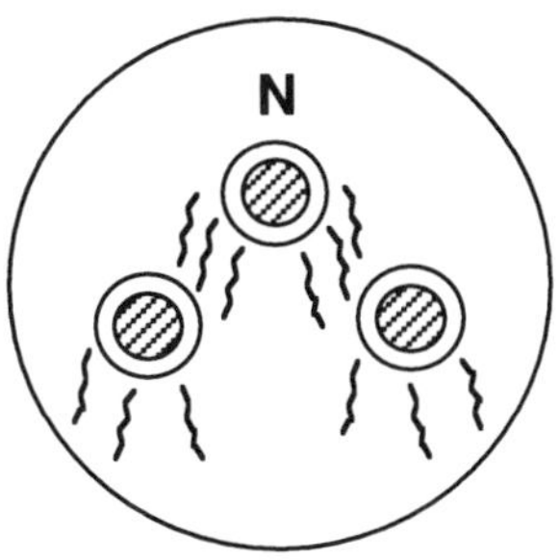

Example: If the load is balanced, the current in the neutral is zero and the two ungrounded conductors will run cooler because the neutral conductor is in intimate contact with them throughout the circuit and will serve to carry off part of the heat.

Using #4 THW copper conductors, 100 feet in length each. From Table 8, Chapter 9 the resistance of a #4 uncoated copper conductor is 0.308 ohms per thousand feet. 0.308 x .100 feet = 0.0308 ohm per conductor.

Let's say the circuit is balanced with each ungrounded conductor carrying a load of 60 amps. Heat = I^2R 60a x 60a x 0.0308 x 2 conductors = 221.76 watts.

Now assume the load is *not* balanced, the ungrounded conductor Line 1 carries 80 amps, Line 2 carries 40 amps, and the neutral carries the *unbalanced* of 40 amps.

The heat loss would be:

Line 1 =	80a x 80a x 0.0308 ohm =	197.12 watts
Line 2 =	40a x 40a x 0.0308 ohm =	49.28 watts
Neutral =	40a x 40a x 0.0308 ohm =	49.28 watts
		295.68 watts

The total of 295.68 watts is *more* than the loss of 221.76 watts in the case of the balanced load. Therefore the balanced load is the best condition for heat. Any amount of load *imbalance* will only make heat conditions worse.

Neutral conductors which carry only unbalanced current from phase conductors, as in the case of normally balanced 3-wire, single-phase or 4-wire, 3-phase circuits are *not* counted as current carrying conductors.

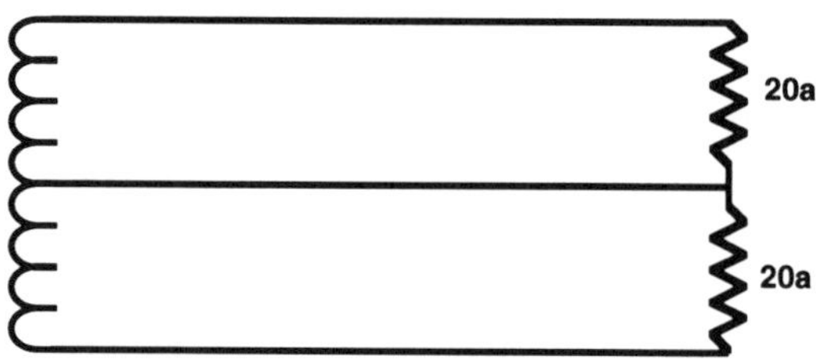

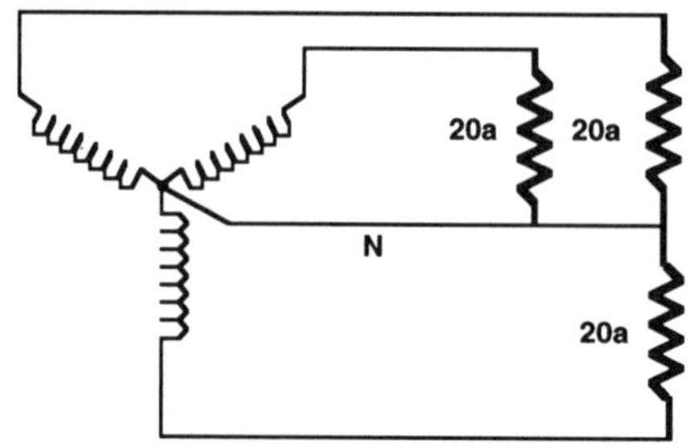

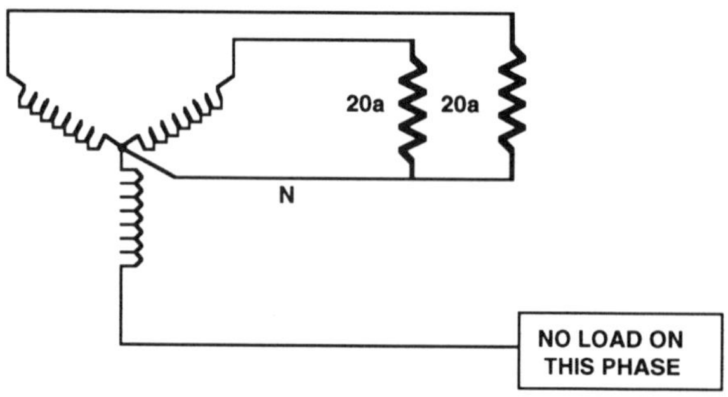

A neutral conductor used with two phase legs of a 4-wire, 3-phase system to make up a 3-wire feeder is *not* actually a *true neutral* since the neutral carries the same current as the other two conductors under balanced load conditions, it must be counted as a phase conductor. This circuit has three current carrying conductors. Note 8 does *not* apply.

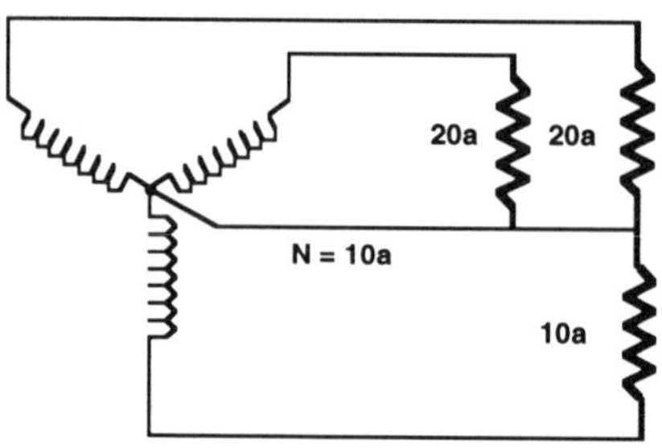

With two phases fully loaded and the third phase partially loaded, the neutral carries the difference between the full phase load and the partial phase load. The heating effect is still only three conductors. Note 8 does *not* apply.

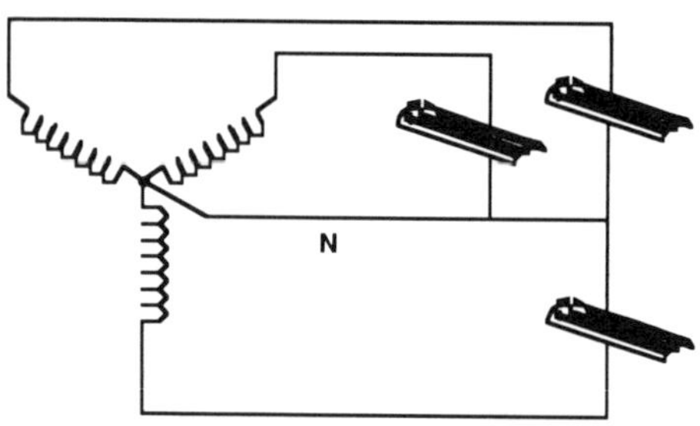

With a balanced load of fluorescent ballasts, third-harmonic generation causes a neutral current approximating the phase current. Now there is four current carrying conductors. Note 8 applies.

HARMONICS

Examples of *non-linear* loads which contain harmonics are computers, static power converters used in adjustable speed drives, solid-state frequency converters, uninterruptible power supplies (UPS), electric welders, discharge lighting, etc.

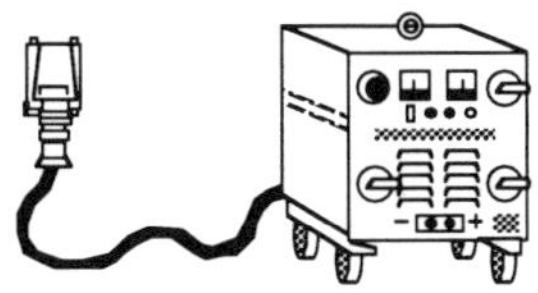
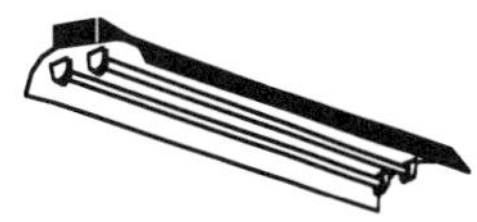

When a linear load is turned on, the voltage and current start and turn off together. When a non-linear load is turned on, the voltage starts but the current is purposely delayed.

Harmonic simply indicates that the current waveform is distorted. The closer the waveform is to a fundamental sine wave, the lower the harmonic content. With a fundamental sine wave, there are no high order harmonics.

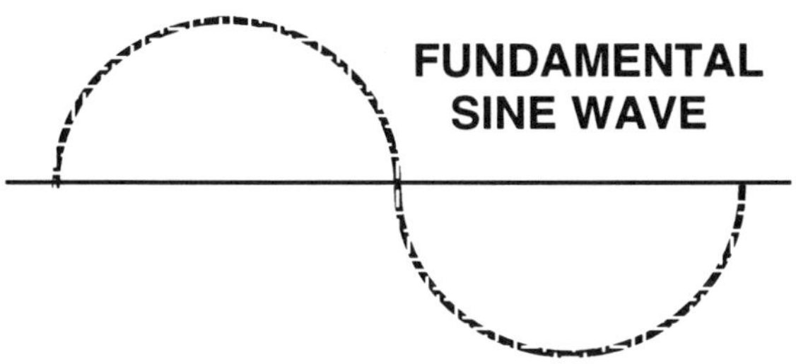

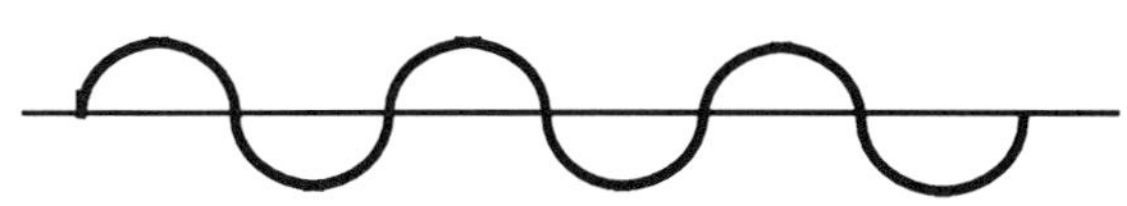

A 3rd harmonic makes 3 alternations in one alternation of the fundamental wave form. A 5th harmonic makes 5 alternations in one alternation of the fundamental wave. The 7th harmonic makes 7 alternations, and so on.

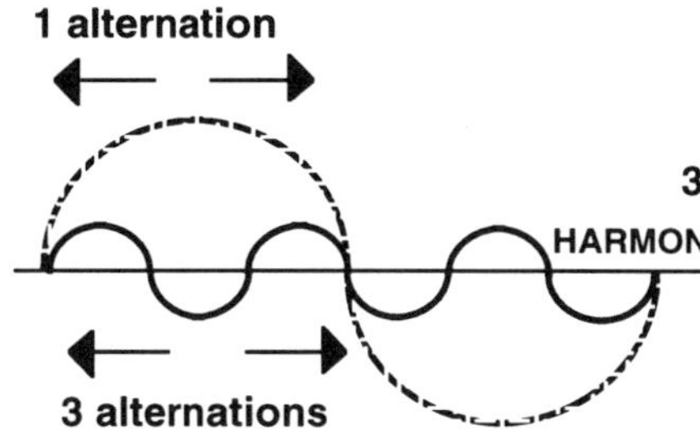

The highest peak of the wave is determined by *adding* all of the odd harmonics together. The frequency is determined by the number of complete cycles per second, measured in Hertz. 60 cycles per second equals 60 Hertz, or 60 Hz.

On a 60 Hz AC system the 3rd harmonic is 180 Hz, the 5th harmonic is 300 Hz, the 7th harmonic is 420 Hz, etc.

HARMONICS ARE ADDITIVE

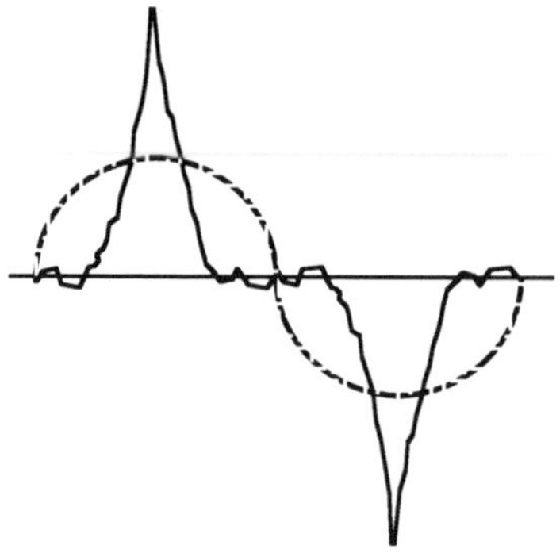

A *harmonic wave* is a *distorted wave* pattern consisting of the fundamental wave and other *higher frequency waves* that are superimposed on the fundamental wave.

The harmonic currents are delayed then *added* together and then burst into action causing a high peak wave form and a higher frequency.

Heat is increased by an increase in current frequency. The heating effect on transformers, circuit breakers and conductors supplying *non-linear* loads is a function of I^2R.

The higher I^2R heating where harmonic currents are supplied is caused by what is known as *skin effect*. Skin effect is an increase in resistance due to the fact that *higher-frequency* currents flow on the *skin* of the conductor, rather than throughout the entire conductor.

AC tends to flow along the surface of a conductor. DC acts through the entire cross-sectional area of the conductor in a uniform manner. The name skin effect is given to the action whereby AC is forced toward the surface of the conductor. Because of skin effect, there is less useful copper conductive area with AC. As a result, there is an increase in resistance.

CROSS-SECTIONAL AREA DENSITY

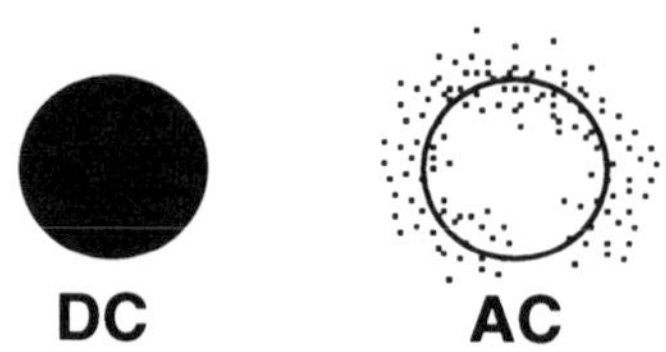

For the 3rd harmonic, instead of 60 Hz it's now 180 Hz. There is a 42% change in the skin effect concentration point. For the 5th harmonic there is a 55% difference and a 62% difference for the 7th harmonic. The result is a proportionate increase of resistance.

As you can see, the harmonic currents raised the frequency, thus raising the resistance and the I^2R heating effect.

In multiwire branch circuits, the odd numbered harmonics do not cancel out on the neutral, but are additive, resulting in a high neutral load. Because the heating effect on the conductor is proportional to I^2R and because the conductor resistance will be greater at 180 Hz than at 60 Hz, the heating effect in the neutral conductor will be greater than that of the phase conductors.

This is the reason in Note 10a of Table 310-16 of the Code it requires counting the neutral conductor in a 3ø, 4-wire circuit supplying *electric discharge lighting*. The neutral conductor is actually *hotter* in temperature than the phase conductors.

An electrician is familiar with the fluorescent light fixture containing a ballast. This fixture is referred to as discharge lighting since it contains a ballast (winding).

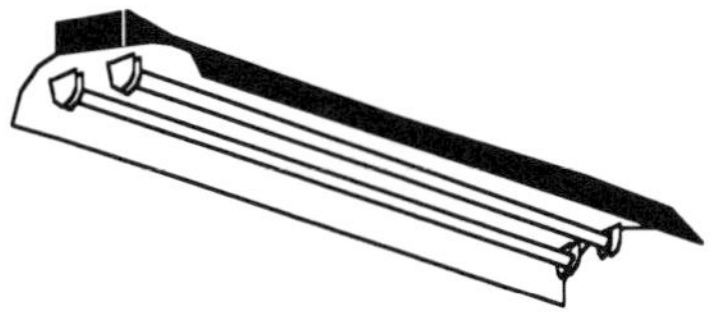

Harmonics cause problems to transformers and neutral conductors in multiwire feeders and branch circuits. The largest contributor to harmonic distortion is the static power converter used in adjustable speed drives.

One of the components for which there is great concern is the the neutral conductor in a multiwire branch circuit. There will be no current flow on the circuit neutral conductor if the circuit supplies the same amount of *resistive* loads from each phase to neutral. With three single-phase *non-linear* loads fed from a 208/120v three-phase feeder due to the phase relationship there is *no* cancellation of the harmonic currents in the neutral conductor.

The neutral will carry the current drawn by each phase and where more than one such load is fed from a given phase, the current drawn by each load on that phase will *add* in the neutral. The heating effect (rms current value) on the neutral could be equal to 1.732 times the current of one phase.

By using a *separate neutral* for each branch circuit phase conductor you could eliminate the *effect* of harmonic loading on the *single* neutral conductor of a multiwire branch circuit. This would not eliminate the harmonics. There would still be harmonic currents in the feeder conductors, which will be carried by the transformer.

Transformers are overheating and burning up and don't appear to be overloaded. Standard dry-type transformers were never designed to handle the harmonic currents that cause added heating. Heat determines the life of the insulation.

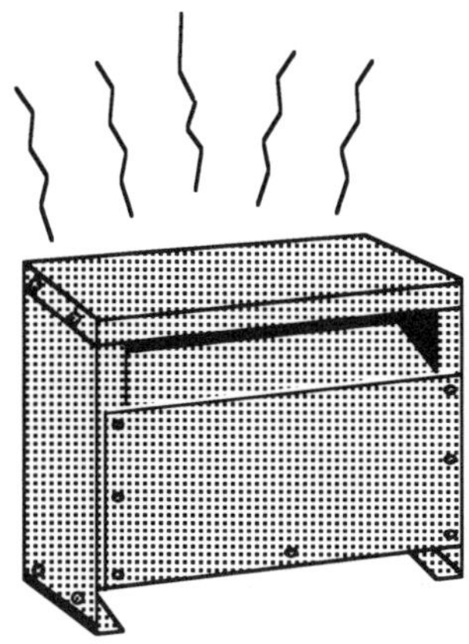

Transformers are now available to handle the harmonic currents. These transformers are *K-Factor rated.* The conventional dry-type transformer has a K-Factor of one. Today with the non-linear loading, a *minimum* K-Factor of 4 is recommended for minimum design.

K-Factor K4 means the transformer is built to handle 4 times the eddy current loss and will allow the transformer to operate within the temperature limit with these harmonic currents.

The K-Factor transformer cost more than the conventional dry-type, but the cost of the total installation is less. A conventional K1 transformer would require a much larger kva rating thus requiring larger conductors, equipment, circuit breakers, etc. As an electrical designer we must remember we have *long term* obligations.

The transformer must be "K-Factor" rated for harmonic loading or they must be derated in kva to assure that the rated temperature rise is not exceeded.

K-Factor rated transformers do *not* get rid of the harmonic currents. They are designed to handle the extra heat generated by these currents. Nobody even proposes that it is advisable to try to get rid of the harmonic currents.

In many cases harmonic currents are an invisible problem for the electrician. A non-rms amp clamp meter shows the transformer is only *half loaded*, but the transformer feels *really hot*. You should trust the heat and not the meter.

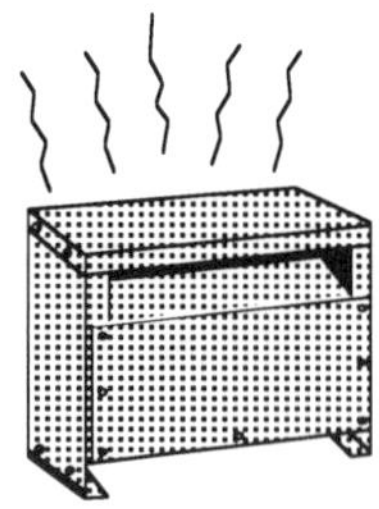

Average-rms ammeters are *not* reliable instruments for evaluating an electrical system's harmonic content. For accurate indication of an electrical system's content, use only a *true-rms meter*.

Phase conductors have been measured showing a balanced load 125 amps per phase and the *neutral* showing *155 amps*. Now you can see how the neutral becomes *a current-carrying conductor* with the added harmonic current heat.

It is quite apparent that feeders and branch circuits serving the *average office space* can have *greater* currents in the neutral than the phase conductors as a result of these *additive* harmonics.

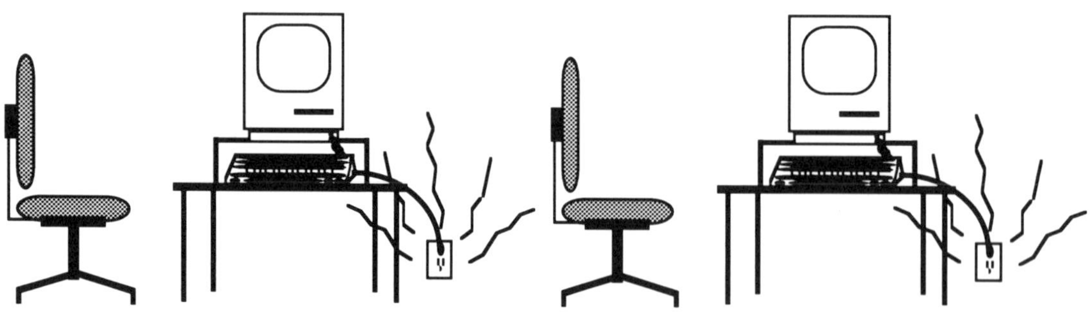

The estimate is that the 1990 load on the utility is 15% harmonic and by the year 2000 harmonic currents will constitute 50% of the loading on the utility transformer.

This message on harmonic currents needs to reach the electrician so the electrician fully understands this problem in our industry. It has been estimated that 70% of the electrical designing is done by the electrician and electrical contractor.

Remember, conductor sizing from the Code is based on *60 Hz*. The system must be designed for the extra heat produced by these non-linear loads.

One might consider the use of nonferrous raceways such as aluminum or PVC. This is a standard practice in 400 Hz applications such as aircraft and large computer rooms to reduce heating (eddy currents - skin effect) from the adjacent steel raceways. Remember, 400 Hz is very nearly the 7th harmonic of a 60 Hz system.

Higher harmonics are worse, percentage wise, than the lower 3rd harmonic.

An electrician can not calculate the amount or order of harmonics contained in a piece of equipment. A ballast generally contains 3rd harmonics, but another type of ballast may contain even a higher order of harmonic. A computer may have some 5th, 7th, 9th or even higher harmonics with different percentages of each order of harmonics.

There is no perfect way, at this writing, to solve this *cancer* that is eating away at our conductors and transformers.

When computers, data processing equipment and solid state devices are involved in the circuitry, the electrician should consider the K-Factor transformer, derating the kva of an existing transformer, upsizing the conductors, installing separate or oversized neutrals on multiwire circuits.

On a 20 amp receptacle multiwire branch circuit supplying computers, it is recommended to use #12 ungrounded (hot) circuit conductors with a *#8* (white) grounded or neutral conductor.

Cable manufacturers are now making a cable containing #12 circuit conductors with a *#8 "super neutral"* conductor. Cables with an individual neutral per phase conductor are also available.

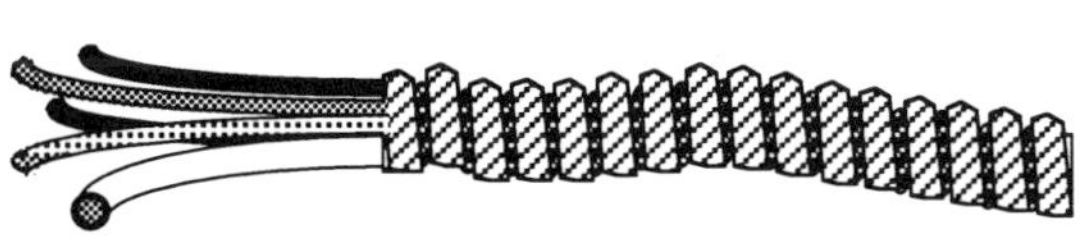

An oversized neutral is not necessary to accommodate the harmonics generated by the 3rd harmonic in ballasts.

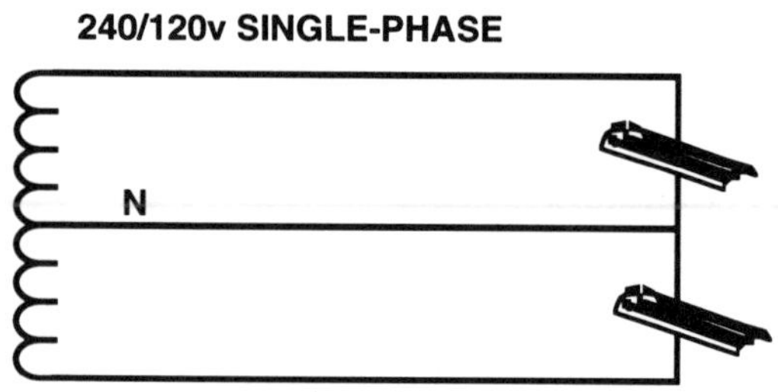

Section 220-22 does *not* permit a reduction on the neutral for a feeder supplying fluorescent lighting over 200 amps even if it is a single-phase 240/120v, 3-wire feeder.

This could be argued as the third-harmonic currents in the ungrounded (hot) conductors are 180° out of phase with each other and therefore, would *not* be additive in the neutral.

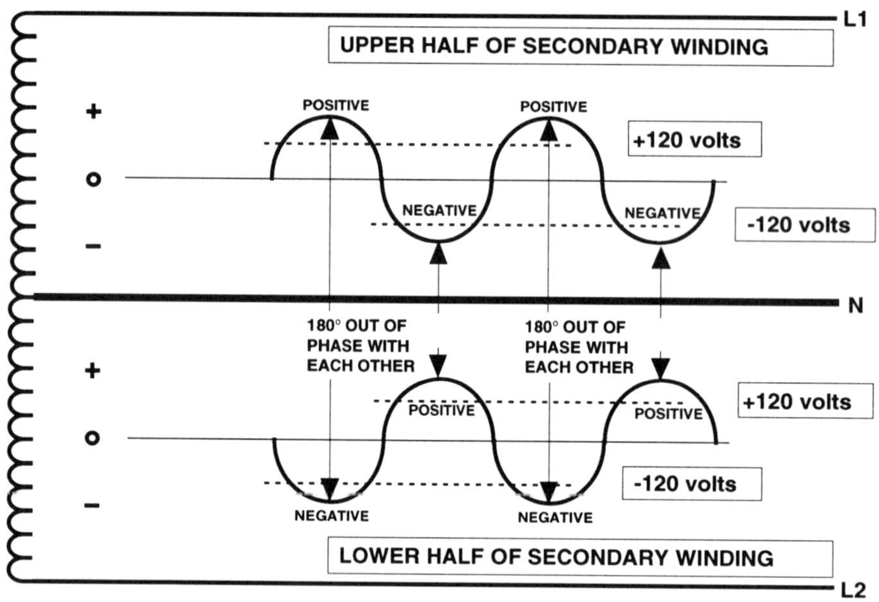

In a 3-phase, 4-wire circuit the third-harmonic components of the phase currents are in phase with each other and *add* together in the neutral instead of cancelling out.

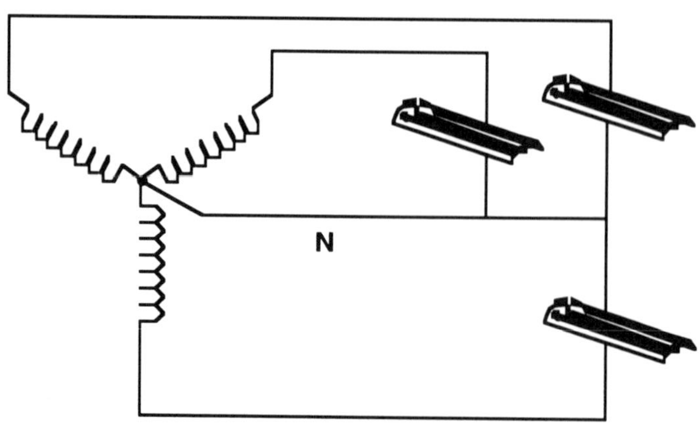

COMMON NEUTRAL

A common neutral is a *single* neutral conductor used as the neutral for more than one set of feeder conductors as permitted in Section 215-4a.

It requires all conductors of feeder circuits employing a common neutral feeder to be within the same enclosure when it is metal.

A common neutral is *not* used for *services* or *branch circuits.* A single neutral of a multiwire branch circuit is *not* a common neutral. It is the neutral of only a single circuit, even though the circuit may contain three or four wires. A common neutral is used only with *more than one feeder circuit.*

NO NEUTRAL LOADS

Whenever a service is derived from a grounded-neutral system, the grounded neutral conductor must be brought into the service entrance equipment, even if there are no neutral loads connected to the panel. The neutral is required to provide a low-impedance path for ground-fault current to return to the neutral to ensure operation of the fuse or circuit breaker.

Neutral conductor provides a low resistance path for fault current to trip breaker

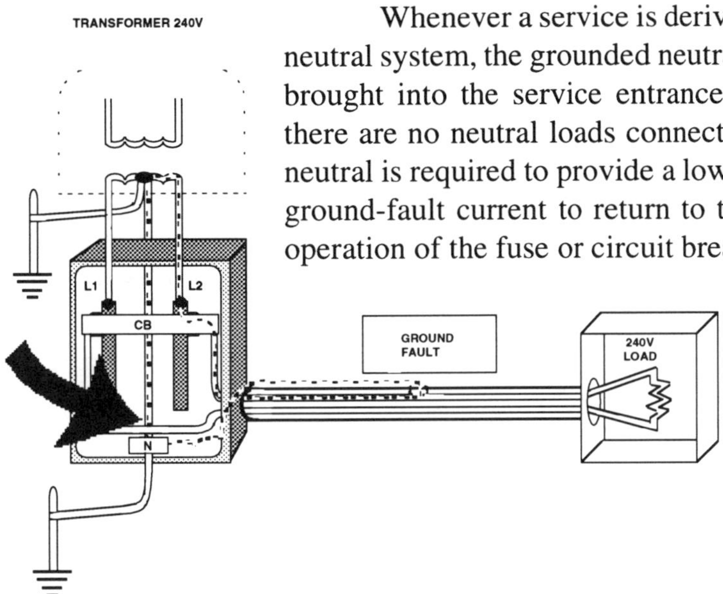

The neutral functions just like an equipment grounding conductor, to provide a path back to the transformer and to the overcurrent device to open the circuit that faulted. Without the neutral in this case, the fault current would have to flow through earth and would not open the overcurrent device.

Without neutral ground path is high resistance earth

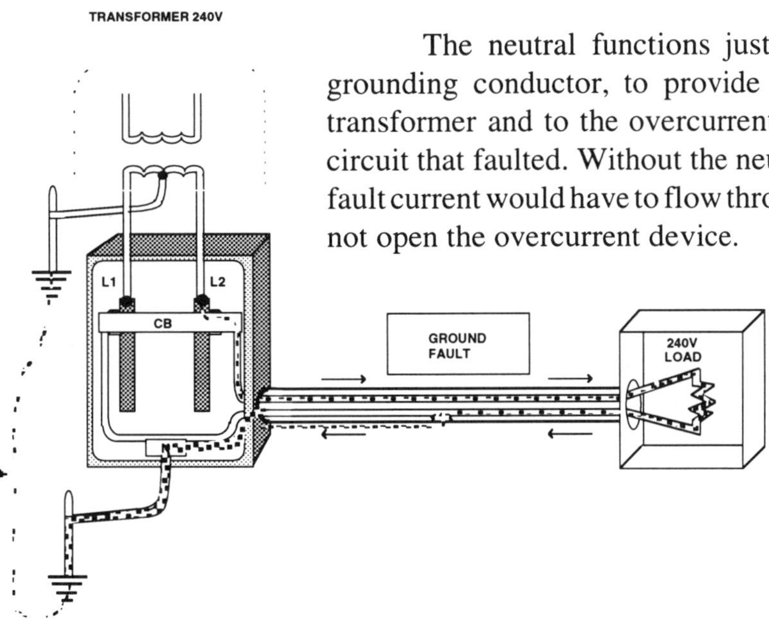

GROUNDED NEUTRAL CONNECTION

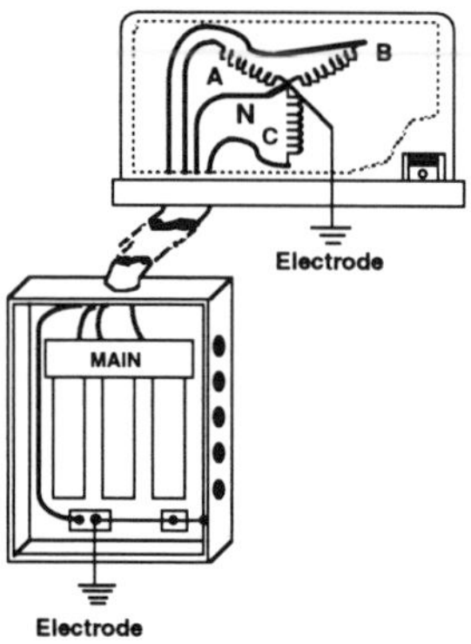

When an electrical system is grounded, a connection to a suitable grounding electrode is required at the service entrance. In addition to the grounding connection for the grounded (neutral) conductor at the point of service entrance to the premises, it is further required that another grounding connection be made to the same grounded conductor at the transformer that supplies the system.

The grounded (neutral) conductor connection to the grounding electrode must be anywhere from the *load end* of the service drop or lateral.

The grounding-electrode conductor may be connected to either the neutral bus (or terminal lug) or to the equipment grounding block or bus in any system that has a conductor or busbar bonding the neutral bus or terminal to the equipment grounding block or bus. Where the neutral is bonded to the enclosure by a bonding *screw*, the grounding-electrode conductor must *not* be connected to the grounding block or bus. It must be connected *only* to the *neutral conductor terminal block or bus*.

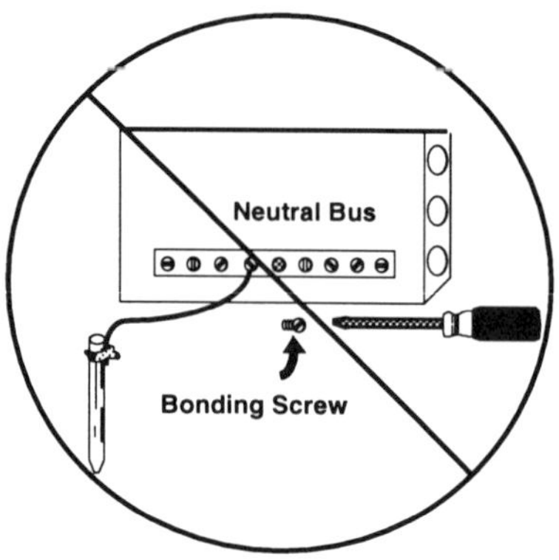

The reason is that the conductivity of a *screw-type* bond to the grounding system is not adequate to pass the high currents of say a lightning strike on the system.

The general rule is: Once the neutral is connected to a grounding electrode at the service equipment, the neutral must be *insulated* from all equipment enclosures and other grounded parts on the *load side* of the service.

The use of a neutral to ground a subpanel, or other equipment on the *load side* of service equipment would be extremely hazardous if the neutral became loose or disconnected.

The bonding of subpanels or any other neutral connection to equipment enclosures is prohibited.

There are a few very specific exceptions to this *general rule*.

Exceptions:

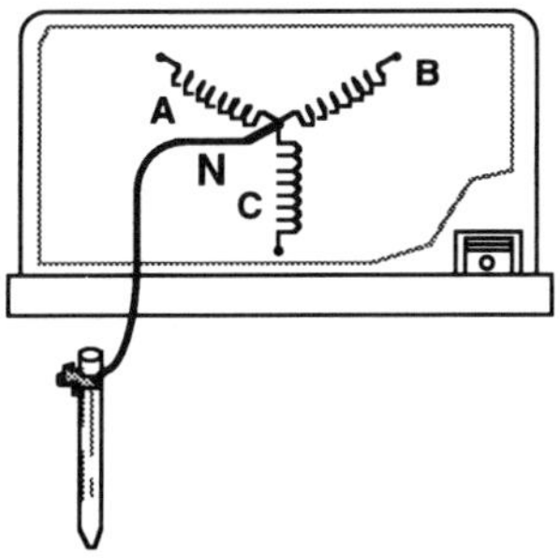

(1) When voltage is stepped down on a system, even though it's on the *load side*, a grounding connection must be made to the secondary neutral.

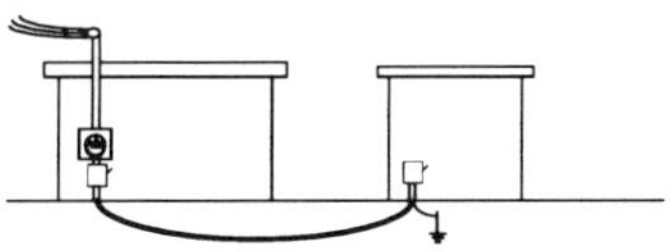

(2) When a circuit is run from one building to a second building and the panel is fed from the first building, it is permissible, in some cases to connect the system grounded (neutral) conductor to a grounding electrode at the second building.

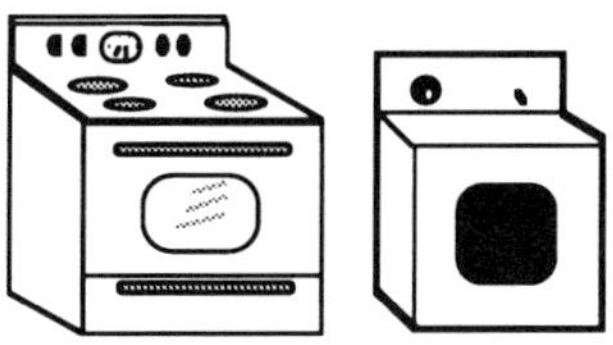

(3) The frames of household cooking equipment and clothes dryers may be grounded by connection to the grounded neutral of their supply circuit as permitted by Section 250-61.

The exception that permits bonding the grounded (neutral) conductor to the metal frame of cooking equipment and clothes dryers has been a very controversial subject over the years.

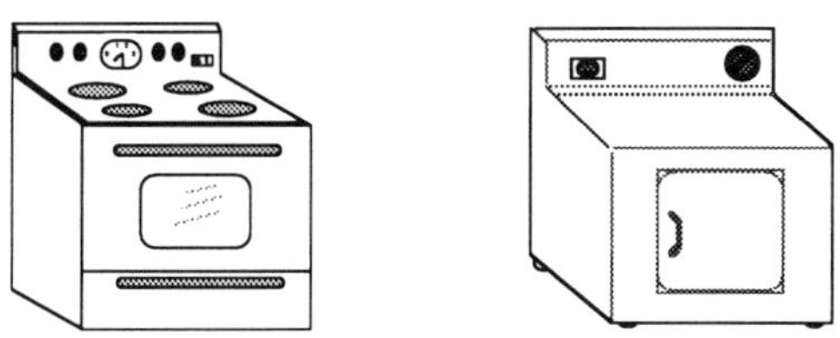

The past Codes permitted connecting the **grounded** circuit conductor to the frame of ranges, wall-mounted ovens, counter-mounted cooktops, clothes dryers, and outlet and junction boxes which are a part of the circuit for these appliances mentioned. When of the following conditions were met:

1. The supply circuit is 120/240v, single-phase, three-wire **or** 208/120v from a three-phase, 4-wire.

2. The grounded conductor is not smaller than #10 copper or #8 aluminum.

3. The grounded conductor is insulated; OR the grounded conductor is uninsulated (bare) and part of a Type SE service-entrance cable and the branch circuit originates at the service equipment.

4. The grounding contacts of receptacles furnished as part of the equipment are bonded to the equipment.

The present rule has a long history. It has been in the Code since 1947. It was estimated in 1983 that there were 40 million electric ranges and 34 million clothes dryers connected to the 3-wire system using the grounded circuit conductor bonded to the frame for equipment grounding, with an excellent safety record.

This book has explained the equipment grounding conductor and its function in the grounding circuit. All receptacles are now required to have a grounding pin for this added protection, but yet we still allow a 3-wire receptacle to ranges and dryers using the grounded conductor for the equipment grounding conductor.

Questions asked are, why does the washing machine have *both* a grounded conductor and an equipment grounding conductor when the clothes dryer sitting right beside it only requires the grounded conductor? The grounded conductor to a dryer is permitted to serve a dual purpose by being used also as the equipment grounding conductor. The 1996 Code has changed this for new circuits.

Ranges and dryers are now designed and manufactured so that they can be connected to either a 3-wire or a 4-wire system. The 4-wire system having *both* the grounded conductor and the equipment grounding conductor.

3-wire 30 amp

4-wire 30 amp

3-wire 50 amp

4-wire 50 amp

The Code panel comment on allowing the 3-wire system to still be used is, the risk involved is where the range or clothes dryer is moved by a family from a residence where the *4-wire* system is used, to an older residence where the *3-wire* system is used. With the range or dryer connected to a 4-wire system, it is likely the bonding strap (connects grounded wire to frame) has been thrown away since it's not needed for the 4-wire connection. Now comes the risk factor; the appliance is connected to the 3-wire system without installing the bonding strap. Now there is *no ground to the frame at all.* If a fault were to develop between a live part in the appliance and the frame, there would be *no protection against electrical shock.*

Since *unqualified* people often install ranges and dryers, the equipment grounding connection will often be made *improperly*, if at all. With a 3-wire system instead of a 4-wire system there is a better chance it will be connected properly. Ask yourself this question, will a buyer of a replacement range or dryer have the existing 3-wire branch circuit rewired to provide for the 4-wire system on his new appliance? The buyer should, but I doubt that it will happen.

With the 3-wire system using the grounded conductor as the grounding conductor, a break or disconnection of the grounded (neutral) conductor becomes evident due to the malfunctioning of the 120 volt loads on the appliance. Lights, timers, and drive motors would be inoperative in this condition. It would be necessary to have this situation repaired before the appliance can function properly.

SE service entrance cable is *the only cable permitted* to be used *with an uninsulated bare grounded conductor.* This is *not* permitted for NM cable (romex). #10-2 w/grd romex, using the bare conductor as a neutral is a VIOLATION! The neutral must be *insulated* in a romex cable, only SE cable is permitted to use bare conductor as a neutral.

The *bare* conductor in SE cable is called the *grounded conductor.* The *bare* conductor in NM cable (romex) is called the *grounding conductor.*

The Code panel does not feel that the present rule should be expanded to permit the *equipment grounding conductor* to serve as the *grounded conductor.*

The sketch below shows a clothers dryer connected SE cable 3-wire with the bare grounded (neutral) conductor connected to the frame of the dryer. The bare neutral is connected to the neutral bus at the *service panel.* The bare neutral could make contact with the metallic service panel as the conductor enters the panel. This is *not* a problem because the service panel is also *bonded* to the neutral bus which is connected to the electrode.

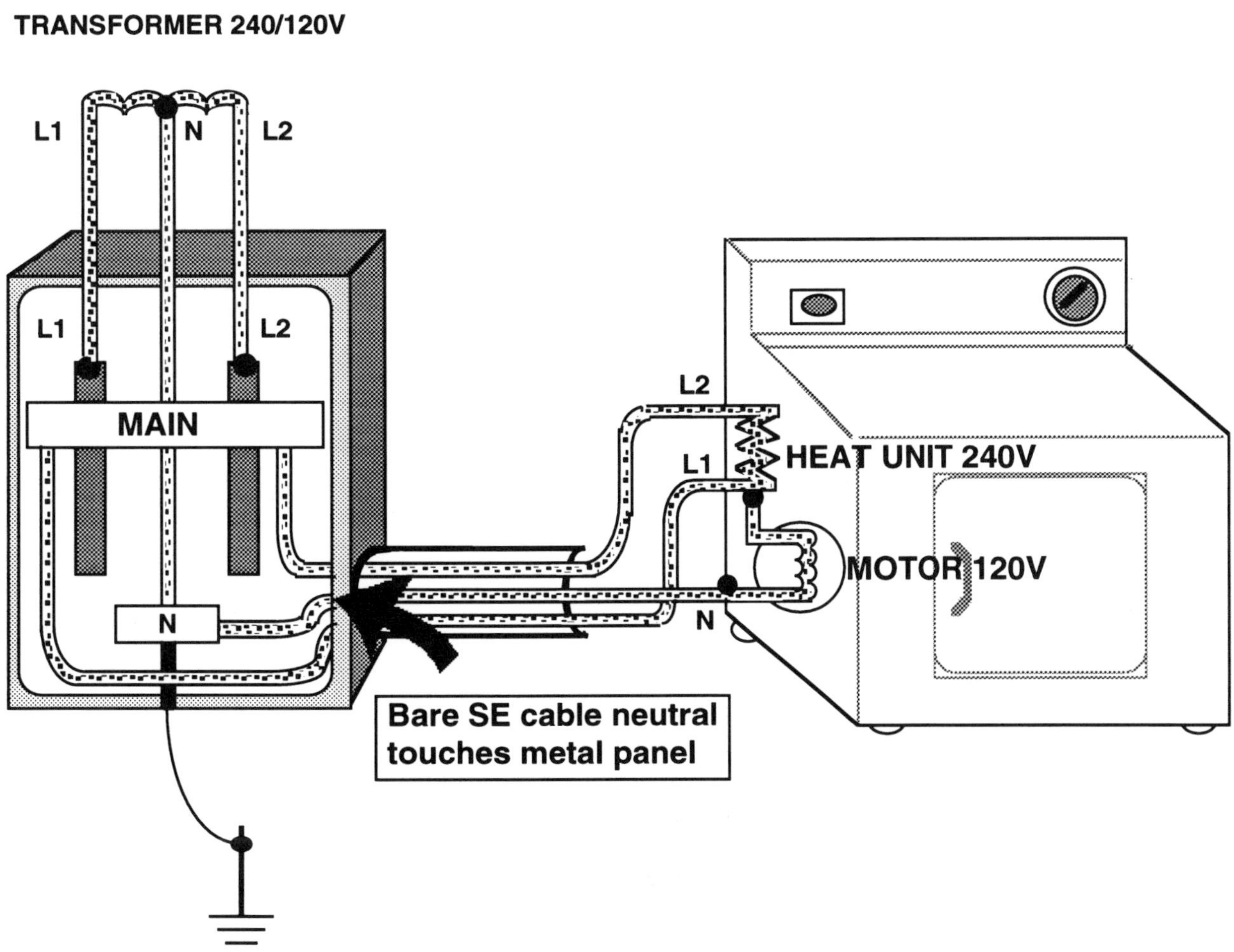

When using the SE service entrance cable with a *bare* neutral conductor, the rule states the branch circuit must originate from the *service panel* and not a *sub-panel* as shown below.

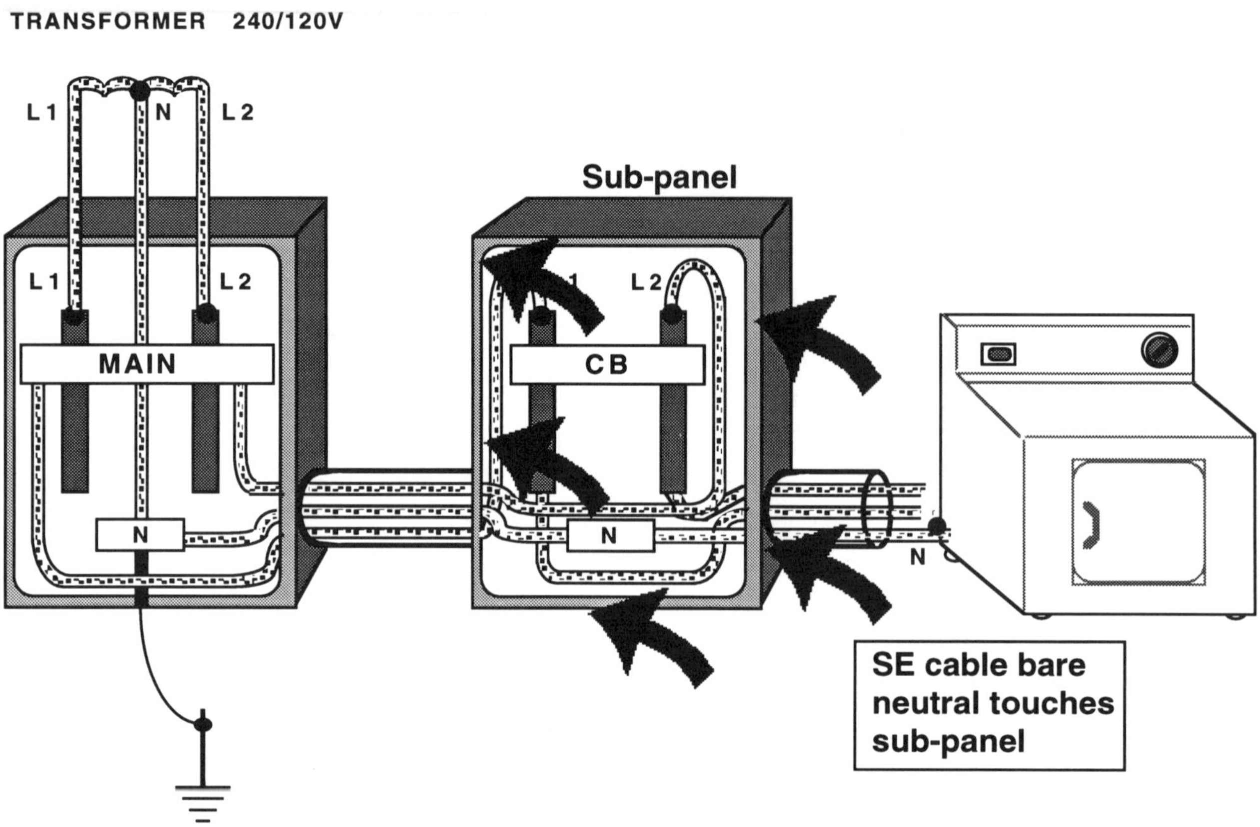

With the condition shown above we have neutral current flowing over the sub-panel which is to be *isolated from the neutral.* The only connection point for the neutral and the equipment grounding is at the *service.* This is the reason the rule requires that the branch circuit must originate from the service panel and not a sub-panel.

The condition above places the bare neutral in parallel with the equipment grounding conductors and conduits, if any, installed in the sub-panel. With an *insulated* neutral you would *not* encounter this condition.

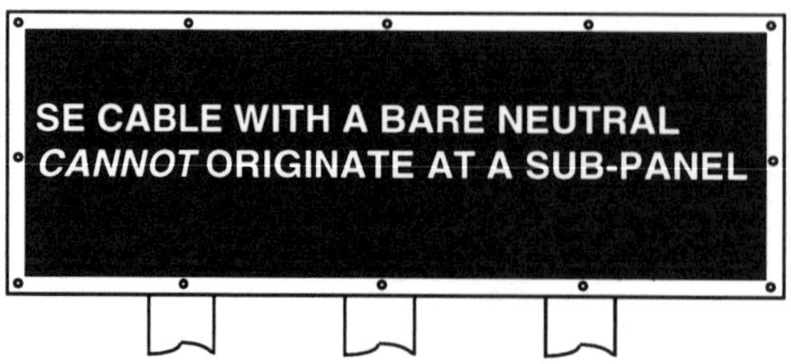

THE GROUNDING CONDUCTOR

THE GROUNDING CONDUCTOR

The Code requires that metal boxes, cabinets, and equipment enclosures be grounded. The equipment grounding connection can be made by the *metal* raceway or by an equipment grounding conductor. An equipment grounding conductor *must* be installed in PVC raceway. When an equipment grounding conductor is used, it must be installed within the *same* raceway, cable or cord with the circuit conductors.

External grounding of equipment housings, frames or enclosures is a violation. The equipment grounding conductor must be run with the circuit conductors.

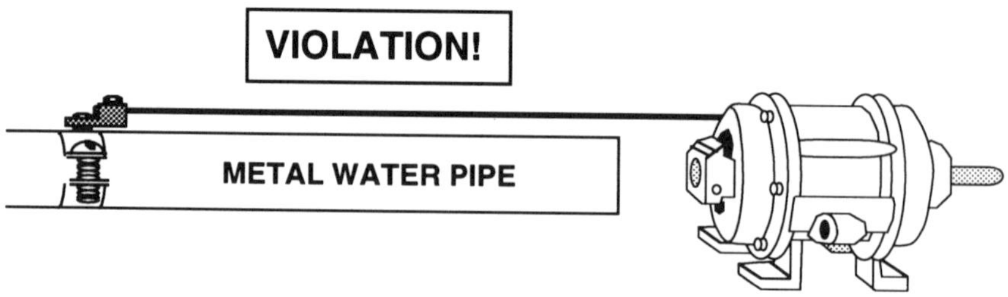

The equipment grounding conductor covered or insulated shall have a continuous outer finish that is *green* or green with one or more *yellow stripes*. An equipment grounding conductor can also be *bare*.

A black or other colored conductor #4 and larger can be used as the *equipment grounding conductor* by stripping the insulation off and making it bare, by coloring the exposed insulation green, or by using green marking tape.

Building steel cannot be used as the equipment grounding conductor for AC equipment.

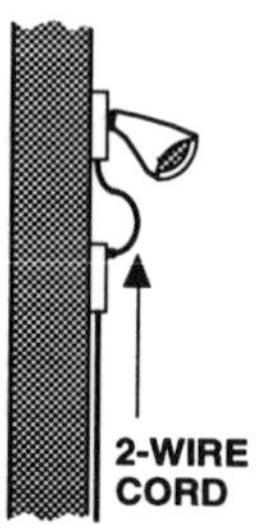

The steel column shown to the left is supporting the box and light. The lower junction box attached to the column is fed with a steel conduit. The light is plugged into the junction box receptacle with a 2-wire cord which has no equipment grounding conductor. The Code permits the *short* section of the steel column to be used as an equipment grounding conductor as it is providing the *support* for the equipment.

EFFECTIVE GROUNDING PATH

The Code states the path to ground from circuits, equipment, and metal enclosures for conductors shall:

(1) Be permanent and continuous.

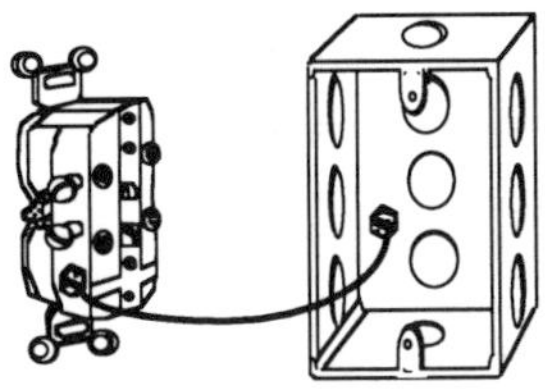

The equipment grounding conductor must be connected at the box where the removal of a receptacle, fixture or other device would not break the continuity path of the grounding conductor (connect with pigtail).

(2) Have capacity to conduct safely any fault current likely to be imposed on it.

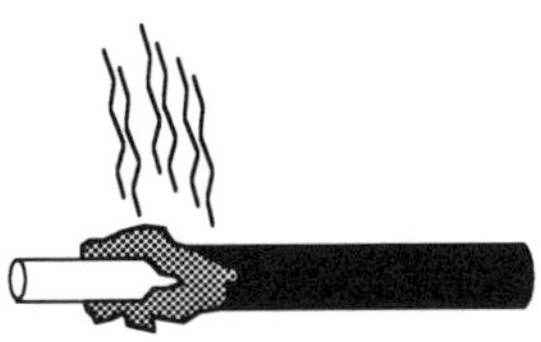

Code Table 250-95 is the *minimum* size equipment grounding conductor allowed. The designer must design the conductor to carry the maximum fault current it would carry.

Example: Let's assume a 200 amp feeder is run in PVC conduit from a 200 amp breaker. The phase-to-ground fault current at the panel is 26,000 amps and the opening of the breaker is one cycle. The circuit conductors are #3/0 THHN copper. Table 250-95 requires a minimum #6 copper for a 200 amp breaker. A #6 THHN will only withstand 14,624 amps for one cycle without insulation damage. This book contains a table which shows a #3 THHN would be required for this example. A #3 THHN has a withstand rating of 29,326 amps for one cycle.

(3) Have sufficiently low impedance to limit the voltage to ground and to facilitate the operation of the fuse or circuit breaker.

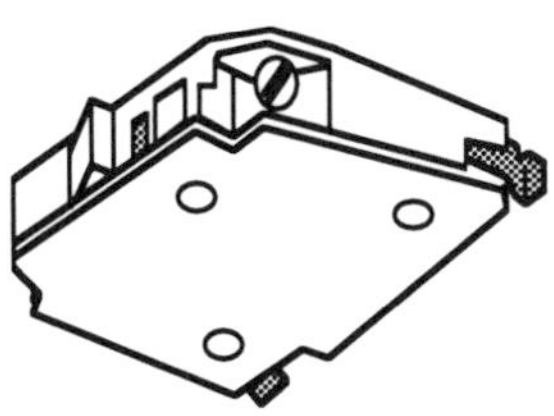

The Code gives no maximum value of impedance. If we assume that 5 times the rating of a breaker or fuse is the *minimum* ground-fault current that will be required to open the breaker or blow the fuse in a reasonable time of about one second, then we can calculate the *maximum* permitted impedance of the circuit.

An effective equipment grounding system should, under conditions of *maximum* ground fault current flow, maintain a low potential difference on metallic enclosures and carry the maximum current without damage to the conductor or equipment. The grounding conductors must not burn off during a fault condition leaving the equipment "energized" creating a shock hazard.

METAL

METAL CONDUIT AS EQUIPMENT GROUND

Many electricians assume that the conduit is used essentially for mechanical protection and do not realize the likelihood of *sparking* when the conduit is called upon to carry fault current.

Metal conduit is part of the electrical circuit, therefore the joints must be tight and clean to carry the large current under accidental fault conditions.

The Code places no restriction on the *size* or *length* of metal conduit. It must have continuity as well as the required conductivity to pass enough current to open the breaker or blow the fuse under a fault condition.

Joints in conduits and raceways require *special* consideration to avoid a shower of sparks and accompanies a fire hazard during a fault condition.

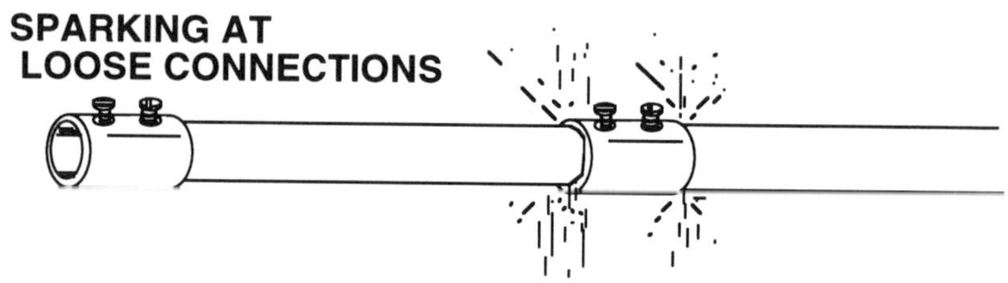

Some electricians know better but are careless in their work and this results in loosely assembled conduit systems which will definitely throw sparks if they are called upon to carry fault currents.

LOOSE LOCKNUTS

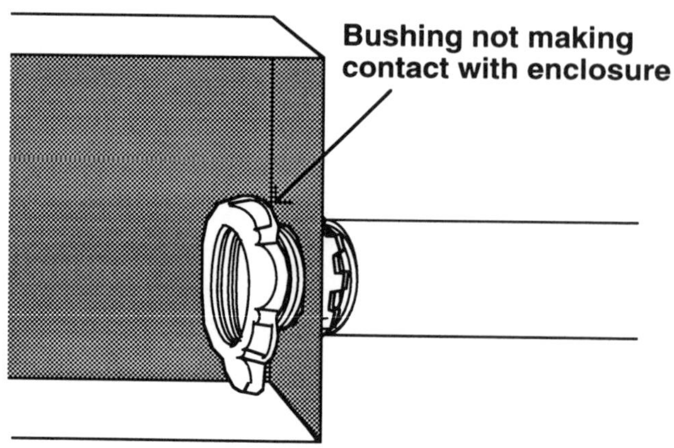

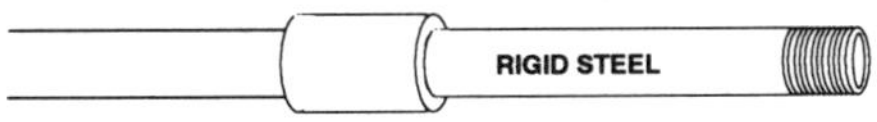

Rigid steel conduit must have the threads securely tightened.

A high-current test on rigid steel conduit produced a shower of sparks from about half the couplings in the conduit run. From one came a blowtorch stream of sparks which burned out many of the threads. Several small fires set in near-by combustible material would have been serious if not properly extinguished. The conduit had been installed by a crew regularly engaged in such work and they gave assurance that the joints had been pulled up to normal tightness and perhaps even a little more.

The importance of proper equipment grounding becomes greater with the larger size feeder circuits and the availability of higher short-circuit currents.

There is no practical substitute for a well designed and well assembled conduit system.

The calculation of impedance for metal conduit is laborious and inaccurate. When metal becomes part of an electric circuit, increases in both resistance and reactance will be large. Moreover, the resistance and reactance of the circuit will vary, depending upon the amount of current being carried.

Tests have shown that a metal conduit properly sized for the enclosed conductors will provide adequate capacity as an equipment grounding conductor. Although steel has a relatively high resistivity, it provides a large cross-sectional area. A 1/2" rigid steel conduit has a cross section of over 300 kcmil. The tubular shape and the fact that it is concentric with respect to the supply conductors it encloses contribute to reducing the overall impedance of conduit. The equipment grounding conductor is *not* designed to carry current continuously, but to provide a low impedance path for fault current for a *short time* until the breaker opens or the fuse blows.

Even though the Code does *not* require an equipment grounding conductor installed within the metal conduit, it still is the best design for a number of reasons.

The safety of the grounding system will depend on how long we can expect the equipment grounding conductor to remain *permanent* and *continuous*.

Corrosion alone can open the continuous path on steel conduit and often a steel coupling is replaced with a PVC coupling. The replacement of equipment and not tightning the locknuts or fittings properly can interrupt the effective grounding path. As you can see, over a period of time, once which was a properly installed equipment grounding circuit can result in an ineffective path of current flow.

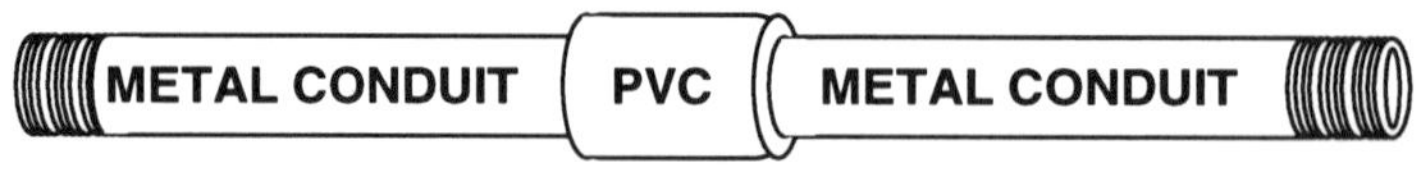

The sketch below shows a test circuit which has a single conductor installed in a steel conduit. The first ammeter reads 100 amps. The single conductor is *bonded* to the steel conduit at both ends providing a parallel path for current flow.

The steel conduit is actually carrying 97 amps and the single conductor only 3 amps. AC current flows to the outer edge of a conductor due to "skin effect". The conduit is the *outer edge* since they are connected in parallel. The conduit is actually more of a current-carrying conductor than the conductor itself.

This points out the importance of making the conduit couplings, fittings, locknuts, etc. *tight* to carry the current effectively in the grounding path when required.

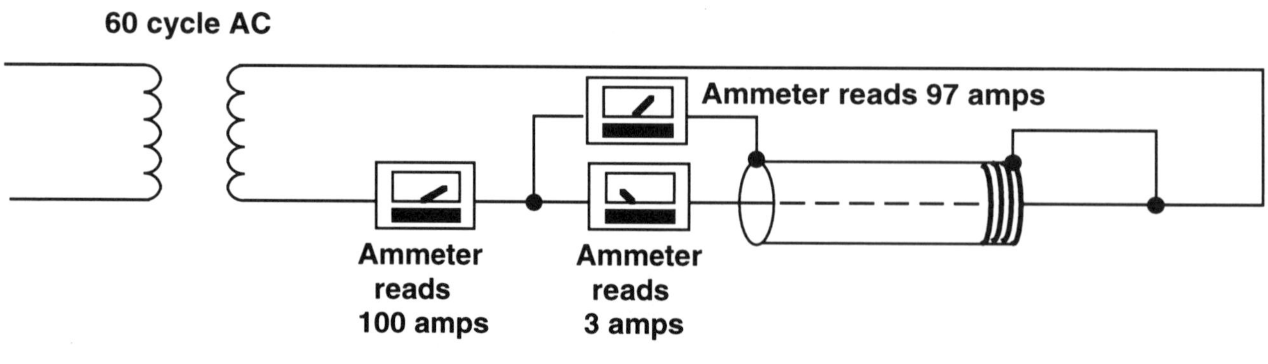

Thus, the steel conduit functions as a magnetic shield preventing the conductor within the conduit from carrying substantial current.

SIZING THE EQUIPMENT GROUNDING CONDUCTOR

Obviously PVC conduit requires an equipment grounding conductor to be installed since the conduit is nonmetallic. Code Table 250-95 gives the *minimum* size conductor permitted. Conductors are sized to the breaker or fuse size.

Where a single equipment grounding conductor is run in a conduit with multiple circuits, it shall be sized to the largest breaker or fuse protecting the conductors in the conduit.

A single metal conduit can be used as an equipment grounding conductor for all the circuits contained within the conduit. A single equipment grounding conductor can also be used for all the circuits contained within the conduit.

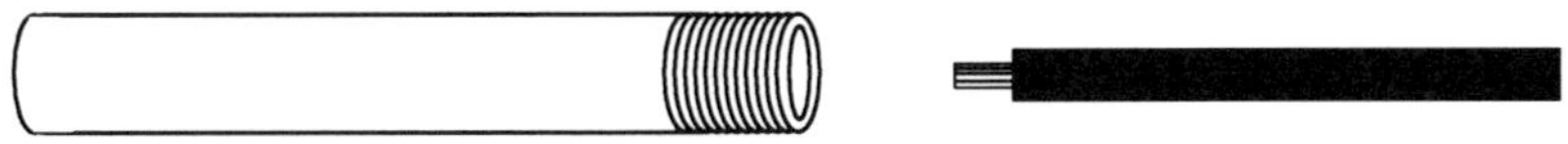

The sketch below shows a PVC conduit with two branch circuits, a 15 amp circuit with #14 conductors and a 30 amp circuit with #10 conductors. Only one equipment grounding conductor can be used for both circuits. The single equipment grounding conductor must be sized to the largest overcurrent device protecting conductors in the raceway, the 30 amp breaker. Table 250-95 requires for a 30 amp breaker a #10 copper equipment grounding conductor.

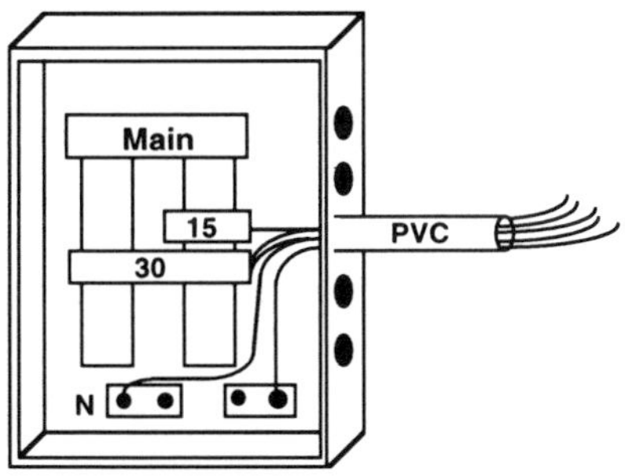

When conductors are increased to compensate for voltage drop, equipment grounding conductors shall also have their circular mil area increased proportionately.

A feeder to a 200 amp load normally has #3/0 THHN conductors and a #6 copper equipment grounding conductor per Table 250-95 for the 200 amp breaker. Because of the distance of the feeder, #250 kcmil conductors were used for voltage drop. To size the equipment grounding conductor proportionately:

A #3/0 has 167,800 circular mils, a #250 has 250,000 circular mils. The circular mil increase ratio is 250,000/167,800 = 1.49 increase of circular mils. The #6 minimum equipment grounding conductor for a 200 amp breaker has 26,240 circular mils. The 26,240 cm would be increased by 1.49 = 39,098 circular mils required. Table 8 requires a #4 equipment grounding conductor to compensate for voltage drop.

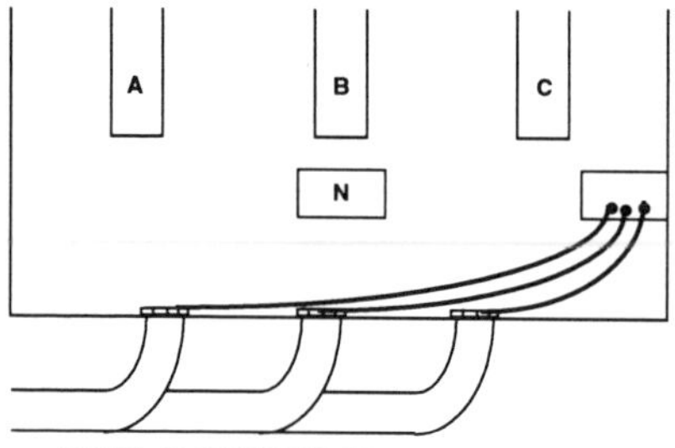

When conductors are paralleled in multiple conduits, an equipment grounding conductor is required in *each* conduit. The grounding conductor is sized based on the overcurrent device (fuse or breaker) protecting the circuit conductors in the conduit. The #1/0 minimum from 310-4 does not apply to equipment grounding conductors in parallel. 310-4 states equipment grounding conductors in parallel are still *sized* from table 250-95.

An exception states that the equipment grounding conductor is not required to be larger than the circuit conductors supplying the equipment. This situation occurs in motor circuits where a larger breaker or fuse is permitted to start the motor and carry the load.

A 28 amp F.L.C. motor would require #10 branch circuit conductors. An inverse time circuit breaker at 250% would allow a 70 amp breaker. If the equipment grounding conductor was sized from Table 250-95 for a 70 amp breaker it would require a #8 copper. It doesn't seem logical to use #10 circuit conductors with a #8 equipment grounding conductor.

The Code says an equipment grounding conductor can be insulated or bare. If a *bare* conductor is used in a *metal* conduit there can be arcing damage during a ground-fault condition. The arcing could cause damage to the other insulated conductors in the conduit. Using an insulated conductor in a metal conduit is the best choice, even if not required.

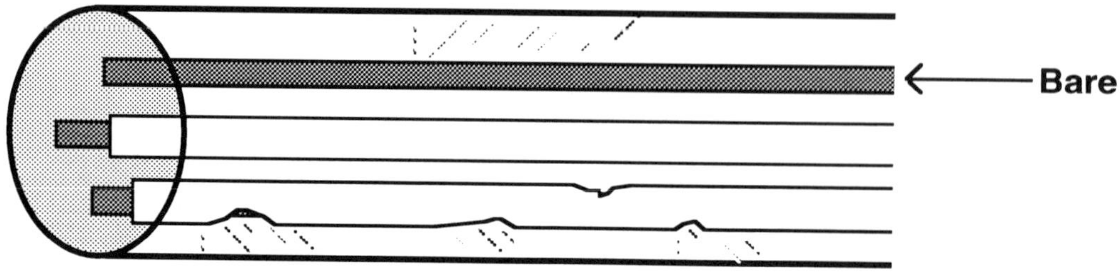

In some cases a designer will require a *bare* equipment grounding conductor to be installed in PVC conduit when the conduit is installed below grade.

Almost all underground PVC conduits will have water seep into them over a period of time. Because of this, the conductors will have a direct connection to ground. This contact to ground is not a problem in itself, but the problem is this: If the insulation in one of the ungrounded (hot) conductors breaks down, cracks or gets nicked during installation, there will be no return path to the panel in which the circuit originates if the grounding wire is *insulated*. This means the leakage current from the hot wire to the water will run through the ground rather than returning through the grounding conductor to open the breaker. The *bare* equipment grounding conductor will make contact with the water in the conduit and carry the fault current to trip the breaker.

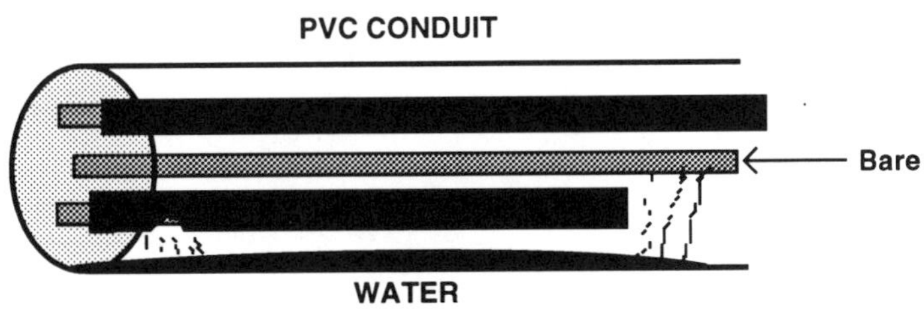

The equipment grounding conductor must always provide a low impedance path to open the breaker or fuse supplying the circuit conductors.

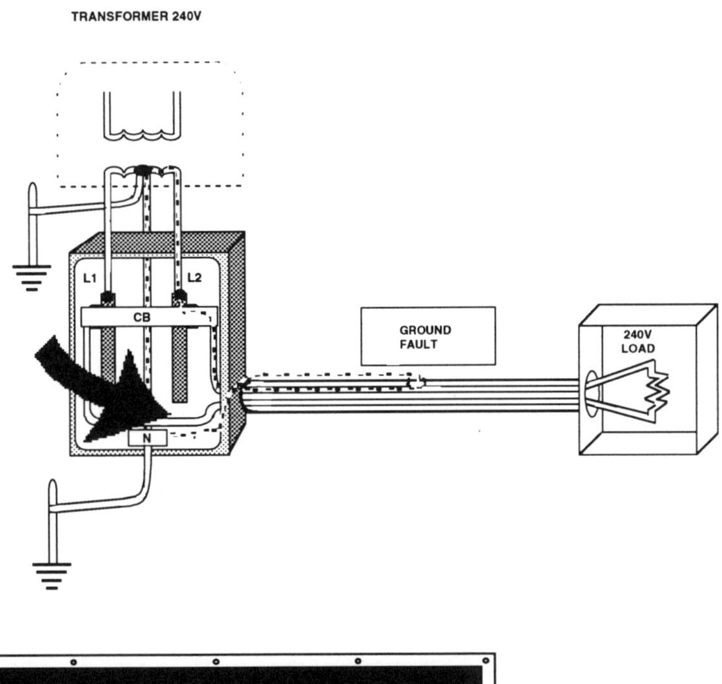

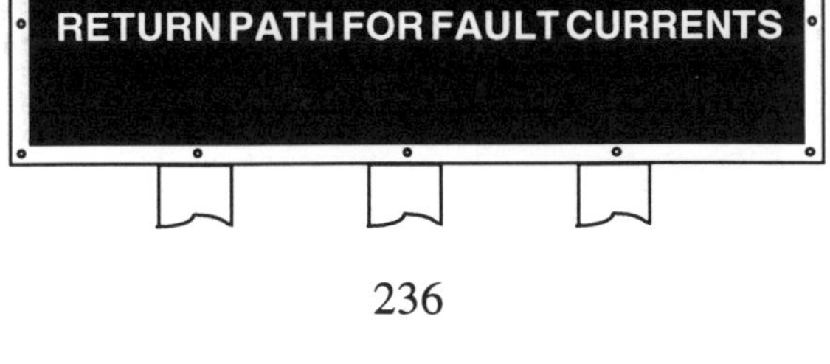

GROUNDING VIOLATIONS

Section 250-91c states earth shall not be used as the sole equipment grounding conductor. This is an area often violated and can cause a dangerous hazard, as planet earth can not carry the fault current to trip the circuit breaker.

The circuit *must* contain an equipment grounding conductor.

The circuit shown below is a definite violation. The metal pole is connected with a jumper to a driven ground rod but the circuit contains *no* equipment grounding conductor. If the light or wiring would fault to the metal pole, there is no path back to the panelboard to trip the circuit breaker.

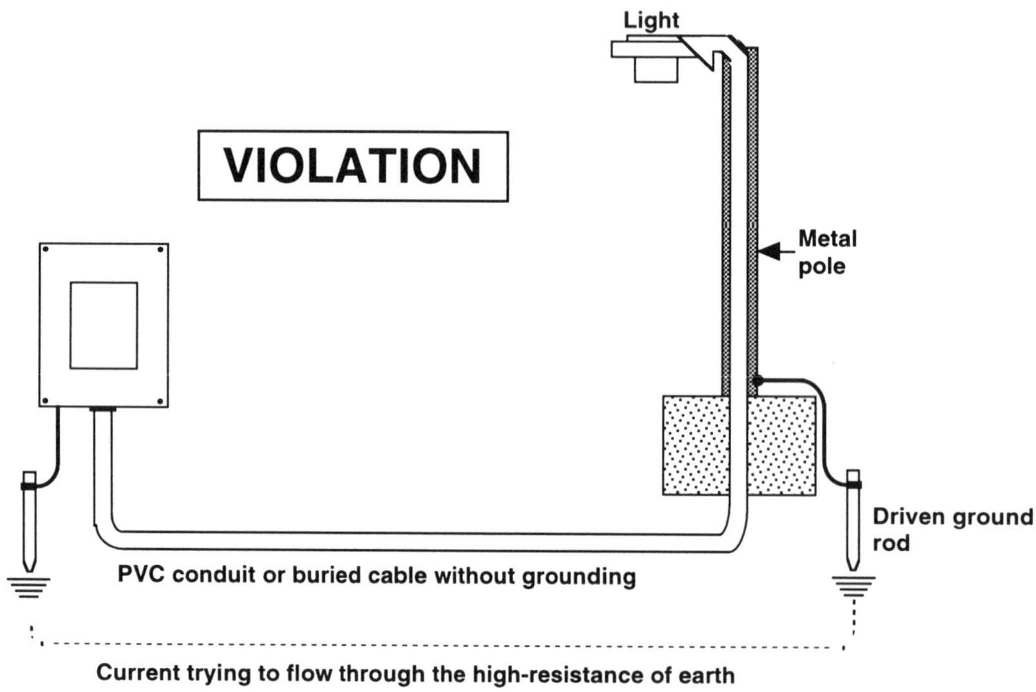

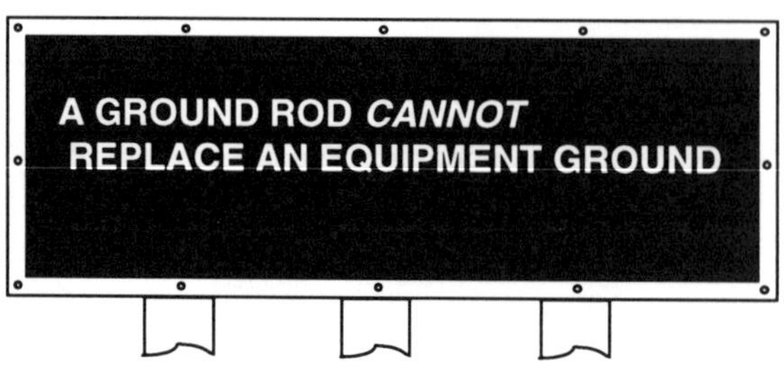

The sketch below shows a common *violation* of bonding. The neutral bus in the *sub-panel* is bonded to the panel. When the neutral is bonded beyond the service it puts the neutral and the conduit in a parallel path, both carrying current. Some current flows in the neutral conductor and some current flows over the metal conduit and panel. The resistance of the conduit is approximately three times greater than the neutral conductor. This allows 1/4 of the current (5 amps) to flow over the conduit with 15 amps flowing in the neutral conductor.

This is a *VIOLATION.* No current except ground-fault current is permitted to flow through metallic equipment.

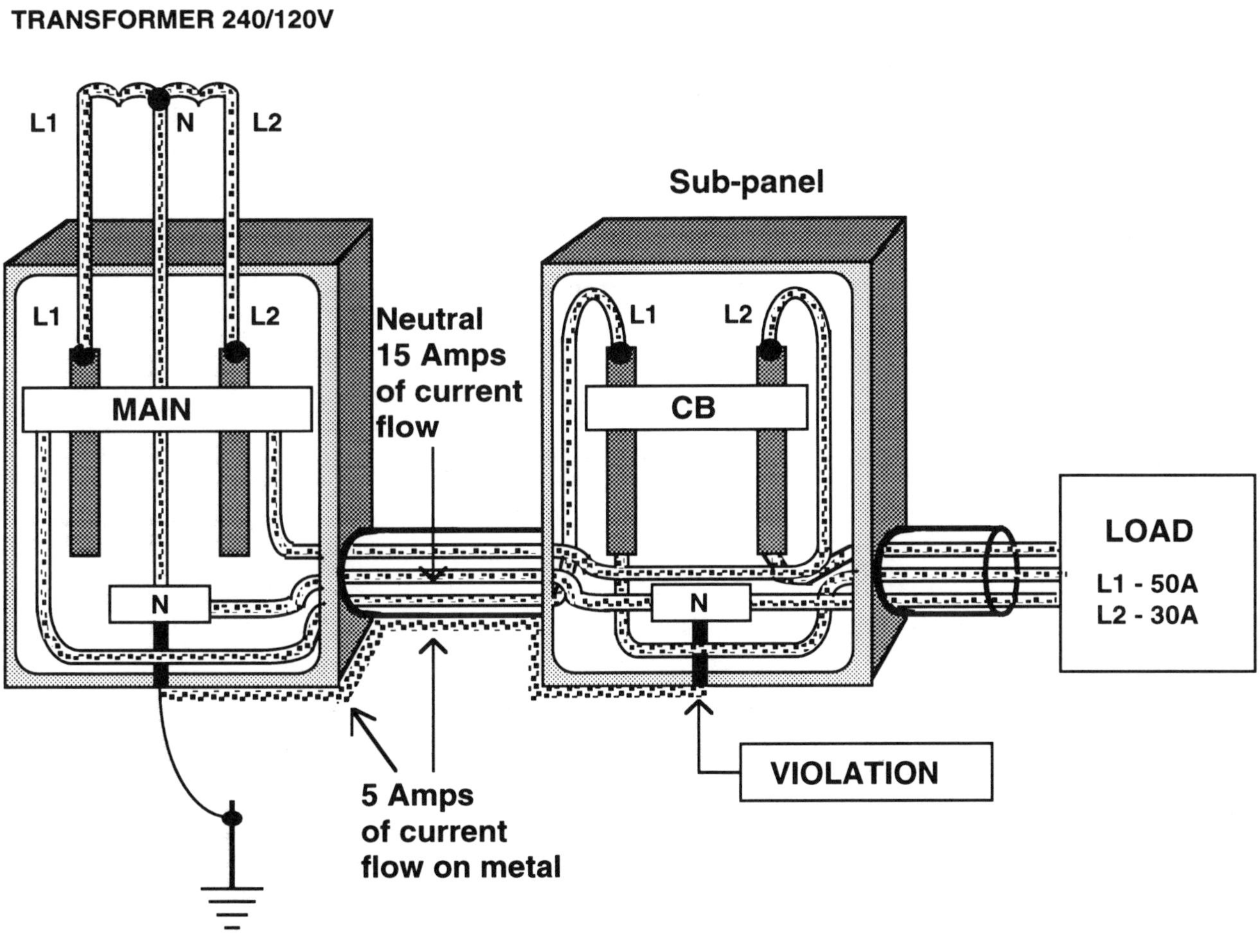

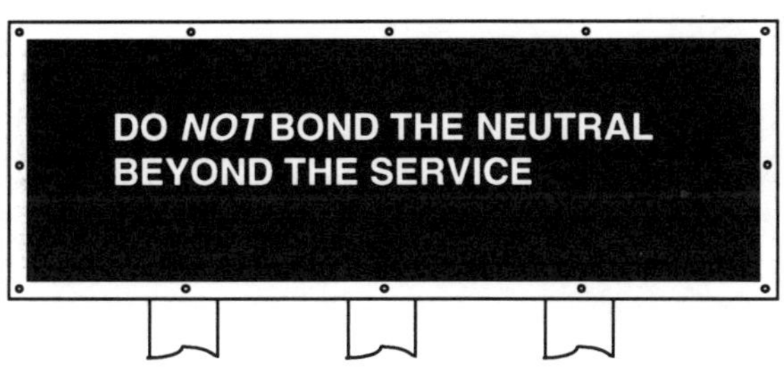

The Code permits in section 250-74 ex.4 an insulated *isolated* grounding conductor to be run without connection to the subpanel grounding terminal for the reduction of electrical noise on the neutral.

The exception permits a receptacle with an *isolating* grounding terminal to be used with no connection between the receptacle grounding terminal and the yoke. The use of an isolated grounding receptacle allows a "clean" path back to the *service grounding bus* in the service equipment.

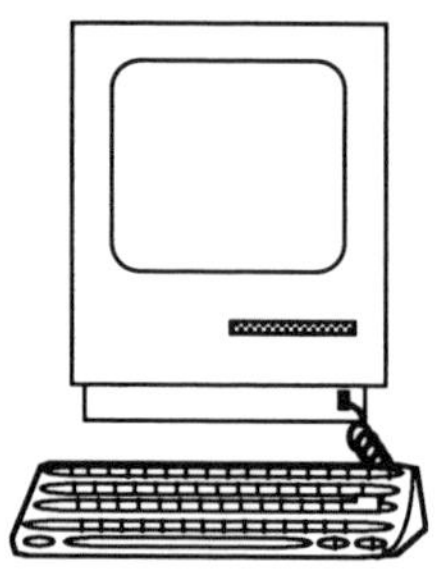

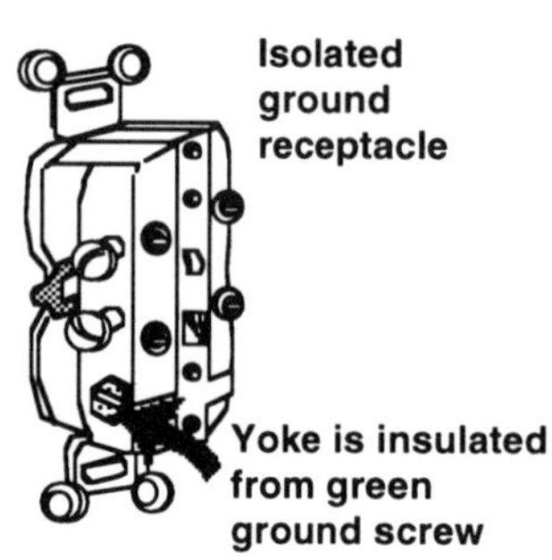

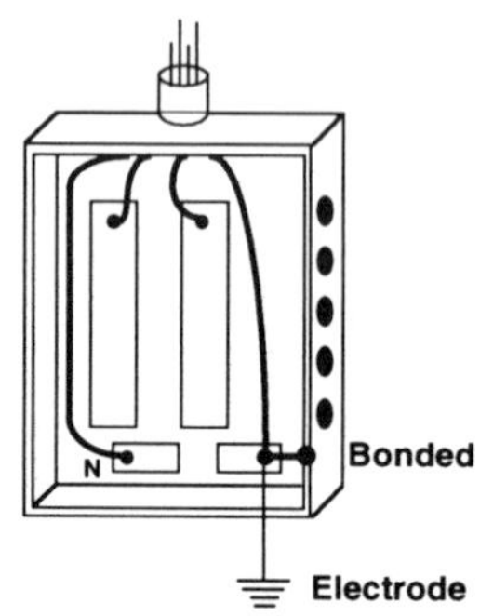

When an isolated equipment grounding conductor is used in a circuit, the circuit is still required to have an equipment grounding conductor connected to the box containing the receptacle. Either the metal conduit can be used or a separate equipment grounding conductor must be installed.

Code section 384-20 ex. states: Where an isolated equipment grounding conductor is provided it shall be permitted to pass through the subpanel without being connected to the panelboard's equipment grounding terminal bar. It is connected at the service equipment neutral ground bus.

The *violation* generally occurs when the designer requires the circuit to have its own individual grounding electrode. The isolated grounding conductor must be connected to the *same* grounding electrode at the service. It is a violation of the Code to provide a separate grounding electrode. *All* grounding ties together at the service to prevent a dangerous voltage potential.

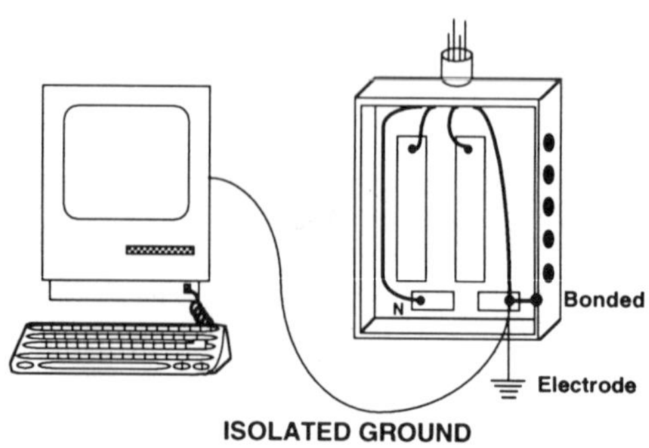

ISOLATED GROUND

GROUNDING ELECTRODE CONDUCTOR

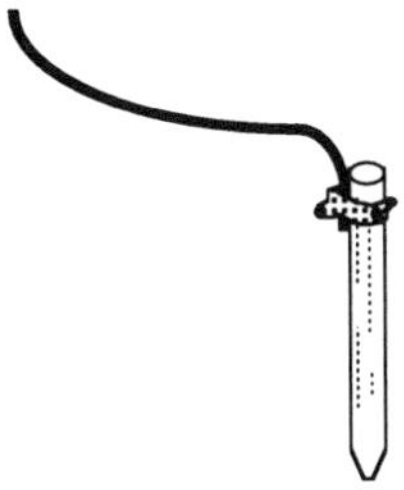

The grounding electrode conductor is sized from Table 250-94 to the largest service entrance conductor or equivalent area for parallel conductors.

A #500 kcmil service conductor would require a #1/0 copper grounding electrode conductor.

There is a difference between an *electrode* and a *made* electrode.

Section 250-81 lists the electrodes in parts a,b,c and d.

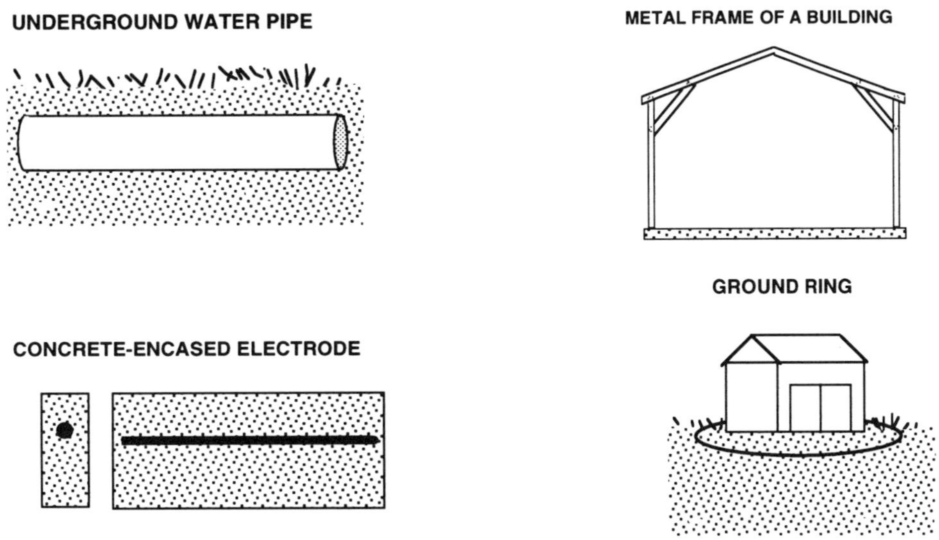

Section 250-83 b,c and d lists *made* electrodes.

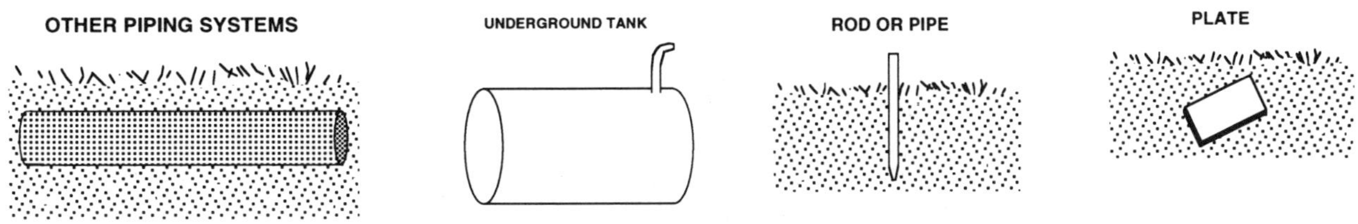

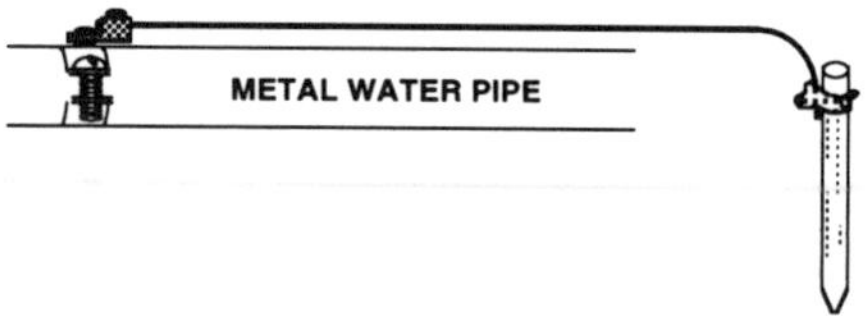

The exceptions for a grounded system are:

(a) A made electrode (rod or plate), that portion of the grounding electrode conductor which is the sole connection to the grounding electrode is not required to be larger than a *#6 copper.*

The #6 copper conductor has been found to be capable of carrying the amount of current that a rod or plate electrode is capable of *dissipating* into the earth. Fortunately, under normal or fault conditions, the grounding electrode will carry *little or none* of the current. It functions primarily to tie the ground voltage of the system to the reference voltage of the earth.

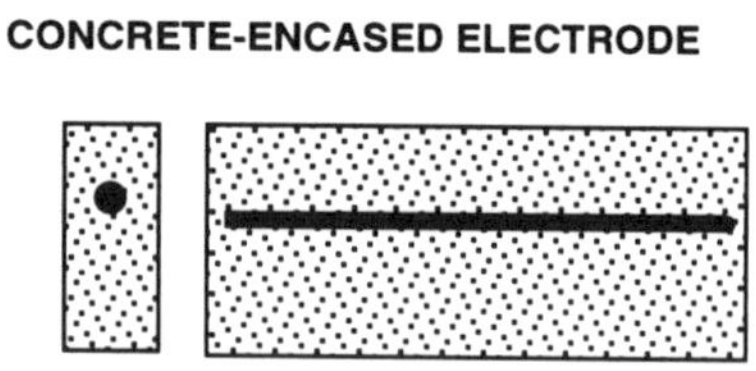

(b) The portion of the grounding electrode conductor to a concrete-encased electrode, which is the sole connection, is not required to be larger than *#4 copper.*

(c) Where connected to a ground ring, that portion of the grounding electrode conductor which is the sole connection to the grounding electrode, is not required to be larger than the ground ring conductor. A *#2 copper* would be the *minimum* permitted.

The reason there are two tables for *sizing* grounding conductors, Table 250-95 is for sizing the *equipment* grounding conductor which carries current only in a fault condition to trip the breaker or blow the fuse. Table 250-94 is for sizing the grounding *electrode* conductor which is used to connect the system to earth to limit the voltage from a lightning strike or a higher impressed voltage from another system.

To understand grounding and to design circuits properly, these two conductors and their functions must be clearly understood.

The main service overcurrent device is what separates *line* from *load.*

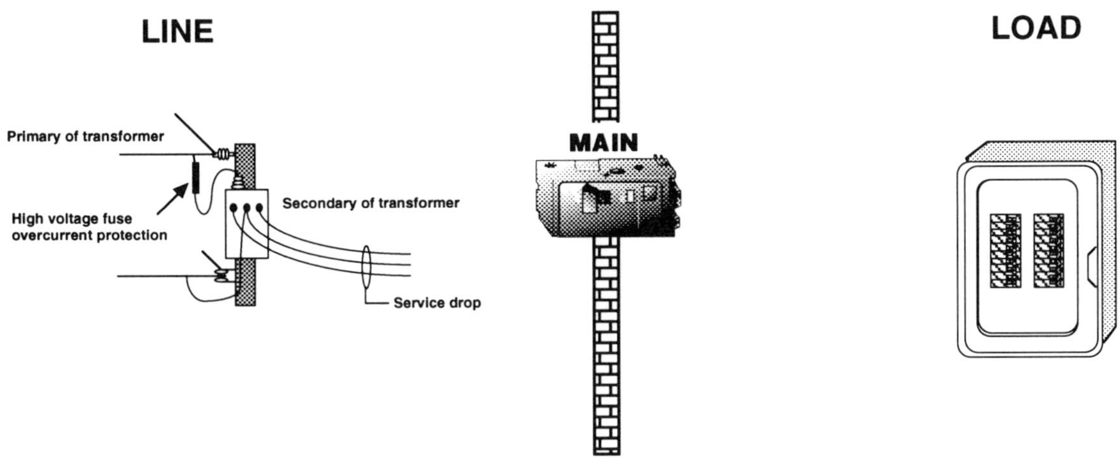

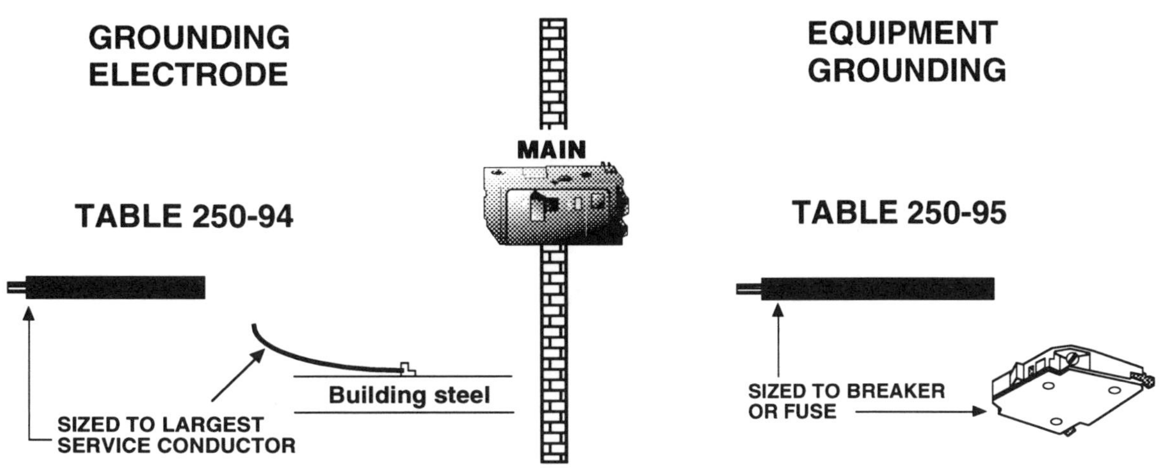

SIZING BONDING JUMPERS - LINE SIDE

Main and equipment bonding jumpers shall be of copper or other corrosion-resistant material. A main bonding jumper shall be a wire, screw, bus, or similar suitable conductor.

MAIN and EQUIPMENT BONDING JUMPER LINE SIDE SERVICE | **TABLE 250-94**

Main and equipment jumper line side service

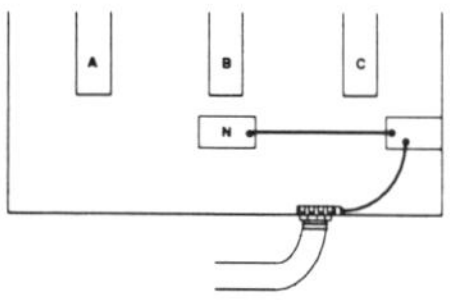

Sized by the largest service entrance conductor using Table 250-94.

Main and equipment bonding jumper line side service, multiple phase conductors in parallel.

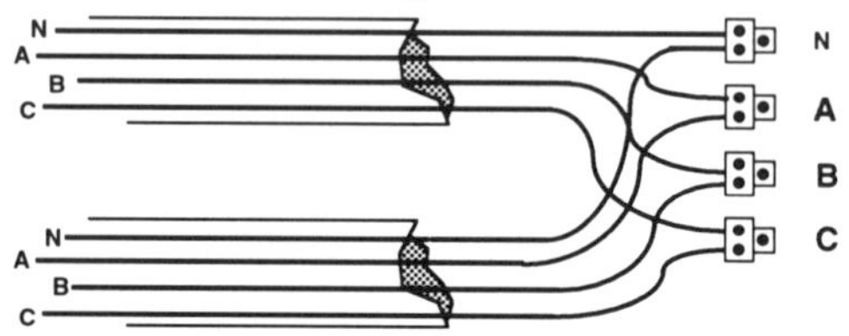

Sized by the equivalent area of one phase. The jumper shall not be less than *12 1/2%* of the area of one phase when larger than 1100 kcmil copper or 1750 kcmil aluminum per section 250-79d.

Main and equipment bonding jumper line side service, different conductor materials (copper-aluminum).

Aluminum phase conductors

Copper bonding jumper

The minimum size bonding jumper shall be based on assumed use of phase conductors of the same material as the bonding jumper with an ampacity equivalent to that of the installed phase conductors.

Example: A service has 4 - #500 kcmil THW *aluminum* conductors per phase. What size *copper* main bonding jumper is required?

Solution: Table 310-16 to find *equivalent ampacity*. A #500 kcmil THW aluminum has an ampacity of 310 amps, equivalent in *copper* would be a #350 kcmil THW copper at 310 amps. Now we assume copper phase conductors at #350 kcmil. 4 per phase x #350 kcmil = 1400 kcmil. The 12 1/2% is applied per 250-79d. 1400 kcmil x 12 1/2% = 175 kcmil or 175,000 cm. Table 8 lists the circular mil area for conductors. 175,000 cm is required. Table 8 requires a *#4/0* bonding jumper at 211,600 cm.

MAIN and EQUIPMENT BONDING JUMPER LINE SIDE SERVICE

TABLE 250-94

One equipment bonding jumper looped to each conduit.

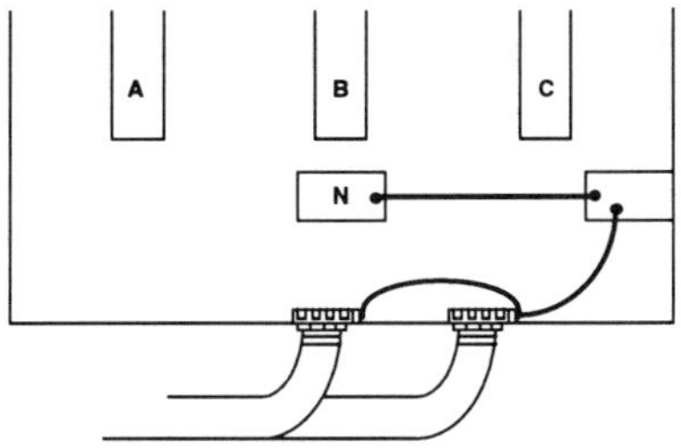

250-79d Sized by the largest service entrance conductor or equivalent area per phase for parallel.

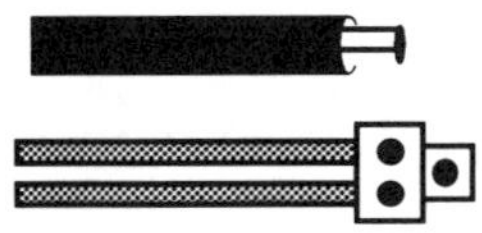

Example: 2 - #500 kcmil THW copper conductors paralleled per phase. What size copper equipment bonding jumper is required when looped through each conduit?

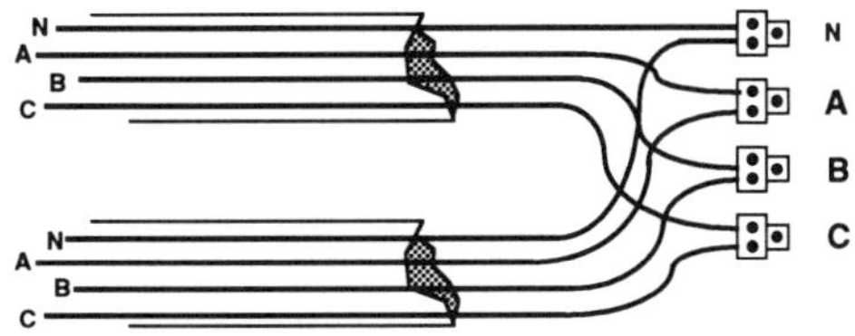

Solution: 2 x 500 = 1000 kcmil copper per phase. Table 250-94 requires a single *#2/0 copper*.

One equipment bonding jumper from each conduit to grounding bus.

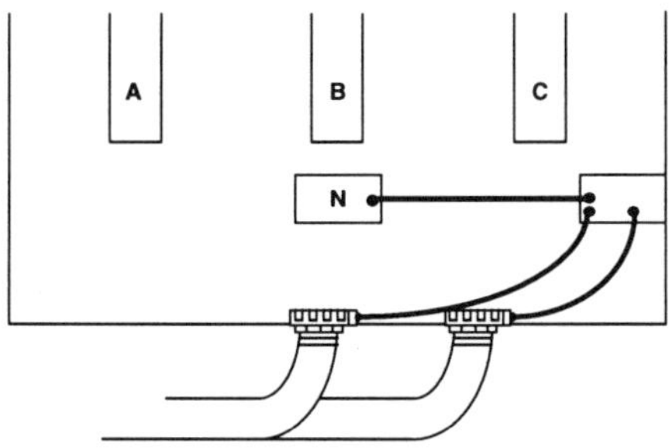

250-79d Sized by the largest service entrance conductor in each raceway.

Example: Each conduit contains 1 - #500 kcmil THW copper conductor. What size copper equipment bonding jumper is required from each conduit to grounding bus?

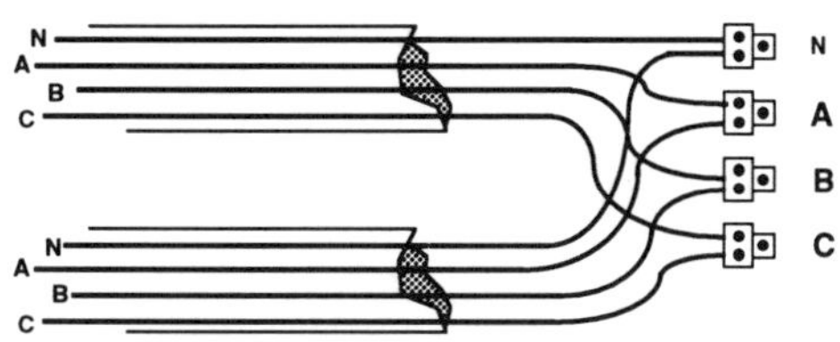

Solution: Table 250-94 requires a *#1/0 copper* jumper for each conduit.

SIZING BONDING JUMPERS - LOAD SIDE

The equipment bonding jumper on the *load side* of the service overcurrent device shall not be smaller than the equipment grounding conductors sized from Table 250-95. The equipment bonding jumper shall be permitted to be installed inside or outside of a raceway or enclosure. Where installed *outside* of a raceway it shall not exceed *6 feet in length*.

BONDING JUMPERS LOAD SIDE

TABLE 250-95

Bonding jumper for expansion joint.

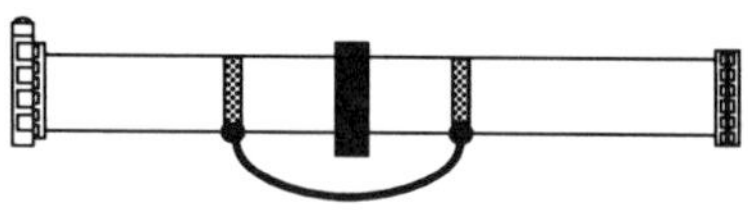

Sized to the largest overcurrent device protecting a circuit in the raceway.

Bonding jumper for flexible conduit.

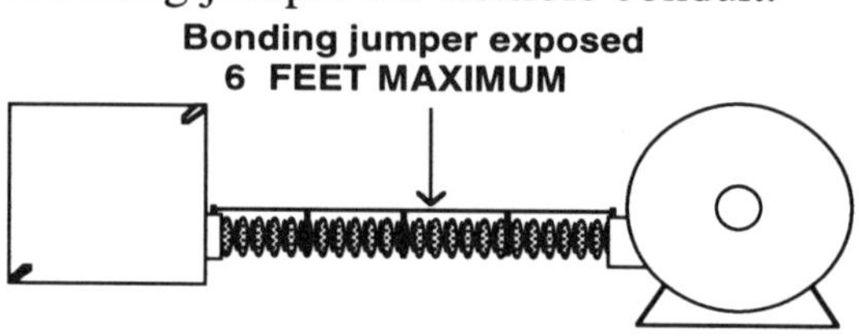

Sized to the largest overcurrent device protecting a circuit in the raceway. The equipment bonding jumper is not required to be larger than the circuit conductors supplying the equipment.

Bonding jumper to bond receptacle to metal box.

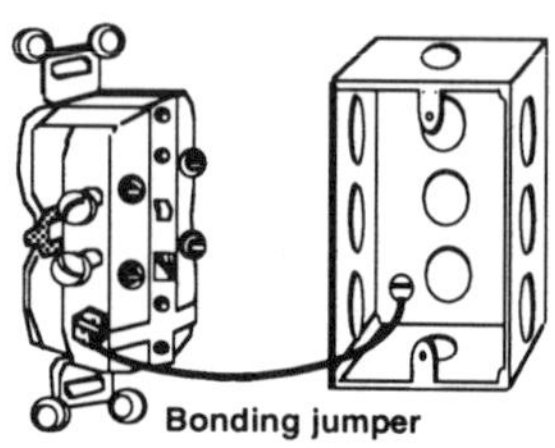

Sized to the overcurrent device (fuse or CB) protecting the circuit.

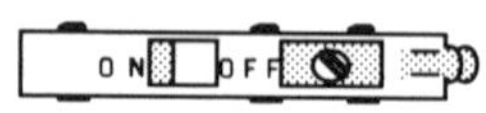

Bonding jumper to interior metal water piping.

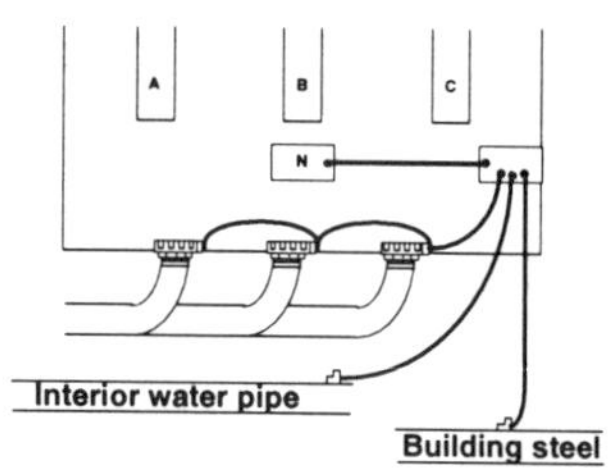

Sized from Table 250-*94* per section 250-80a.

Bonding jumper to interior metal water piping, multiple occupancies.

Sized from Table 250-95 per section 250-80a ex.

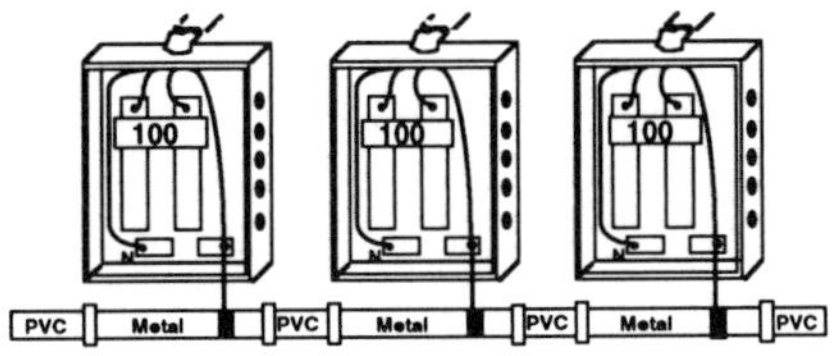

Bonding jumper to other metal piping and metal duct work that could become energized.

Sized from Table 250-95 per section 250-80b. The equipment grounding conductor for the circuit which may energize the piping may serve as the bonding means.

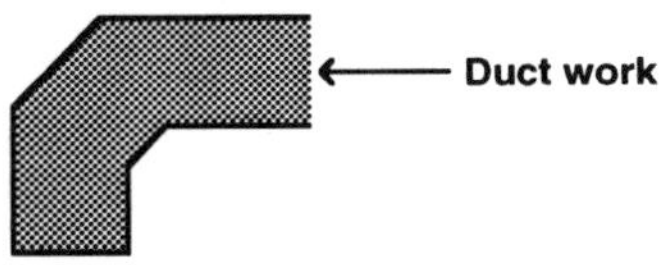

It is important to understand which is the "supply side" and which is the "load side" when making connections. The supply side (line side) are the service conductors from the weatherhead to the connection to the disconnecting means (main breaker) and neutral bus in the loadcenter.

On the supply side a standard locknut can be used on the *outside* of the enclosures and a *bonding bushing* on the inside of the enclosures (bonding bushing on the threaded end of the conduit or fitting).

On the *load side* standard locknuts can be used on both sides of an enclosure.

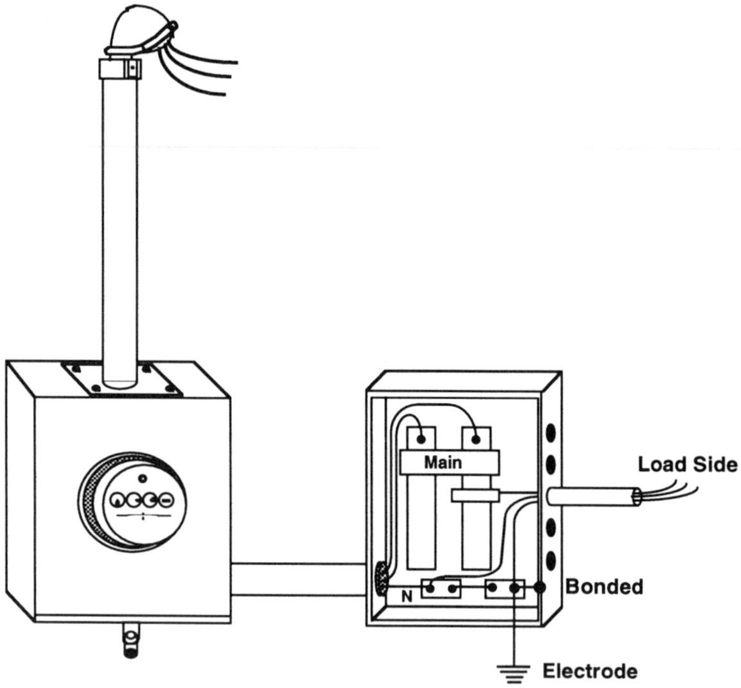

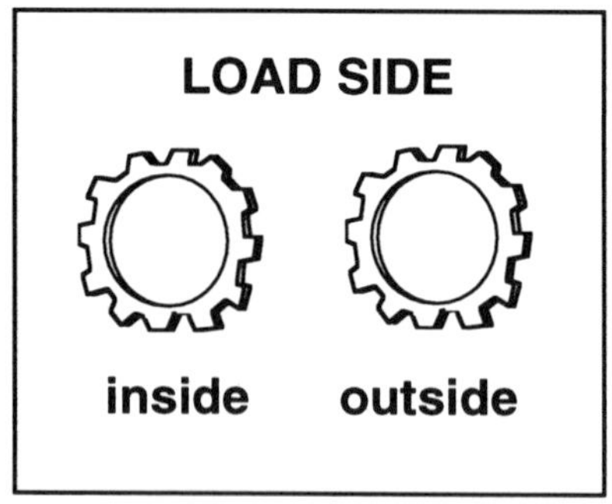

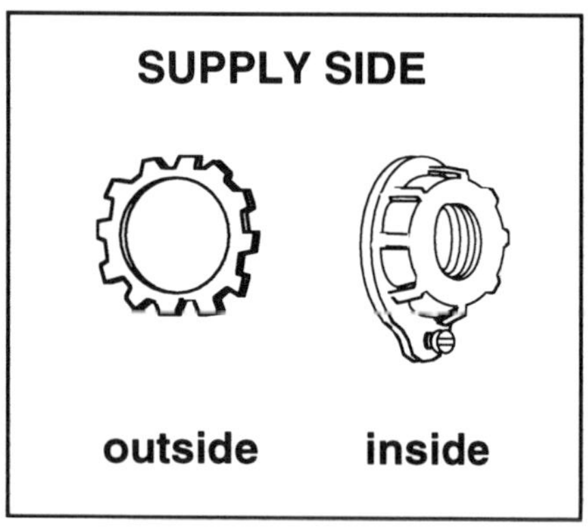

A conduit connection to a meter base may be made by connecting a threaded conduit end into a threaded hub on the meter base.

Bonding of service equipment requires all noncurrent-carrying parts of the service equipment to be bonded together. Service bonding (line-supply side) must assure burn-clear on shorts and grounds in service conductors as these conductors are not properly protected against faults and shorts on the *supply side* of the service disconnecting means. The primary function of equipment bonding ahead of the service disconnect is to limit the voltage to ground on the enclosures. If a fault occurs ahead of the service overcurrent protection it is likely to burn clear, the only protection is on the utility company's transformer on the oversized primary side.

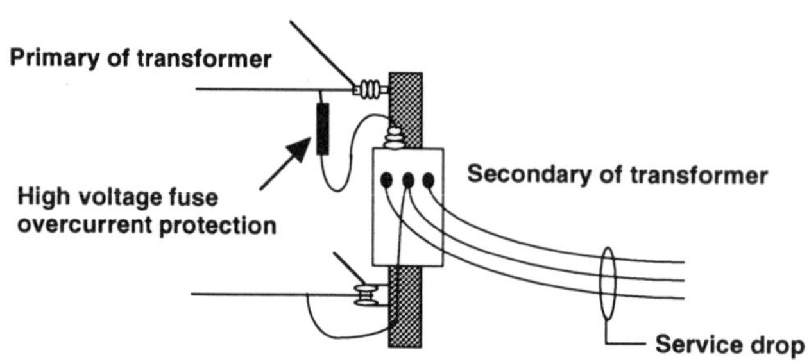

LOAD BALANCING

LOAD BALANCING - SINGLE PHASE

Code section 220-4d states the load shall be *evenly proportioned* among tne branch circuits within the panelboard.

EXAMPLE 1

Balance the following loads as closely as possible. 3-wire, 1 ø, 230/115v.

1 - 5000 watt 1 ø, 230v load
2 - 3000 watt 1 ø, 230v loads
6 - 400 watt 1 ø, 115v loads

1 ø 230 / 115v

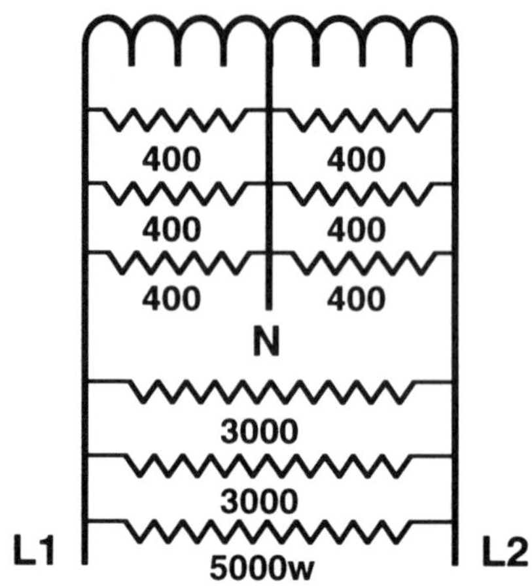

The 5000 watt load is 230v so it would connect between L1 and L2. The two 3000 watt loads are also 230v so they connect between L1 and L2.

The six 400 watt loads are 115v so they connect from line to neutral and are balanced as closely as possible. In this case we can obtain a balance by connecting three 400 watt loads on line one to neutral and the other three 400 watt loads line two to neutral.

If the loads are not 115 volt they do not connect to the neutral. Only the 115 volt loads connect to the neutral.

Calculate the current in each of the conductors L1, L2, and the neutral with all loads on:

CURRENT IN LINE ONE:

Line one would carry half of the 5000 watt load	=	2500 watt
Line one would carry half of the two 3000 watt loads	=	3000 watt
Line one would carry three 400 watt loads	=	1200 watt
		6700 watt

$$\frac{6700 \text{ watt}}{115 \text{ volt}} = 58.26 \text{ amperes current in line one}$$

CURRENT IN LINE TWO:

Line two would carry half of the 5000 watt load	= 2500 watt
Line two would carry half of the two 3000 watt loads	= 3000 watt
Line two would carry three 400 watt loads	= 1200 watt
	6700 watt

$$\frac{6700 \text{ watt}}{115 \text{ volt}} = 58.26 \text{ amperes current in line two}$$

The current in the neutral conductor *with all loads on would be zero.*

The *maximum current* the neutral conductor would carry in this circuit connected in a balance would be when all the loads on either line one or line two were shut off.

Example: With line one loads shut off the neutral would carry the loads connected line two to neutral.

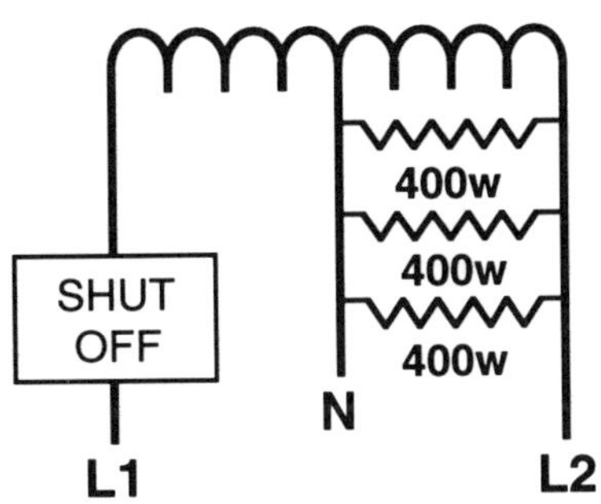

$$\frac{1200\text{w}}{115\text{v}} = 10.43 \text{ AMPERES } \textit{MAXIMUM} \text{ NEUTRAL CURRENT!}$$

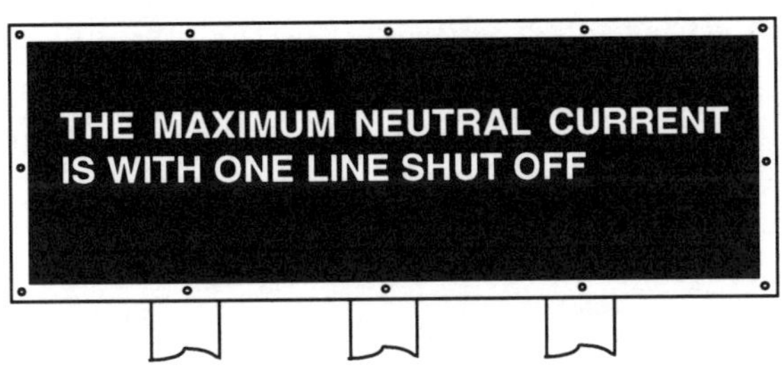

The electrician can perfectly balance these loads as they are connected in the panel. It's almost impossible to maintain a perfect balance as all loads will not always be on nor will all the loads always be off.

So as the 400 watt, 115 volt loads are switched on and off, the current flowing in the neutral conductor varies from zero current with all loads on to a maximum current of 10.43 amps for three loads on one line at the same time with all loads off on the other line.

Total power required by all the loads in EXAMPLE ONE would be:

1 - 5000 watt = 5000 watt
2 - 3000 watt = 6000 watt
6 - 400 watt = 2400 watt
13,400 watt total power

$$\text{TOTAL CURRENT} = \frac{\text{TOTAL WATTAGE}}{\text{TOTAL VOLTAGE}} = \frac{13{,}400 \text{ watt}}{230 \text{ volt}} = 58.26 \text{ amperes current}$$

EXAMPLE 2

An unbalanced calculation:

1 - 6000 watt 1 ø 230 volt load
1 - 4000 watt 1 ø 230 volt load
5 - 1500 watt 1 ø 115 volt loads

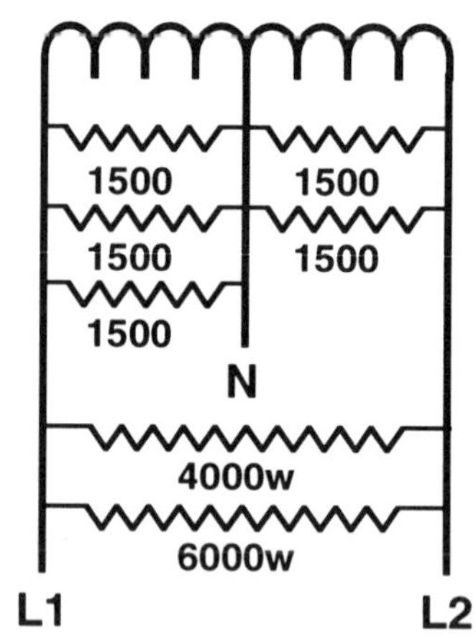

The 6000 watt load is 230 volt. It would connect between line one and line two. The 4000 watt load is also 230 volt so it would also connect between line one and line two.

The five 1500 watt loads are 115 volt, they would connect line to neutral. They are balanced as closely as possible. In this case connect three 1500 watt loads from line one to neutral and two 1500 watt loads from line two to neutral.

Calculate the current in each of the conductors L1, L2, and the neutral with all loads on:

CURRENT IN LINE ONE:

Line one would carry half of the 6000 watt load	= 3000 watt
Line one would carry half of the 4000 watt load	= 2000 watt
Line one would carry three 1500 watt loads	= 4500 watt
	9500 watt

$$\frac{9500 \text{ watt}}{115 \text{ volt}} = 82.6 \text{ amperes current in line one}$$

CURRENT IN LINE TWO:

Line two would carry half of the 6000 watt load	= 3000 watt
Line two would carry half of the 4000 watt load	= 2000 watt
Line two would carry two 1500 watt loads	= 3000 watt
	8000 watt

$$\frac{8000 \text{ watt}}{115 \text{ volt}} = 69.56 \text{ amperes current in line two}$$

Current flowing in the neutral with all the loads on would be 13.04 amperes; the unbalanced current. 1500 watt the unbalanced divided by 115 volt = 13.04 amperes.

The maximum current the neutral conductor would carry connected as shown, would be when the line two loads are both off, the three loads line one to neutral would be the maximum unbalance 4500w/115v = 39.13 amperes maximum current.

Total power required by all the loads in EXAMPLE 2 would be:

1 - 6000 watt = 6000 watt
1 - 4000 watt = 4000 watt
5 - 1500 watt = 7500 watt
17,500 watt total power

$$\text{TOTAL CURRENT} = \frac{\text{TOTAL WATTAGE}}{\text{TOTAL VOLTAGE}} = \frac{17{,}500 \text{ watt}}{230 \text{ volt}} = 76.08 \text{ amperes current}$$

EXAMPLE 3

Determining the neutral current in an unbalanced single-phase 3-wire, 230/115 volt system.

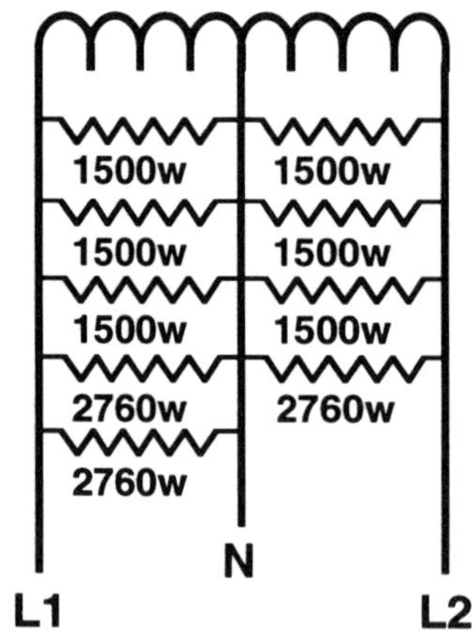

L1	**L2**
1500 watt	1500 watt
1500 watt	1500 watt
1500 watt	1500 watt
2760 watt	2760 watt
2760 watt	------------
10,020 watt	7,260 watt

Total power would be 10,020 watt plus 7,260 watt = 17,280 watt

$$\frac{17{,}280 \text{ watt}}{230 \text{ volt}} = 75.13 \text{ amperes current}$$

By assuming the loads to be balanced and dividing the total power by the line voltage, the total current equals 75.13 amperes. For more exact results for an *unbalanced* system, calculate the neutral load by the largest line to neutral total 10,020 watts divided by the line to neutral voltage of 115, equals 87.13 amperes or 12 amperes higher.

So the maximum neutral current as connected would be if the line two loads were off. The line one to neutral current would be the maximum 87.13 amperes.

CALCULATING THE NEUTRAL LOAD

When a 3-wire 230/115 volt circuit serves a total load that is balanced from each hot leg to neutral, that is, half the total load is connected from one hot leg to neutral and the other half of the total load from the other hot leg to neutral, the condition of maximum unbalance occurs when all the load fed by one hot leg is operating and all the load fed by the other hot leg is off. Under that condition, the neutral current and hot leg current are equal to half the total watts divided by 115 volts (half the voltage between hot legs).

But that current is exactly the same as the current that results from dividing the total load (connected hot leg to hot leg) by 230 volt, (which is twice the voltage from hot leg to neutral) because of this relationship, it is easy to determine the neutral current by simply calculating hotleg load (the total load from hot leg to hot leg) and dividing by 230 volts.

COMPUTATION ACCURACY

When 115 volt loads are exactly balanced on a 230/115 volt circuit, it is correct to add the load in watts and then divide by 230 volts to determine the line and neutral load in amperes. The neutral load represents the maximum unbalanced load at 115 volts. For balanced 115 volt loads, the neutral is one-half the total load; therefore, dividing the total load in watts by 230 volts yields the correct value for the neutral in amperes.

If a 115 volt load is *not balanced* by an identical load, the neutral load should *not* be determined by dividing by the 230 volt circuit voltage if an *exact* result is desired.

Standard practice allows the neutral load in amperes to be determined by summing the loads that require a neutral and dividing the total watts by 230 volts. It should be remembered that this method is slightly *inaccurate* when *unbalanced* loads are present on 3-wire circuits.

EXAMPLE 3 shows the *exact* calculation at 87.13 amperes.

LOAD BALANCING - THREE PHASE

The three-phase 4-wire secondary can be connected only two ways, either WYE or DELTA.

Let's start with the *wye* 4-wire 208 / 120v three-phase secondary.

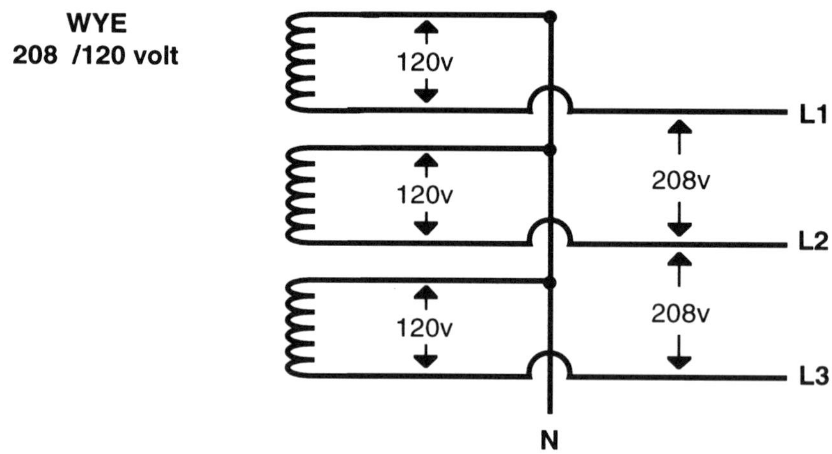

We have three possible voltage connections:

- 208 volt 3ø loads connect L1 - L2 - L3
- 208 volt 1ø loads connect L1-L2 or L2-L3 or L3-L1
- 120 volt 1ø loads connect L1-N L2-N L3-N

Since in either of the three-phase systems, delta or wye, the voltages are vectors, the currents derived from loads are vector quantities. In order to add these currents, they must be added vectorially.

The easiest way to handle transformer loads is to convert all the loads to volt-amps and distribute the volt-amps to the phase to which they are connected.

Division of Loads:

208 volt three-phase loads	- 1/3 on phase A
	- 1/3 on phase B
	- 1/3 on phase C
208 volt single-phase loads	- 1/2 on any one phase
	- 1/2 on any other phase
120 volt single-phase loads	- 100% on any one phase

EXAMPLE 1

Balance the following loads on a 208 / 120v 4-wire three-phase secondary.

1 - 6000va 3ø 208v motor
3 - 5000va 1ø 208v heaters
3 - 150va 1ø 120v lights

Solution: The 6000va three-phase 208v motor divides 1/3 on each phase. The three 5000va heaters would divide 1/2 per phase since the single-phase voltage is 208v, the heaters would connect from L1-L2, L2-L3, and L3-L1. The three 150va lights are single-phase 120 volt, 120v connects from line to neutral. Since we have three 150va loads, we can equally balance them one per phase.

PHASE A	**PHASE B**	**PHASE C**
2000va	2000va	2000va
2500va	2500va	2500va
2500va	2500va	2500va
150va	150va	150va
7150va	7150va	7150va

L1 amps $\frac{7150va}{120v} = 59.5$ **L2 amps** $\frac{7150va}{120v} = 59.5$ **L3 amps** $\frac{7150va}{120v} = 59.5$

- *Checkpoint: WYE-connected phase amps = line amps,*
  Total va = 7150va x 3 phases = $\frac{21{,}450va}{208v \times 1.732}$ *= 59.5 line amps.*

NEUTRAL:	**PHASE A**	**PHASE B**	**PHASE C**
	150va	150va	150va

With all loads on, the neutral would carry ZERO current as we have a balance. Maximum neutral current would occur when two 150va loads are OFF, then the neutral would carry $\frac{150va}{120v}$ = 1.25 amps.

No three-phase loads appear on the neutral, single-phase 208v loads connected across two phases will *not* appear on the neutral. Only 120 volt single-phase loads are connected to the neutral.

Maximum neutral current:	**PHASE A**	~~**PHASE B**~~	~~**PHASE C**~~	
	150va	~~150va~~	~~150va~~	*LOSS OF TWO PHASES*

If any one of the three 150va loads shut off, with two loads on, the neutral will carry the unbalance which would be 150va.

PHASE A	**PHASE B**	**PHASE C**	
150va	off - - - - -	150va	*NEUTRAL WOULD CARRY 150va*

Example One is shown below in a picture form.

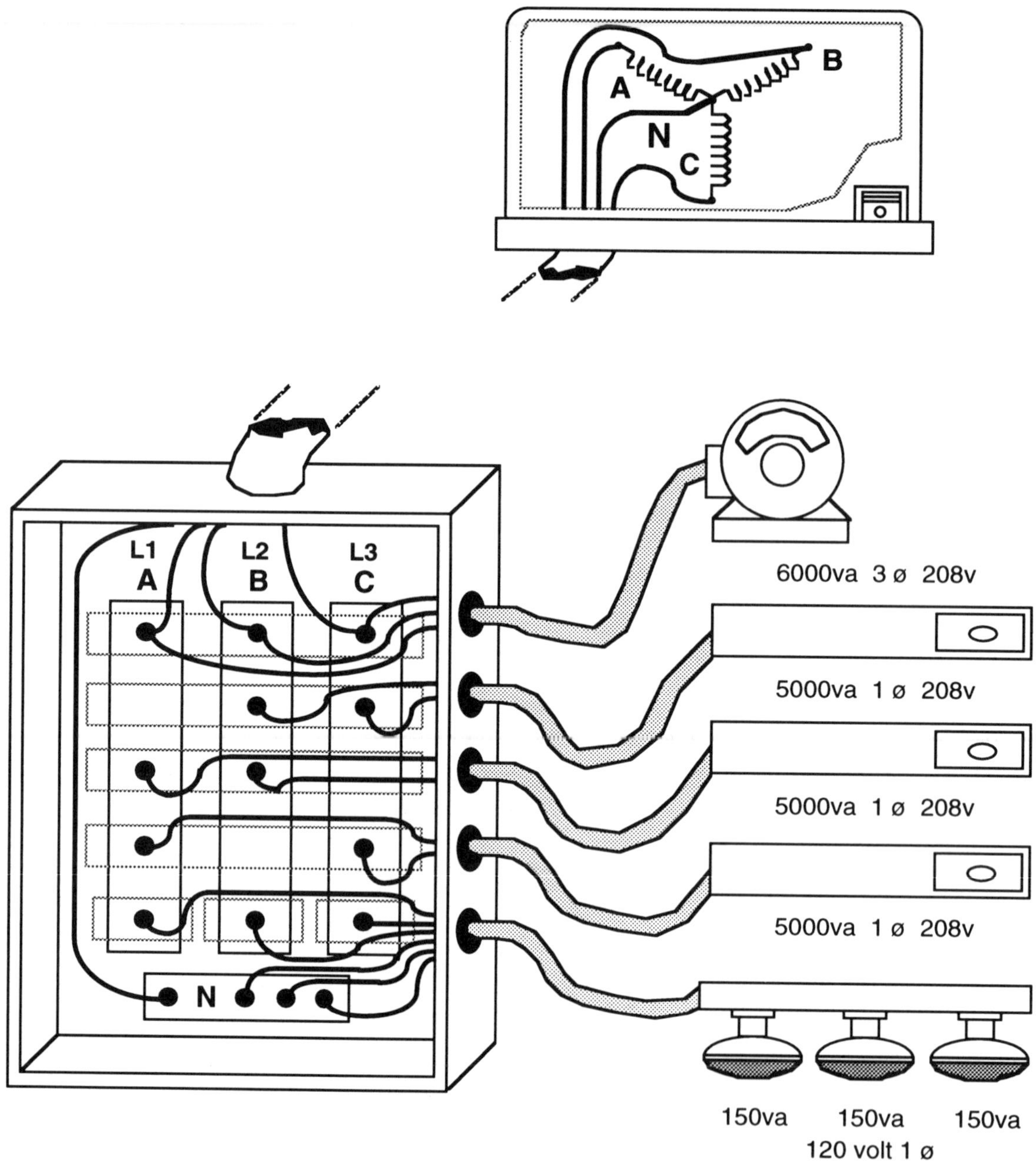

Delta 4-wire, 240 / 120v three-phase secondary:

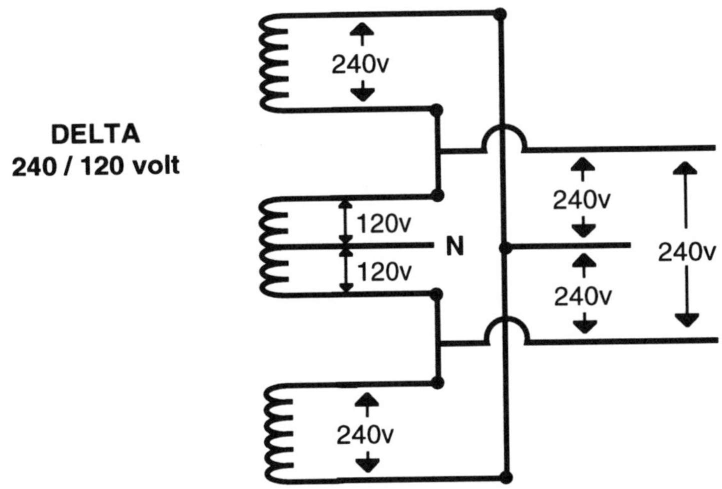

We have three possible voltage connections:

- 240 volt 3 ø loads connect L1 - L2 - L3
- 240 volt 1 ø loads connect L1-L2 or L2-L3 or L3-L1
- 120 volt 1 ø loads connect L1-N or L3-N

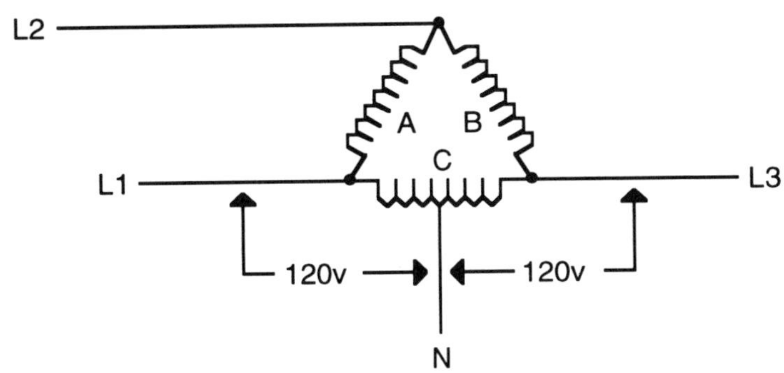

Division of loads:

240 volt 3 ø loads - 1/3 on phase A
- 1/3 on phase B
- 1/3 on phase C

240 volt 1 ø loads - 100% on any one phase
120 volt 1 ø loads - 100% on phase C only

EXAMPLE 2

Let's connect the same loads on a delta 240 / 120v, 4-wire, 3 ø secondary:

1 - 6000va 3 ø 240v motor
3 - 5000va 1 ø 240v heaters
3 - 150va 1 ø 120v lights

Solution: The 6000va three-phase 240v motor divides 1/3 on each phase. The three 5000va heaters 240v single-phase would balance 5000va on each phase. The three 150va lights are single-phase 120 volt so they connect Line One or Line Three to neutral. They can only connect on phase C.

PHASE A	PHASE B	PHASE C	
2000va	2000va	2000va	
5000va	5000va	5000va	
		150va - - -	150va
		150va	
7000va	7000va	7450va	

$$7000va + 7000va + 7450va = \frac{21{,}450va}{240v \times 1.732} = 51.6 \text{ amps line current}$$

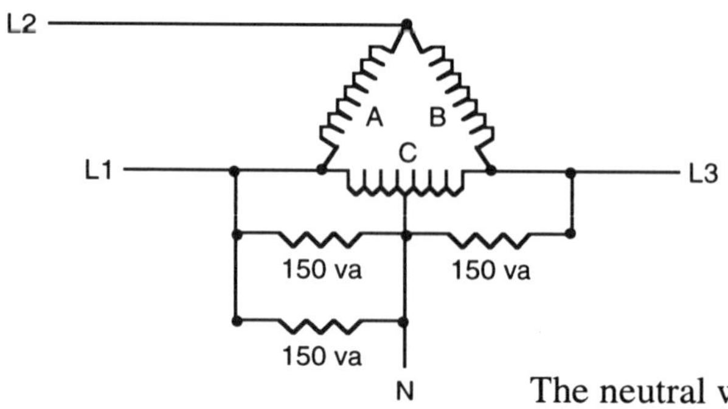

The neutral with all loads on would carry $\frac{150va}{120v} = 1.25$ amp.

Maximum neutral current would occur when the L3 to N load is shut off, then the neutral would have to carry the two L1 to N loads $\frac{300va}{120v} = 2.5$ amps.

By now you can see the differences between a WYE or a DELTA secondary. With a WYE, the 120 volt loads balanced, the 5000va 208v 1 ø loads divided half on each phase. With the DELTA secondary, the 120 volt loads *did not* balance and the 5000va 240v 1 ø loads did balance with one on each phase.

Since the delta shows an *unbalance* in the phases, let's calculate the current in each line.

DELTA UNBALANCED

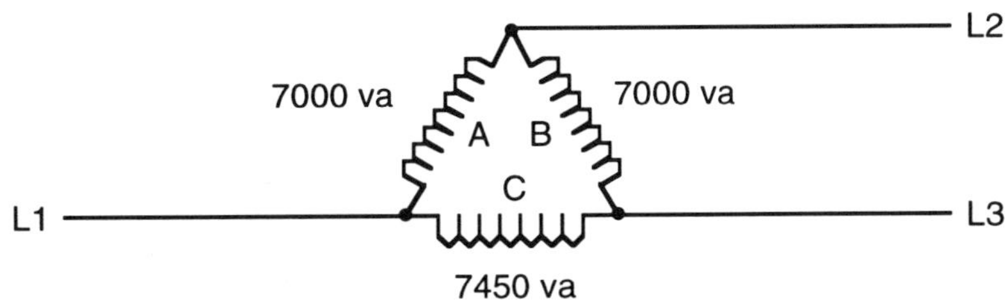

Change to current:

Phase A = 7000va/240v = 29.16 amp
Phase B = 7000va/240v = 29.16 amp
Phase C = 7450va/240v = 31.04 amp

FORMULA FOR UNBALANCED DELTA CURRENT:

L1 current = Phase A + Phase C, divided by 2, times 1.732
L2 current = Phase A + Phase B, divided by 2, times 1.732
L3 current = Phase B + Phase C, divided by 2, times 1.732

Using the above formula calculate the current in the three lines:

L1 = 29.16a + 31.04a = 60.2a/2 = 30.1a x 1.732 = **52.13 amperes in L1**

L2 = 29.16a + 29.16a = 58.32a/2 = 29.16a x 1.732 = **50.5 amperes in L2**

L3 = 29.16a + 31.04a = 60.2a/2 = 30.1a x 1.732 = **52.13 amperes in L3**

EXAMPLE 3

This example shows the difference in current flow in the line conductors using a 45 kva three phase *balanced* and a 45 kva with three phases *unbalanced.*

BALANCED 45 kva 480v

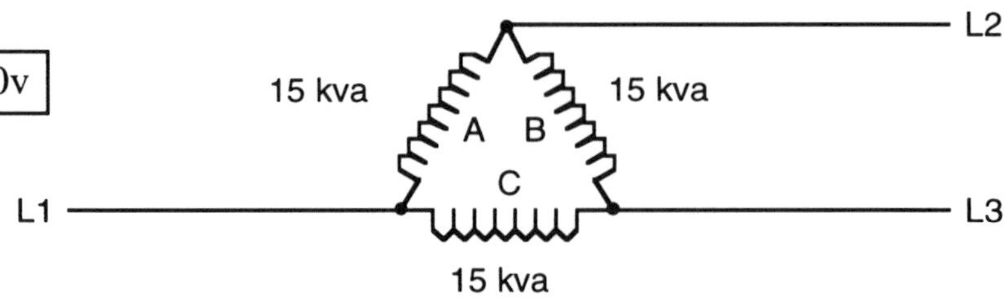

L1 current = 15000va/480v = 31.25 phase amps x 1.732 = **54.125 amperes L1**
L2 current = 15000va/480v = 31.25 phase amps x 1.732 = **54.125 amperes L2**
L3 current = 15000va/480v = 31.25 phase amps x 1.732 = **54.125 amperes L3**

In a balanced delta the line current is the *same* in each line 54.125 amperes.

Now let's calculate the same 45 kva total but unbalanced in three phases.

UNBALANCED 45 kva 480v

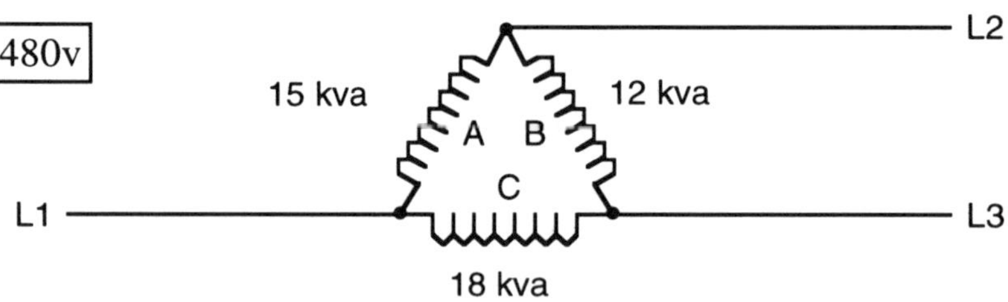

Phase A current = 15000va/480v = 31.25 phase amps
Phase B current = 12000va/480v = 25 phase amps
Phase C current = 18000va/480v = 37.5 phase amps

L1 current = 31.25a + 37.5a = 68.75a/2 = 34.375a x 1.732 = **59.5 amperes in L1**
L2 current = 31.25a + 25a = 56.25a/2 = 28.125a x 1.732 = **48.7 amperes in L2**
L3 current = 25a + 37.5 = 62.5/2 = 31.25a x 1.732 = **54.1 amperes in L3**

• **Checkpoint:** L1 59.5a
L2 + 48.7a
L3 + 54.1a
162.3 amps total divided by 3 = 54.1 per line average
or
$$\frac{\textbf{45000 total va}}{\textbf{480v x 1.732}} = \textbf{54.1 line amps}$$

EXAMPLE 4

DELTA UNBALANCED

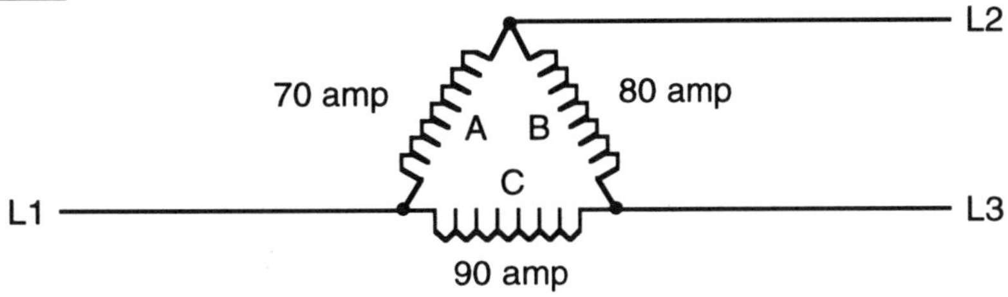

Calculate the current in each line.

L1 current = 70a + 90a = 160a/2 = 80a x 1.732 = **138.6 amperes in L1**

L2 current = 70a + 80a = 150a/2 = 75a x 1.732 = **129.9 amperes in L2**

L3 current = 80a + 90a = 170a/2 = 85a x 1.732 = **147.2 amperes in L3**

• **Checkpoint:** L1 current 138.6
L2 current + 129.9
L3 current + 147.2
415.7 divided by 3 = **138.56 ampere average**

Phase A = 70a x 1.732 = 121.24
Phase B = 80a x 1.732 + 138.56
Phase C = 90a x 1.732 + 155.88
415.68 divided by 3 = **138.56 ampere average**

EXAMPLE 5

A 10 H.P. 230 volt 3 ø motor has a F.L.C. of 28 amps, three-phase power = E x I x 1.732. 230v x 28a x 1.732 = $\frac{\text{11,154 total}}{\text{3 phases}}$ = 3718 per phase.

A 10 H.P. 208v 3 ø motor has a F.L.C. of 30.8 amps,
208v x 30.8a x 1.732 = $\frac{\text{11,096 total}}{\text{3 phases}}$ = 3699 per phase.

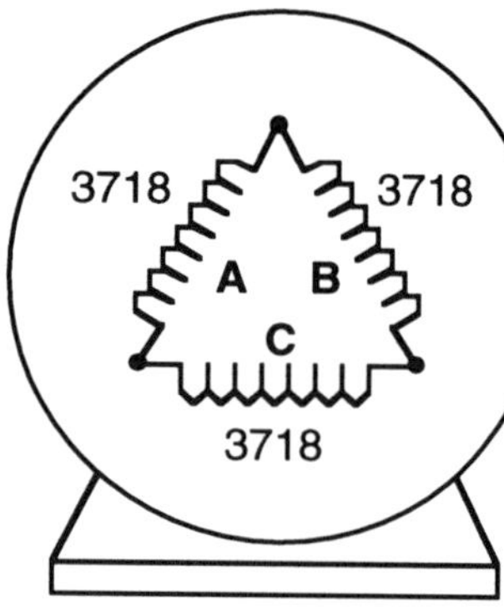

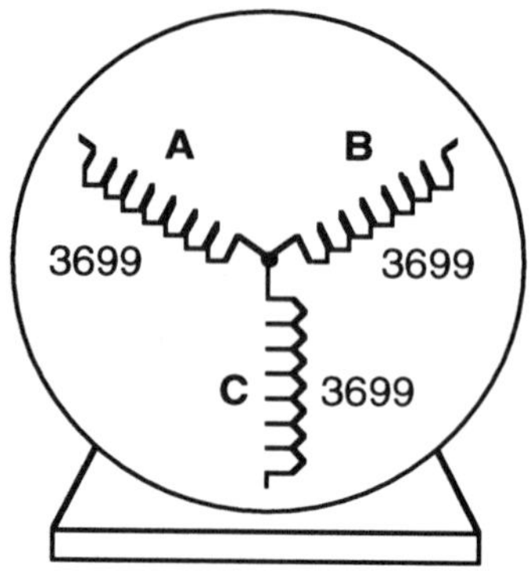

A 5000 watt 1 ø 208v heater connected on a WYE would have 2500w on phase A, and 2500w on phase B.

The same 5000 watt 1 ø heater connected DELTA would have the entire 5000w on phase A.

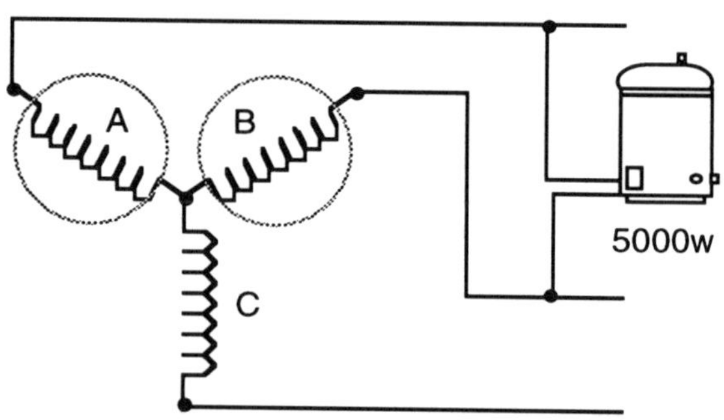

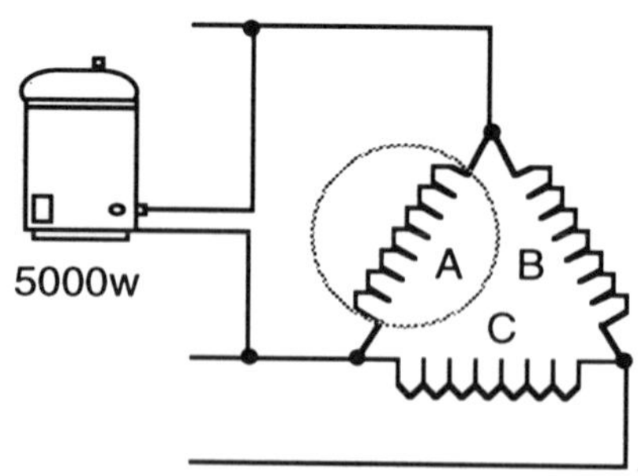

PHASE A	PHASE B	PHASE C
2500w	2500w	--------

PHASE A	PHASE B	PHASE C
5000w	--------	--------

EXAMPLE 6

Calculating the neutral current in the WYE secondary.

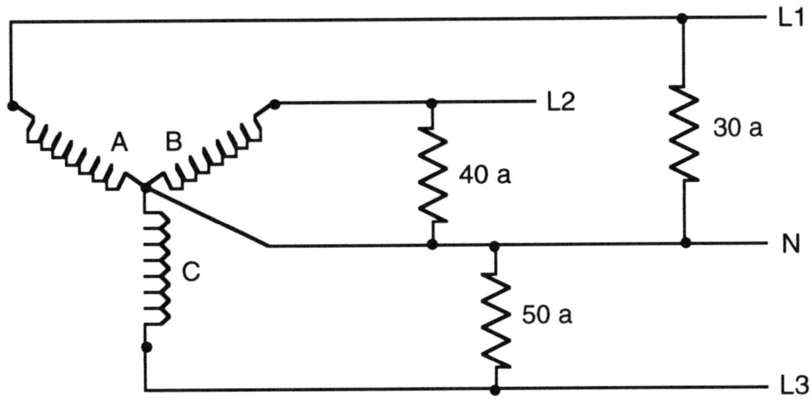

The three loads, 30 amp, 40 amp, and 50 amp, are connected line to neutral with all loads "on". What is the unbalanced current flowing in the neutral?

Solution: In = $\sqrt{I2A + I2B + I2C - (IA\ IB) - (IB\ IC) - (IC\ IA)}$

The formula at first looks very difficult, but really it's not.

Everything under the square root sign must be done first, which means:

current in A squared = 30 x 30 = 900
+ current in B squared = 40 x 40 = 1600
+ current in C squared = 50 x 50 = 2500
5000 total (call this total "X")

Now the right side of the formula shows:

current in A x current in B = 30 x 40 = 1200
current in B x current in C = 40 x 50 = 2000
current in C x current in A = 50 x 30 = 1500
4700 (call this total "Y")

Now subtract total "Y" from total "X" = 5000 total X
— 4700 total Y
300

Now extract the square root by pressing the $\sqrt{}$ button on your calculator. The answer is 17.32 amps is the unbalanced current flowing in the neutral. 17.320508 is the square root of 300. 17.320508 x 17.320508 = 300.

EXAMPLE 7

Five 1500 watt 120 volt 1 ø light circuits are connected to a WYE 4-wire secondary 208 / 120v three-phase. The loads are connected 3000 watt, 3000 watt, and 1500 watt per phases A, B, and C. With all loads on, what is the current flowing in the neutral?

(a) 100 amps (b) 50 amps (c) 25 amps (d) 12.5 amps

Solution:

PHASE A	PHASE B	PHASE C
1500 watt	1500 watt	1500 watt
1500 watt	1500 watt	- - - - - - -
3000 watt	3000 watt	1500 watt

The loads are connected 3000-3000-1500. You can see the unbalance would be 1500. If we added 1500 watt to phase C, we would have a balance. But since we have an unbalanced situation, let's check using the formula to see if the correct answer is 1500 watt or $\frac{1500 \text{ watt}}{120 \text{ volt}} = 12.5$ amp.

First step convert watts to amps since the question asks for current.

PHASE A: $\frac{3000 \text{ watt}}{120 \text{ volt}} = 25$ amp

PHASE B: $\frac{3000 \text{ watt}}{120 \text{ volt}} = 25$ amp

PHASE C: $\frac{1500 \text{ watt}}{120 \text{ volt}} = 12.5$ amp

Using the formula:

current in A squared = 25 x 25 = 625
+ current in B squared = 25 x 25 = 625
+ current in C squared = 12.5 x 12.5 = 156.25
1406.25 amp total ("X")

current in A x current in B = 25 x 25 = 625
current in B x current in C = 25 x 12.5 = 312.5
current in C x current in A = 12.5 x 25 = 312.5
1250 amp total ("Y")

Now subtract total "Y" from total "X" = 1406.25 total "X"
— 1250 total "Y"
156.25 amp

Now extract the square root by pressing the square root button on your calculator. The answer is 12.5 amp. 12.5 amp is the unbalanced current flowing in the neutral.

The maximum neutral current would occur with the loss of either phase A and phase C *or* the loss of phase B and phase C. In either situation, the neutral would have to carry the maximum of 3000 watt or $\frac{3000 \text{ watt}}{120 \text{ volt}} = 25$ amp.

PANELBOARD SCHEDULE - SINGLE PHASE

The actual connected loads are shown in the load balancing schedule. It is very important that the electrician connect the circuits as shown in the panelboard schedule to maintain the best possible balance.

LOAD BALANCE SCHEDULE		H - 6677		
Description	**Load**	**L1**	**L2**	**N**
Air conditioner	6720	3360	3360	
Water heater	5000	2500	2500	
Range	12000	6000	6000	•8400
Dryer	5000	2500	2500	5000
Dining room	2400	2400		2400
Kitchen-SW	2400		2400	2400
Disposal	1176	1176		1176
Kitchen-SE	2400		2400	2400
Dishwasher	1200	1200		1200
Kitchen Lts-fan	1356		1356	1356
Washer	1176	1176		1176
Furnace	1176		1176	1176
Garage Recpts.	2400	2400		2400
Garage Lts.-Etc.	2400		2400	2400
Living room	2400	2400		2400
Bedroom-Bath	2400		2400	2400
TOTAL	51604	25112	26492	36284

PANELBOARD SCHEDULE H- 6677					
MAINS 150 PHASE 1 VOLTS 240/120				AMPERE 150	
Circuit Number	Circuit Breaker		Conductor		Serves
	Poles	Amps	Number of	Size	
1	2	35	2	#8	Air Conditioner
2	2	30	2	#10	Water Heater
3	—	—	—	—	Air Conditioner
4	—	—	—	—	Water Heater
5	2	50	3	#6	Range
6	2	30	3	#10	Dryer
7	—	—	—	—	Range
8	—	—	—	—	Dryer
9	1	20	2	#12	Dining Room
10	1	20	2	#12	Kitchen - SW
11	1	20	2	#12	Garbage Disposal
12	1	20	2	#12	Kitchen - SE
13	1	20	2	#12	Dishwasher
14	1	20	2	#12	Kitchen Lights-Fan
15	1	20	2	#12	Washer
16	1	20	2	#12	Furnace
17	1	20	2	#12	Garage Recpts.
18	1	20	2	#12	Garage Lights, etc
19	1	20	2	#12	Living Room
20	1	20	2	#12	Bedrooms-Bath

All loads will connect to a line wire, but not all loads connect to the neutral. This is why the neutral can generally be reduced in size.

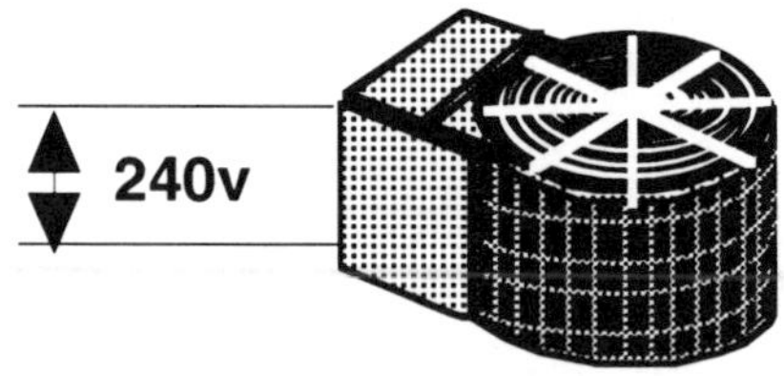

The air conditioner and water heater have no neutral connection so there is no neutral load. The range and dryer have a neutral and the Code requires 70% of the load to be on the neutral. The washer, disposal, dishwasher are 120 volt loads.

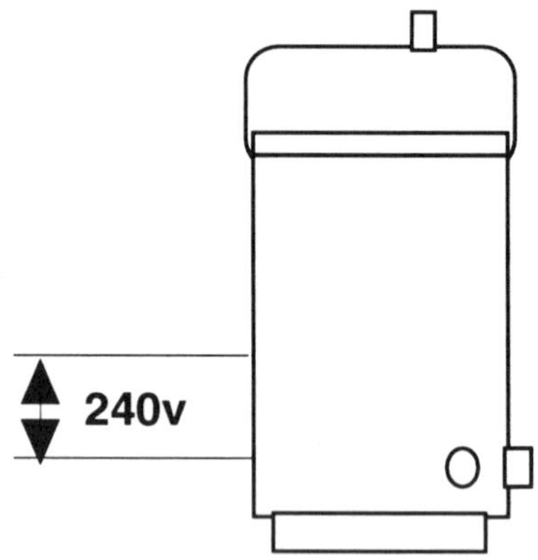

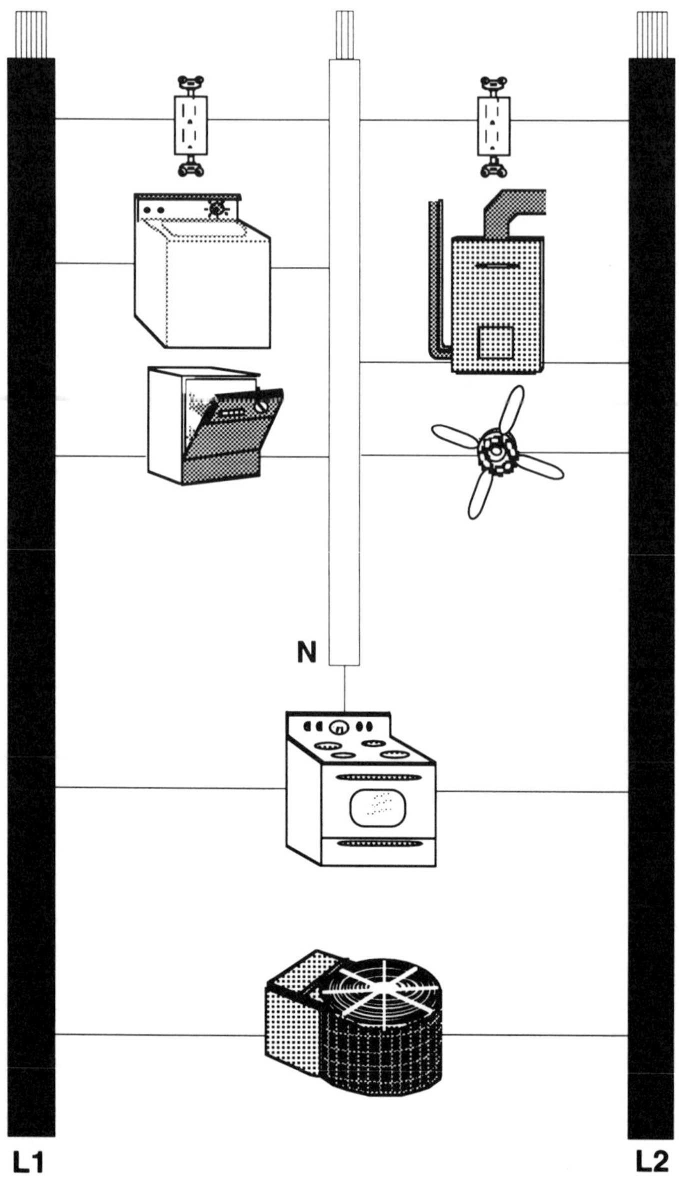

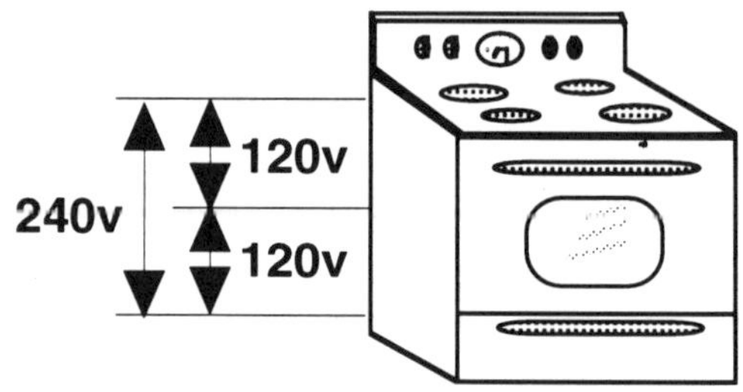

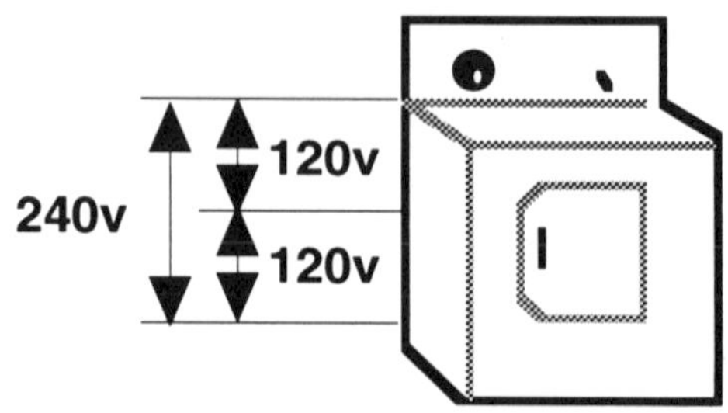

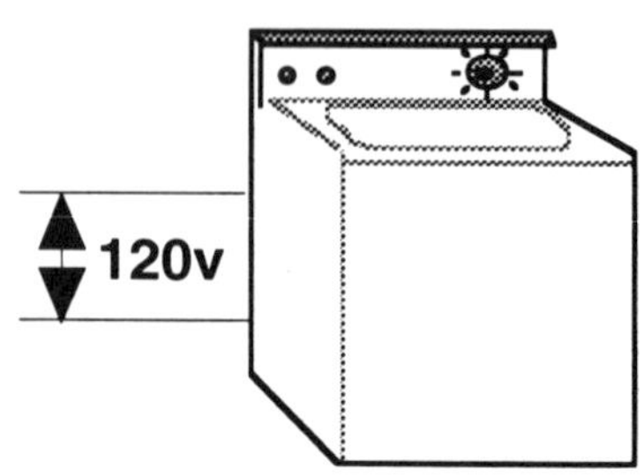

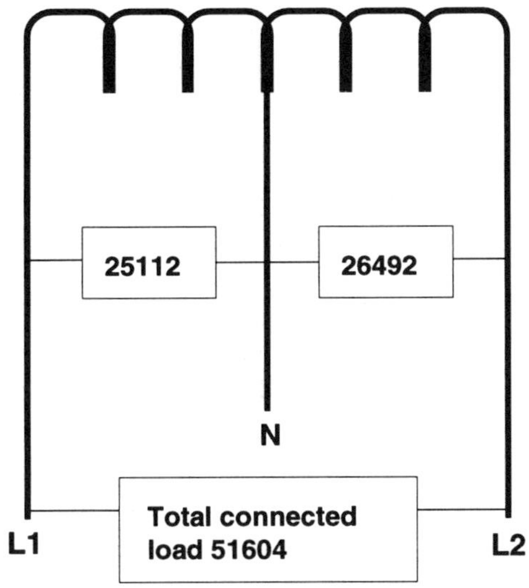

The total connected load is 51,604va/240v = 215 amps.

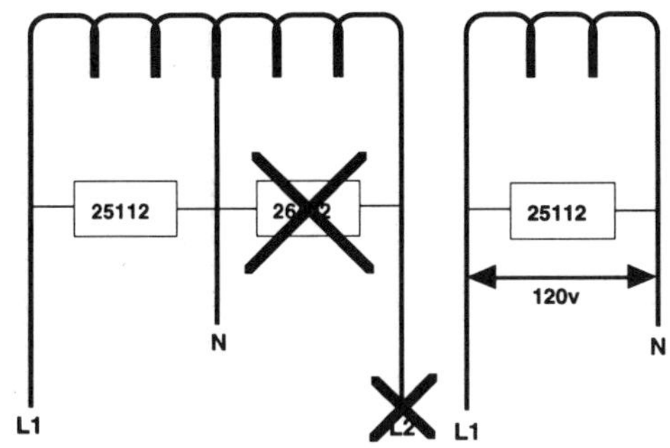

If Line 2 was completely shut off Line 1 would carry 25112va/120v = 209 amps.

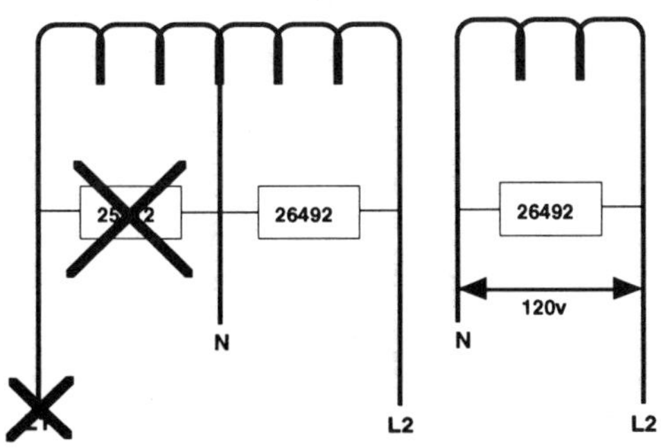

If Line 1 was completely shut off Line 2 would carry 26492va/120v = 221 amps.

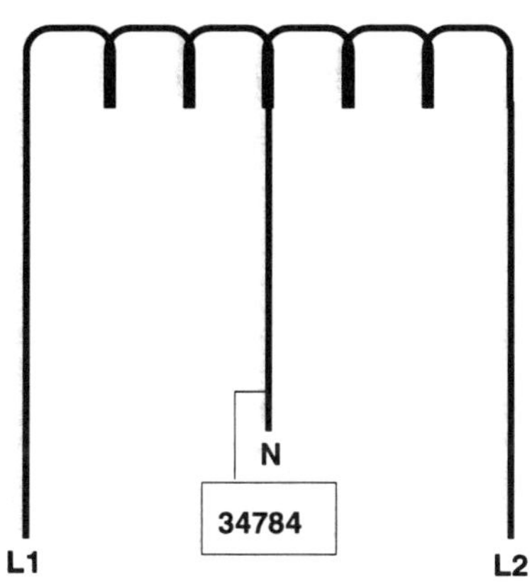

The neutral shows a connected load of 36284/240v = 151 amps. The •8400 @ 70% makes no logic, but the Code requires 70%. Most ranges only have a time clock and light that are connected to the 120v neutral, hardly a 8400/240 = 35 amp load. The worst condition on the neutral would only be 50% (the maximum unbalance) of 12kw = 6000.

There is no 70% on the neutral for a dryer *branch circuit*. If the dryer had a 1/2 hp, 120v drive motor the amperage is 9.8, plus a time clock and a light. 5000/240v = 21 amps would cover the neutral for the dryer branch circuit.

Example: The 12kw range has four cook-top elements which are 240 volts connected with low-medium-high settings. The range has an oven with 240 volt connected elements. The range has a 120 volt connected clock and light.

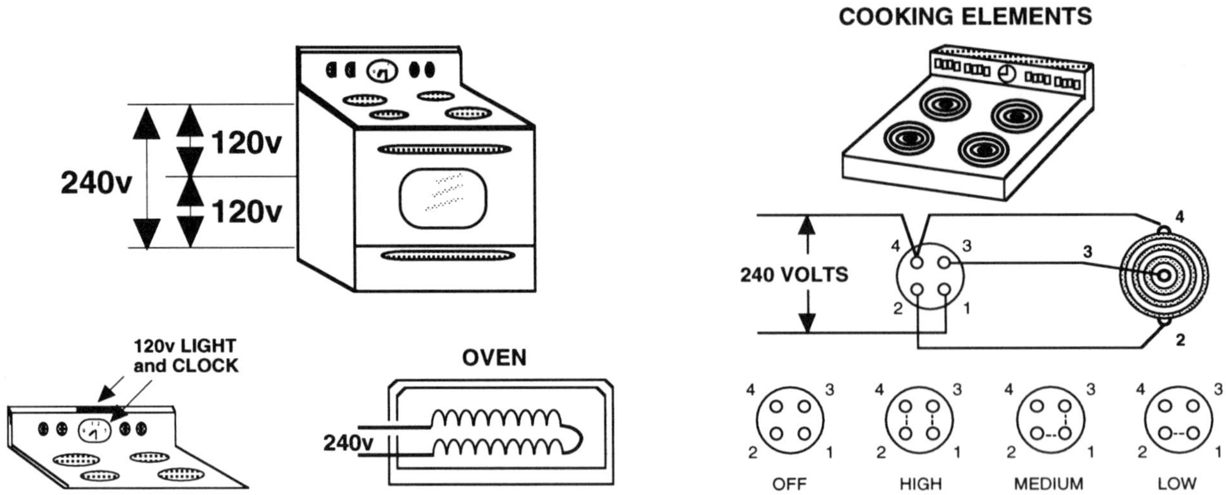

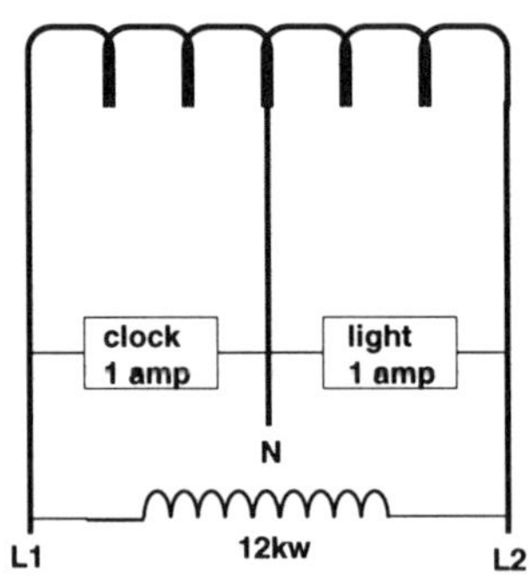

If the clock and light drew one amp each, the highest load on the neutral would be one amp with one shut off and the other one on. If both were on, the load would be zero on the neutral. If the cooktop elements were connected to the neutral the maximum unbalance could only be 6kw plus the one amp for the clock or light. 70% on the neutral makes no logic as 50% could be the highest unbalance.

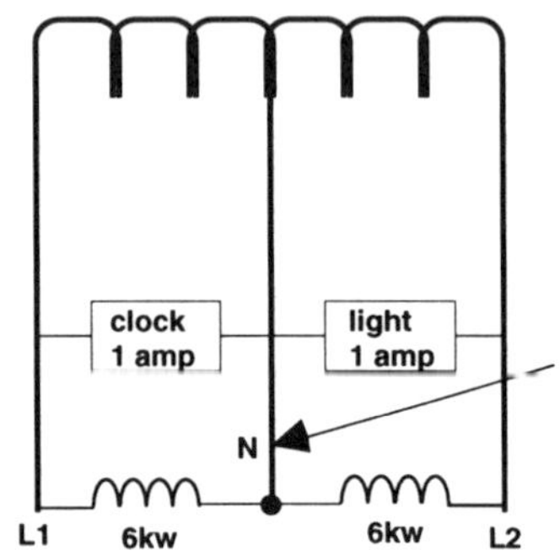

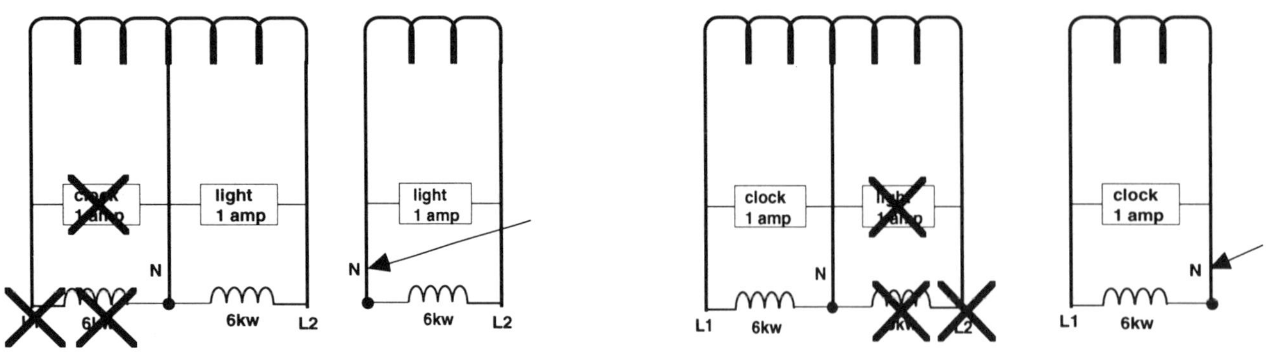

THE NEUTRAL CARRIES THE UNBALANCED CURRENT

Example: The dryer has a 1/2 hp, 120v motor to turn the drum, timer and maybe a 120v control light. The heating element is 240v connected.

It is important for the electrician reading the drawings to have a mental picture of how the loads are connected and balanced.

The reason for balancing loads on the neutral is to ensure that the neutral conductor never has to carry a large amount of current. If the loads are unbalanced, considerable current will flow in the neutral conductor. The neutral conductor is sized by the designer by balancing loads. If the loads are *not* connected as shown on the drawing schedule, overheating can occur causing insulation damage to the neutral conductor.

Shown below is how the AC sine wave looks on a 240 volt secondary winding.

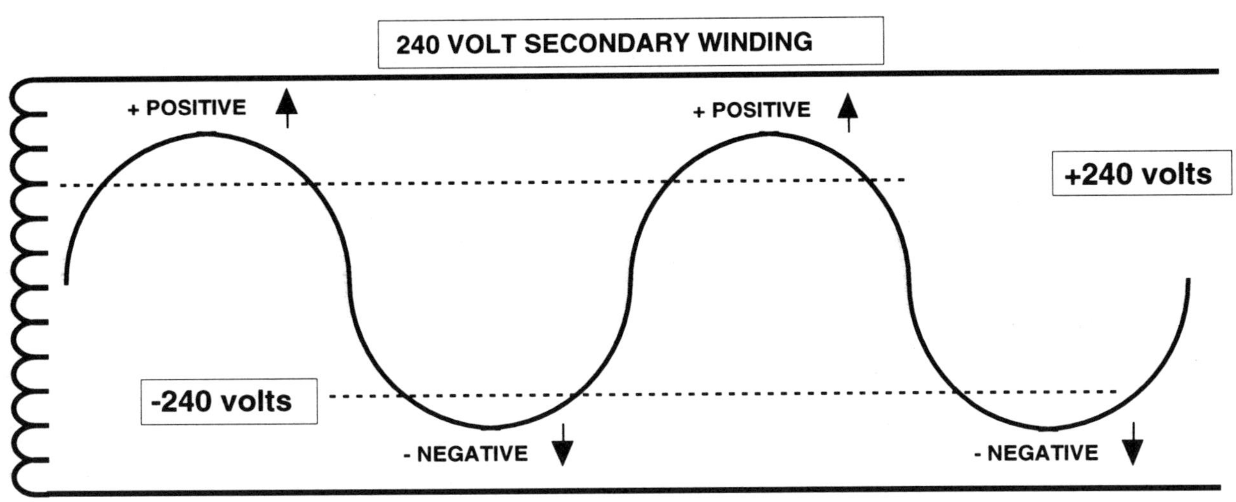

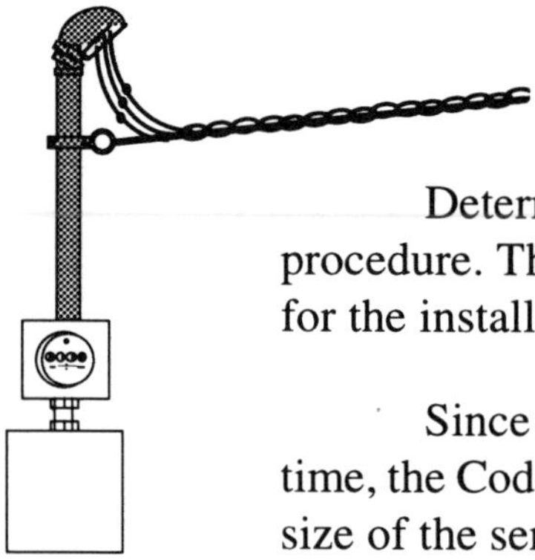

Determining the minimum size service is accomplished by a step-by-step procedure. The designer, inspector, plans examiner, electrician, etc. are responsible for the installation of the service to be within the requirements of the Code.

Since not all the connected loads will be turned on and fully loaded at the same time, the Code permits reducing the size of the connected load when determining the size of the service.

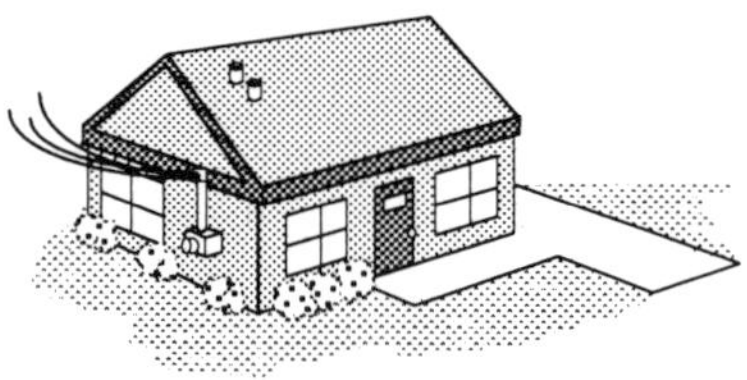

House plan **H -6677**

Example: A dwelling has a total of 2394 sq.ft. of living area (outside dimensions) with the following equipment:

5 hp air conditioner 28a x 240v	=	6720
5 kw water heater	=	5000
12 kw range	=	12000
5 kw dryer	=	5000
1/2 hp disposal 9.8a x 120v	=	1176
1200 va dishwasher	=	1200
1/4 hp kitchen fan 5.8 x 120v	=	696
1/2 hp furnace blower motor	=	1176
1/4 hp garage door opener	=	696

The Code requires a minimum of 3va per square foot of living area be calculated. This load is for the 120v receptacles throughout the house (with the exception of the small appliance and laundry receptacles).

The Code requires a minimum of 1500va each for a minimum of two small appliance circuits. This minimum calculated load is for all the 120v receptacles in the kitchen and dining room. This is the heaviest loaded area of the house; 120 volts.

The Code requires a minimum of 1500va be calculated into the size of the service conductor for the laundry. This is the 120 volt receptacle for the washing machine.

The Code permits a demand (reduction) to be applied to this calculated receptacle load. The first 3000va cannot be reduced, but the next 117,000va can be reduced 35%.

	LINE LOAD	**NEUTRAL LOAD**
2394 sq.ft. x 3va	~~= 7182va~~	~~7182va~~
Small appliance 2 x 1500va	~~= 3000va~~	~~3000va~~
Laundry	~~= 1500va~~	~~1500va~~
Total 120v receptacle load	~~**= 11682va**~~	~~**11682va**~~
The 11682va can be reduced:		
1st 3000va @ 100%	= 3000va	3000va
Next 8682va @ 35%	= 3039va	3039va
Total 120v receptacle demand load	= **6039va**	**6039va**
The 11682va is now reduced to 6039va.		
The air conditioner load is 6720 with no neutral.	= 6720	0
The 5kw water heater is a fastened in place appliance and has a 75% demand. There is no neutral.	= 3750	0
The 12kw range is permitted a demand of 8kw with 70% on the neutral.	= 8000	5600
The 5kw dryer requires 70% on the neutral.	= 5000	3500
The 1/2 hp (9.8a x 120v) disposal is a fastened in place appliance and has a 75% demand.	= 882	882
The 1200va dishwasher is a fastened in place appliance and has a 75% demand.	= 900	900
The 1/4 hp (5.8a x 120v) kitchen fan is a fastened in place appliance and has a 75% demand.	= 522	522
The 1/2 hp (9.8a x 120v) furnace blower motor is a fastened in place appliance @ 75%.	= 882	882
The 1/4 hp (5.8a x 120v) door opener is a fastened in place appliance @ 75%.	= 522	522
The Code requires that 25% of the largest motor be added to the service demand. The A/C is the largest motor at 6720 x 25%.	= 1680	0
TOTAL LINE DEMAND	= **34,897**	NEUTRAL = **18,847**

LINE CONDUCTOR SIZE

34,897/240v = 145 amps

NEUTRAL CONDUCTOR SIZE

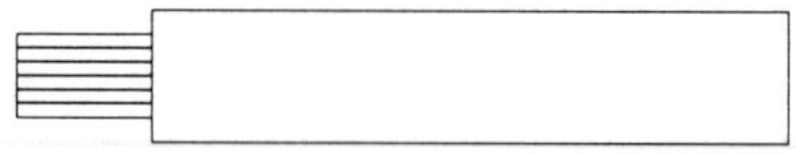

18,847/240v = 79 amps

The selection of a panelboard is first determined by the number of circuits it must supply. Then the panelboard size is determined by the **calculated** load it will supply. The calculated load is 145 amperes minimum. The next higher standard size service is 150 amps.

A panelboard may have a busbar current rating *larger* than the calculated load, but never less than the calculated load.

The overcurrent protection (main) cannot be larger than the *rating* of the panelboard.

THE OVERCURRENT PROTECTION CANNOT BE LARGER THAN THE RATING OF THE PANELBOARD

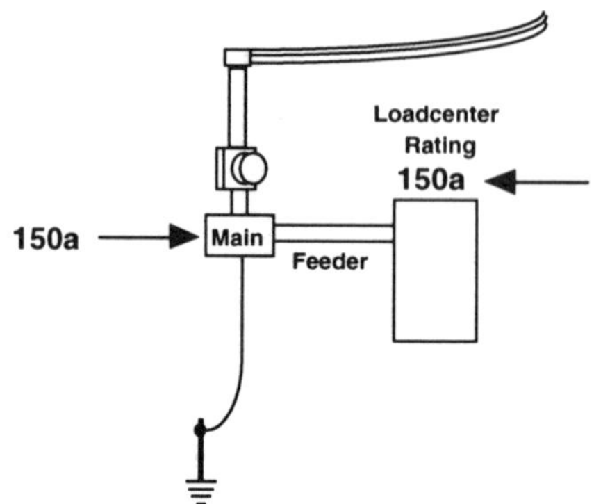

Often the question arises, can you install 150 amp service cable to a 200 amp, 30 circuit main lug panel, with a 150 amp main breaker? The answer is YES!

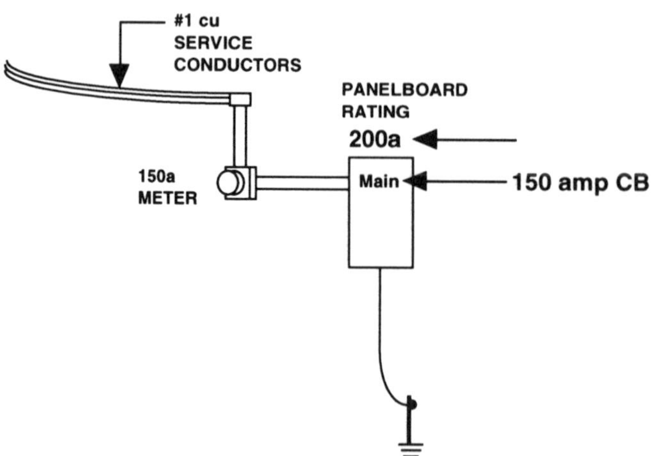

Sometimes you find there is a need for a panelboard with a main breaker that has a current rating LESS than the normally required rating of the panelboard to have the number of branch circuits required.

The Code states each lighting and appliance panelboard shall be individually protected on the supply side by not more than two main circuit breakers or two sets of fuses having a combined rating not greater than that of the panelboard. The exception to this rule is individual protection for a lighting and appliance panelboard shall *not* be required if the panelboard *feeder* has overcurrent protection not greater than the rating of the panelboard.

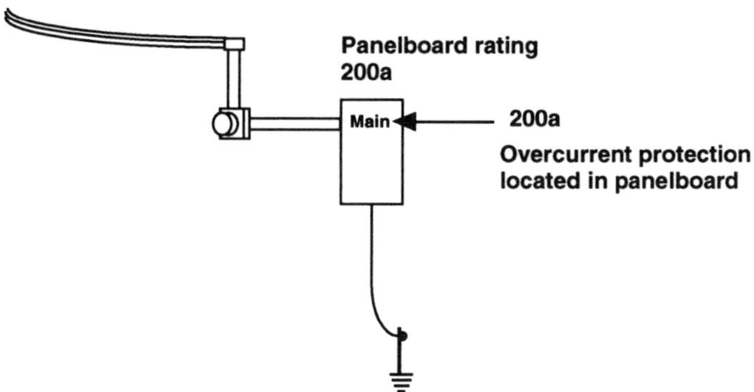

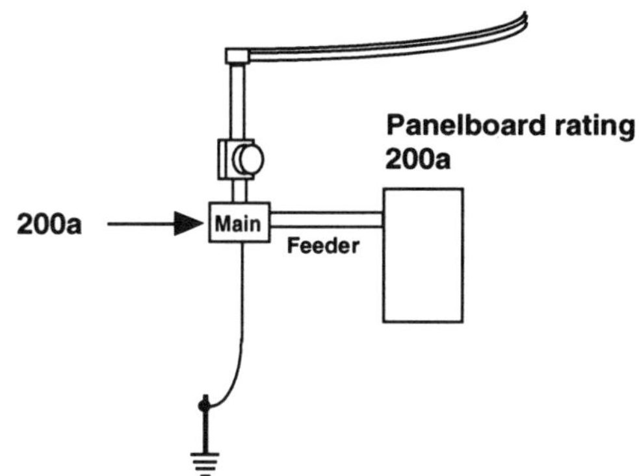

A panelboard might be a service equipment panelboard, distribution panelboard, or a branch-circuit panelboard. A single panelboard could combine two of these functions or all three within a single cabinet.

If the panelboard is used as a service equipment panelboard it must be marked *suitable for service equipment.*

The most common panelboard is the lighting and appliance panelboard. No more than 42 overcurrent devices (in addition to the main) may be installed.

It is very important to understand the definition of a *lighting and appliance* branch circuit panelboard.

The Code defines a lighting and appliance panelboard as one having *more* than 10% of its overcurrent devices rated *30 amps or less*, for which *neutral connections* are provided.

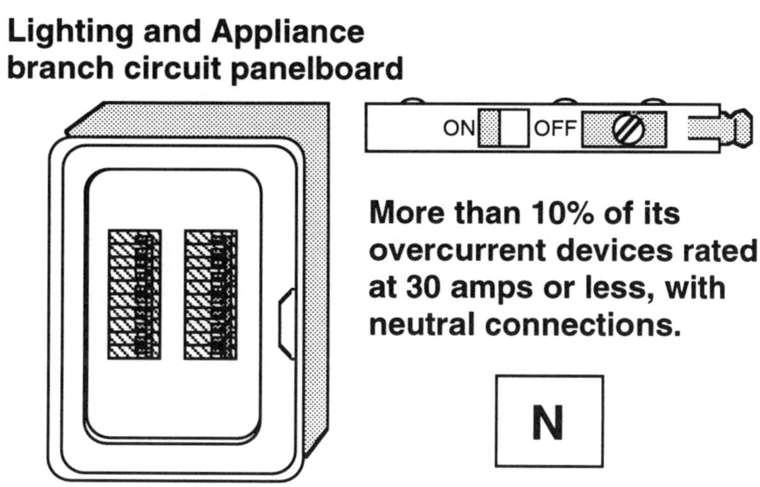

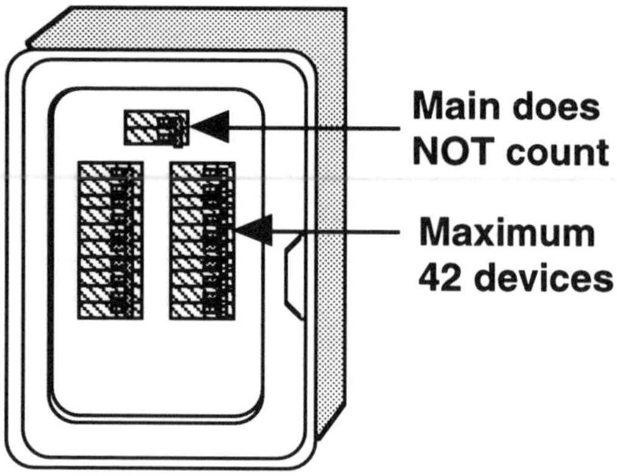

The lighting and appliance panelboard can contain up to 42 overcurrent devices maximum. The main overcurrent device is NOT counted.

1-POLE COUNTS 1

2-POLE COUNTS 2

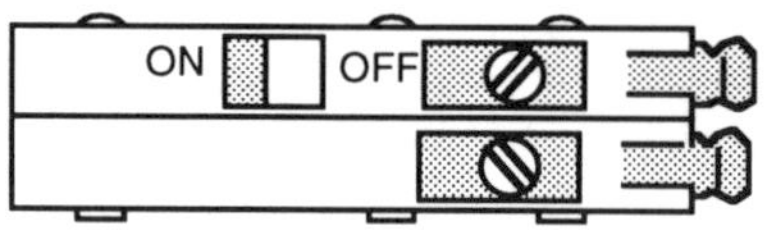

3-POLE COUNTS 3

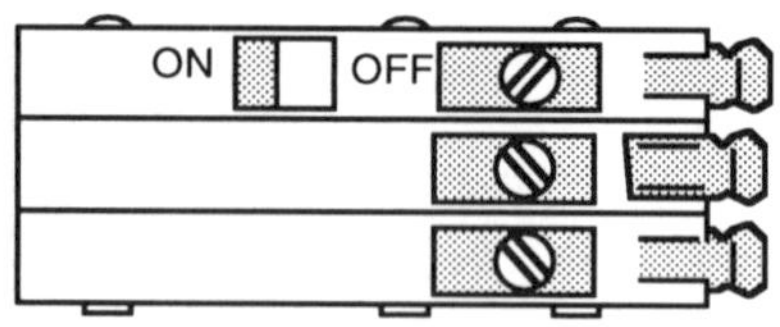

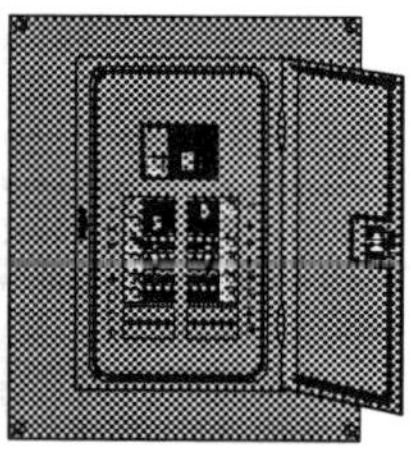

Different manufacturers of panelboards make different sizes of ampere ratings and spaces for single-phase and three-phase panelboards.

Manufacturer "X"

SINGLE-PHASE

Ampere Rating	Spaces
60a	12-18-24
100a	12-18-24
150a	30-36-42
225a	12-18-24-30-36-42

Manufacturer "Y"

SINGLE-PHASE

Ampere Rating	Spaces
100a	12-20
125a	30-42
150a	30-42
175a	30-42
200a	30-42
225a	30-42

PANELBOARD SCHEDULE

The following pages show the panelboard schedules for a restaurant which has a 200 amp single-phase panel "A" and a 225 amp three-phase panel "B".

PANEL "A" Single-phase 240/120 volt

PANELBOARD SCHEDULE R-3344					
MAINS 175	PHASE 1	VOLTS 240/120		AMPERE 200	
Circuit Number	Circuit Breaker		Conductor		Serves
	Poles	Amps	Number of	Size	
1	1	20	2	#12	4-corner roof lights
2	1	20	2	#12	Pole Sign
3	1	20	2	#12	Two center roof lights
4	1	20	2	#12	Pole Sign
5	1	20	2	#12	Exterior bracket lights
6	1	20	2	#12	Freezer
7	1	20	2	#12	Exterior bracket lights
8	1	20	2	#12	Dining Room lights
9	1	20	2	#12	Food warmer & hood
10	1	20	2	#12	Dining Room lights
11	1	20	2	#12	Lights @ bar & recpt.
12	1	20	2	#12	Work area lts. & recpt.
13	1	20	2	#12	Coffee Urn & Tea
14	1	20	2	#12	Kitchen & Storage Lts.
15	2	20	2	#12	Walk-in-cooler
16	1	20	2	#12	Ice machine/glass froster
17	2	20	2	#12	Walk-in-cooler
18	1	20	2	#12	Dough roller
19	1	20	2	#12	Rest room lights & fan
20	1	20	2	#12	Rest room lights & fan
21	1	20	2	#12	Cash register
22	1	20	2	#12	Fans
23					SPARE
24					SPARE

PANEL "B" Three-phase 240 volt

PANELBOARD SCHEDULE R-3344					
MAINS 225 **PHASE** 3 **VOLTS** 240/120 **AMPERE** 225					
Circuit Number	**Circuit Breaker**		**Conductor**		**Serves**
	Poles	**Amps**	**Number of**	**Size**	
1	3	175	3	#2	Air-conditioner
2	3	20	3	#12	Dough mixer
3	3	175	3	#2	Air-conditioner
4	3	20	3	#12	Dough mixer
5	3	175	3	#2	Air-conditioner
6	3	20	3	#12	Dough mixer
7	3	60	3	#6	Electric Heater
8	3	20	3	#12	Electric hot plate
9	3	60	3	#6	Electric Heater
10	3	20	3	#12	Electric hot plate
11	3	60	3	#6	Electric Heater
12	3	20	3	#12	Electric hot plate
13	3	60	3	#6	Electric Heater
14					SPARE
15	3	60	3	#6	Electric Heater
16					SPARE
17	3	60	3	#6	Electric Heater
18					SPARE
19					SPARE
20					SPARE
21					SPARE
22					SPARE
23					SPARE
24					SPARE

LOAD BALANCING - DELTA

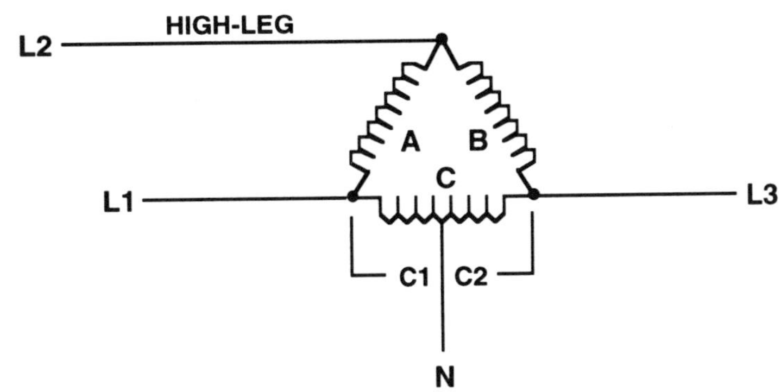

LOAD BALANCE SCHEDULE

TRANSFORMER					
A	**B**	**C**			DESCRIPTION
6000	6000	6000			18 kva heater 3ø 240v
6000	6000	6000			18 kva heater 3ø 240v
8000	8000	8000			24 kva A/C 3ø 240v
2000	2000	2000			6 kva hot plate 3ø 240v
600	600	600			1800va dough machine 3ø 240v
			C1	**C2**	
			6000	6000	Lights and recpts. 1ø 120v
			2000	2000	Walk-in cooler 1ø 240v
			1800	1800	Corner roof lights 1ø 120v
			1700	1700	Pole mounted sign 1ø 120v
			1500		Coffee & tea 1ø 120v
				1000	Center roof sign 1ø 120v
				830	Freezer 1ø 120v
				830	Glass froster 1ø 120v
			830	830	Exhaust hood fans 1ø 120v
			830		Ice machine 1ø 120v
				580	Glass washer 1ø 120v
			530	530	Bath exhaust fans 1ø 120v
			530		Hot plate hood 1ø 120v
22600va	**22600va**	**54420va**			**Load per transformer** **Total load = 99620va**

NEUTRAL CALCULATION - DELTA

6000	6000	Lights and recpts. 1ø 120v
~~2000~~	~~2000~~	~~Walk-in cooler 1ø 240v~~
1800	1800	Corner roof lights 1ø 120v
1700	1700	Pole mounted sign 1ø 120v
1500		Coffee & tea 1ø 120v
	1000	Center roof sign 1ø 120v
	830	Freezer 1ø 120v
	830	Glass froster 1ø 120v
830	830	Exhaust hood fans 1ø 120v
830		Ice machine 1ø 120v
	580	Glass washer 1ø 120v
530	530	Bath exhaust fans 1ø 120v
530		Hot plate hood 1ø 120v
13720 va	**14100va**	

•NOTE: Walk-in cooler is *not* a neutral load at 240v.

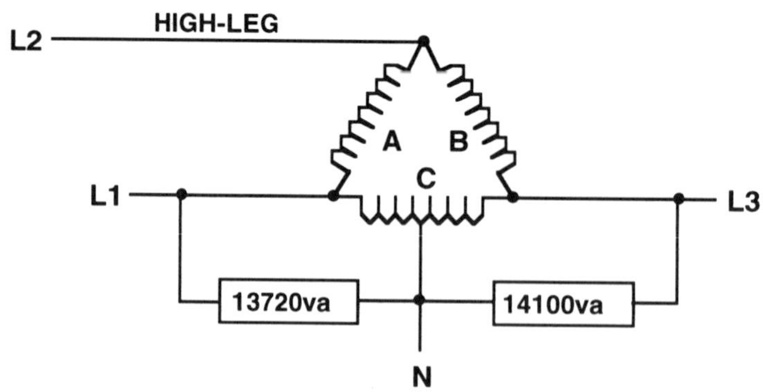

The maximum load on the neutral conductor would be 14100va/120v = **117.5 amperes**.

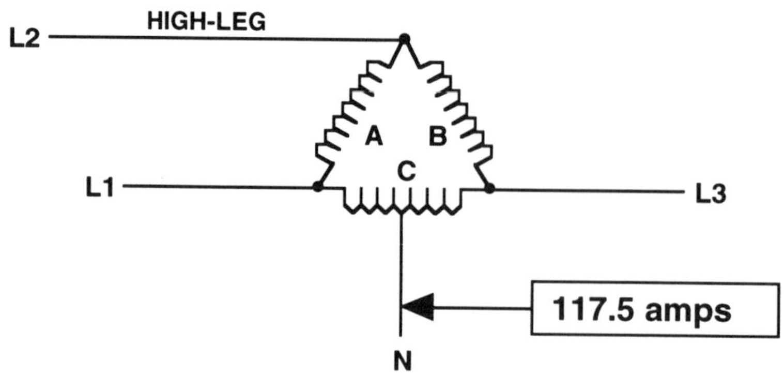

PHASE LOADS

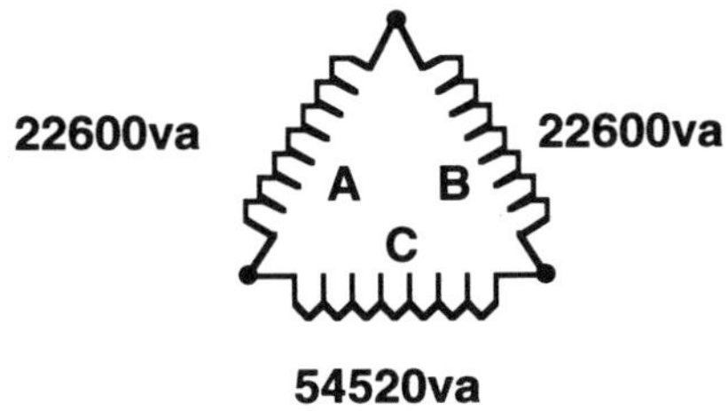

22600va/240v = 94a
54520va/240v = 227a

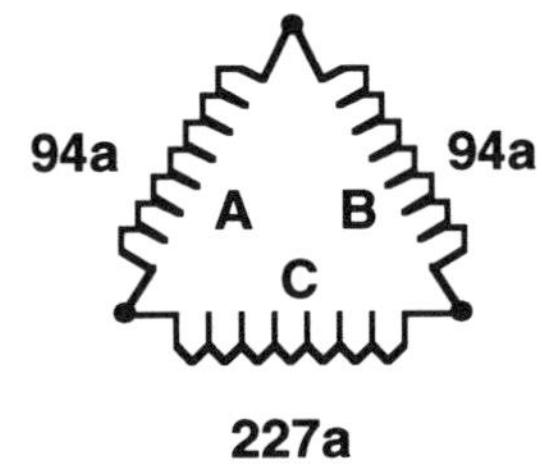

LINE LOADS

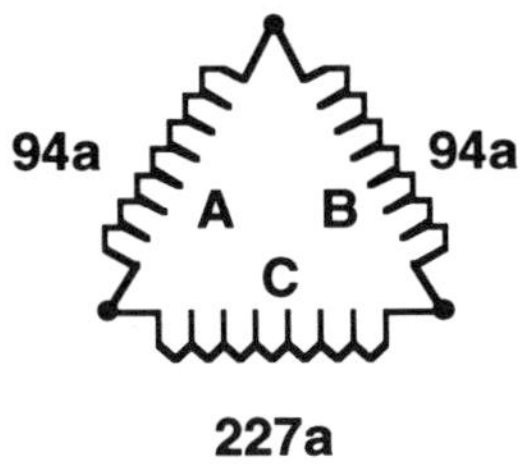

FORMULA FOR UNBALANCED DELTA CURRENT:

L1 current = Phase A + Phase C, divided by 2, times 1.732
L2 current = Phase A + Phase B, divided by 2, times 1.732
L3 current = Phase B + Phase C, divided by 2, times 1.732

L1 current = 94a + 227a = 321a/2 = 160.5a x 1.732 = 278a
L2 current = 94a + 94a = 188a/2 = 94a x 1.732 = 163a
L3 current = 94a + 227a = 321a/2 = 160.5a x 1.732 = 278a

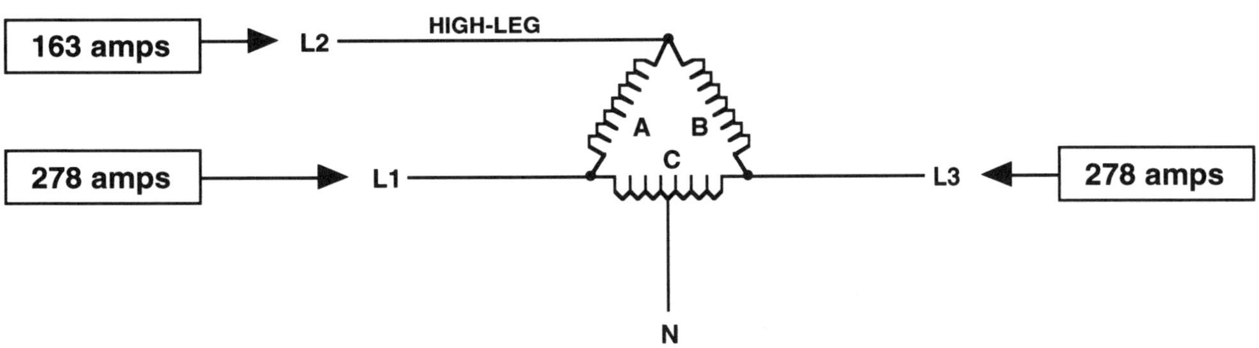

Total load is 99,620va/240v/1.732 = 240 amps.

SIZING THE CONDUCTORS - DELTA

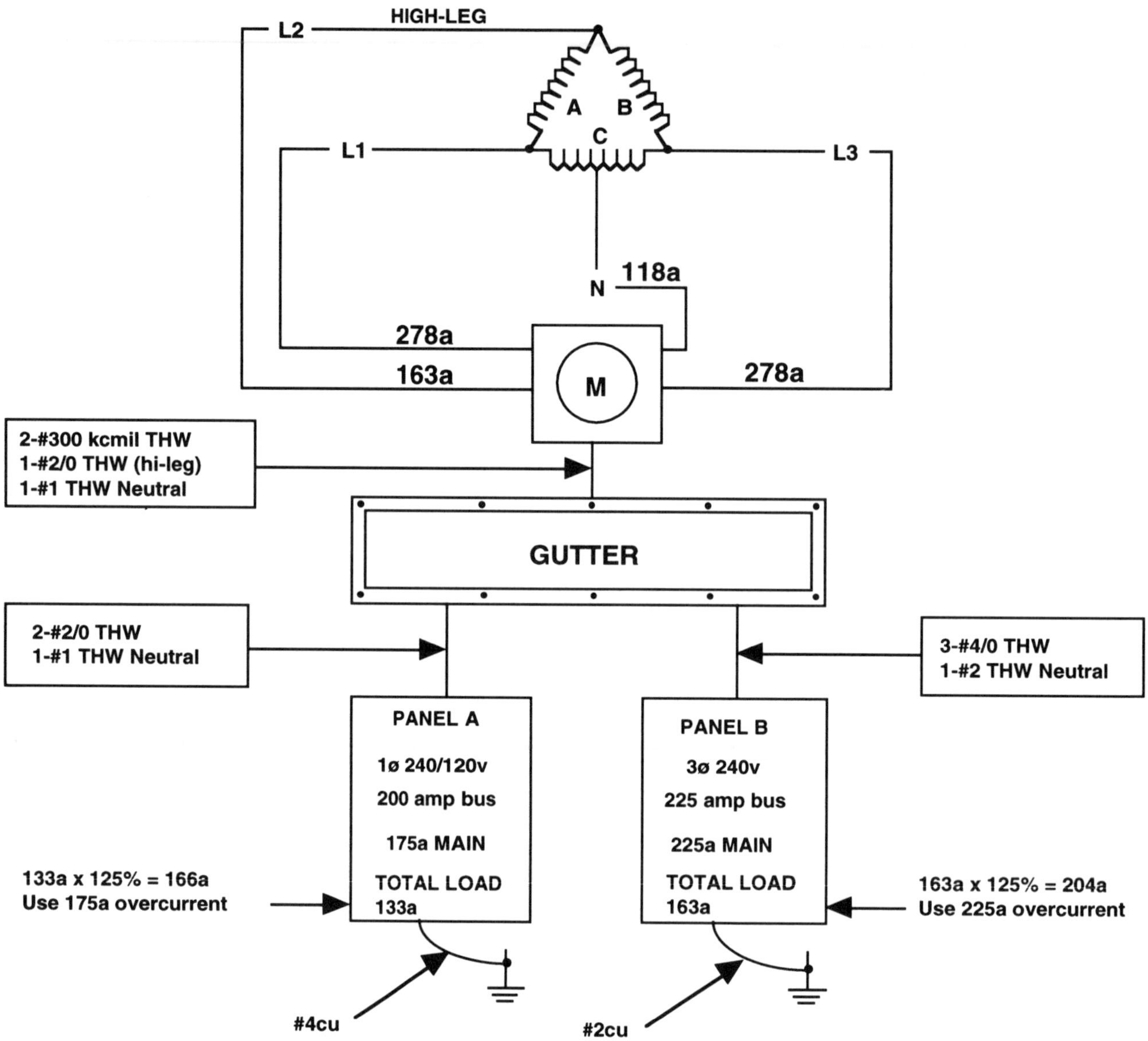

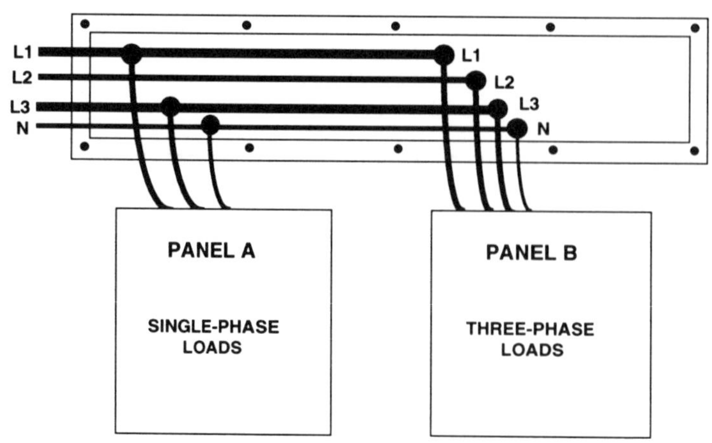

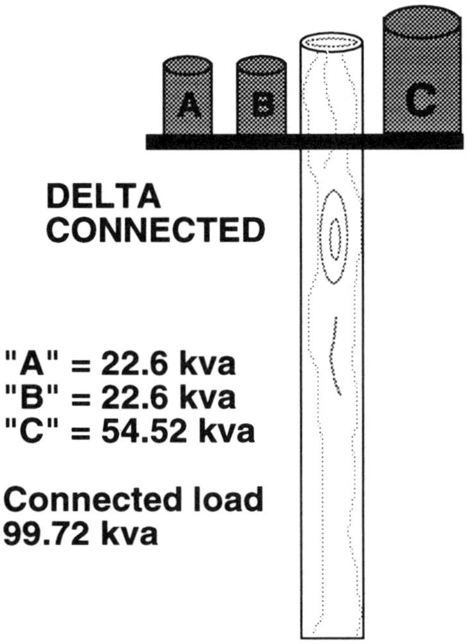

The delta connection will require one larger transformer ("C") which carries one-third of all three-phase loads plus all of the 120 volt loads. Only transformer "C" can provide 120 volts.

The following pages will show the same loads only now they are 208/120v three-phase wye connected. With the wye, all three phases have a neutral connection and will provide a more equal balancing among the three transformers.

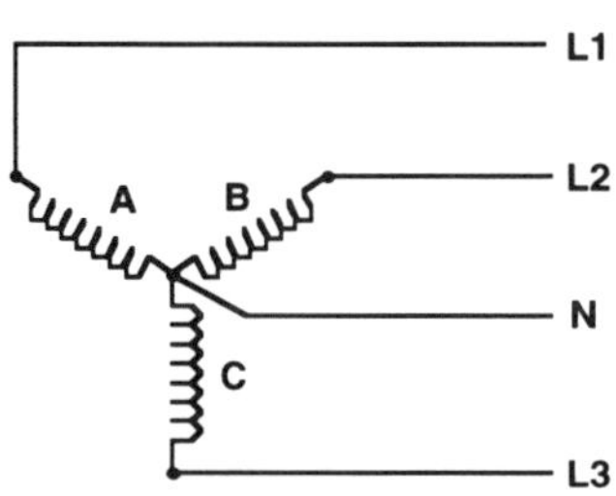

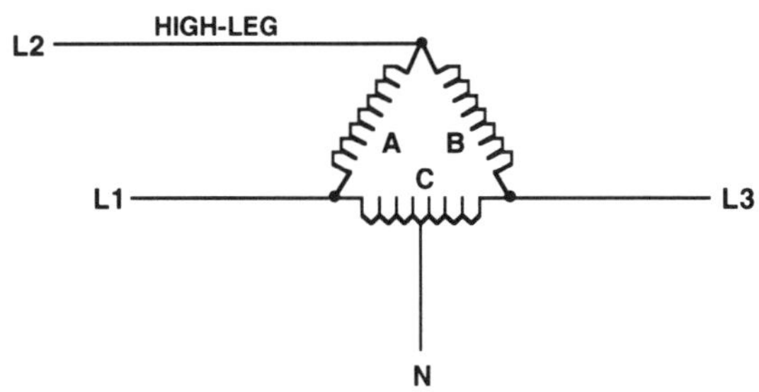

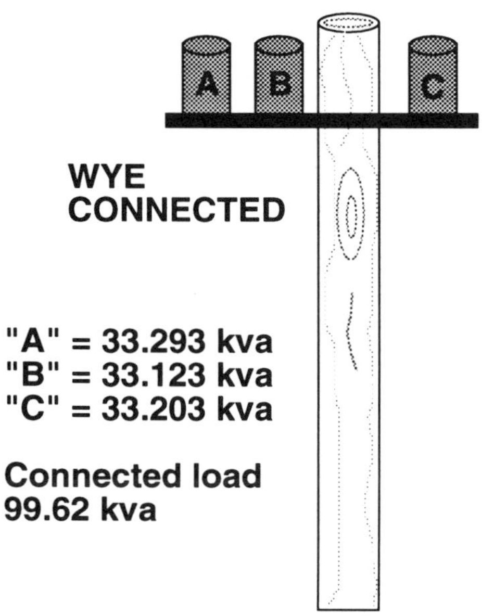

The wye connection provides an equal distribution of the120 volt loads among all three transformers.

LOAD BALANCING - WYE

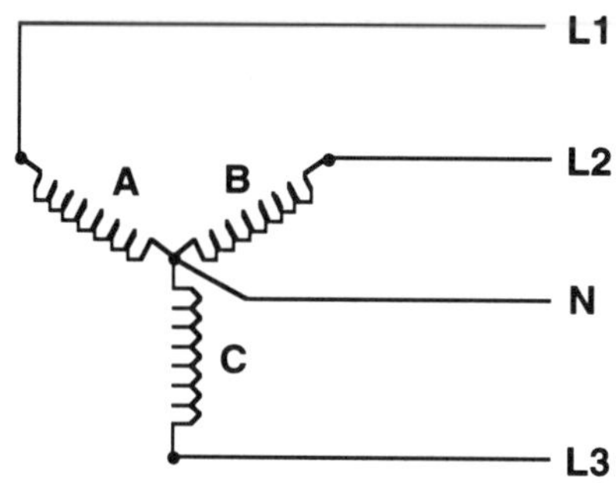

LOAD BALANCE SCHEDULE

TRANSFORMER			
A	B	C	DESCRIPTION
6000	6000	6000	18 kva heater 3ø 208v
6000	6000	6000	18 kva heater 3ø 208v
8000	8000	8000	24 kva A/C 3ø 208v
2000	2000	2000	6 kva hot plate 3ø 208v
600	600	600	1800va dough machine 3ø 208v
4000	4000	4000	Lights and recpts. 1ø 120v
2000	2000		Walk-in cooler 1ø 208v
1200	1200	1200	Corner roof lights 1ø 120v
1133	1133	1133	Pole mounted sign 1ø 120v
		1500	Coffee & tea 1ø 120v
1000			Center roof sign 1ø 120v
	830		Freezer 1ø 120v
		830	Glass froster 1ø 120v
830	830		Exhaust hood fans 1ø 120v
		830	Ice machine 1ø 120v
		580	Glass washer 1ø 120v
	530	530	Bath exhaust fans 1ø 120v
530			Hot plate hood 1ø 120v
33293va	**33123va**	**33203va**	**Load per transformer** **Total load = 99619va**

NEUTRAL CALCULATION - WYE

A	B	C	
4000	4000	4000	Lights and recpts. 1ø 120v
~~2000~~	~~2000~~		~~Walk-in cooler 1ø 208v~~
1200	1200	1200	Corner roof lights 1ø 120v
1133	1133	1133	Pole mounted sign 1ø 120v
		1500	Coffee & tea 1ø 120v
1000			Center roof sign 1ø 120v
	830		Freezer 1ø 120v
		830	Glass froster 1ø 120v
830	830		Exhaust hood fans 1ø 120v
		830	Ice machine 1ø 120v
		580	Glass washer 1ø 120v
	530	530	Bath exhaust fans 1ø 120v
530			Hot plate hood 1ø 120v
8693va	**8523va**	**10603va**	

•NOTE: Walk-in cooler is *not* a neutral load at 208v.

Transformer "A" = 8693va/120v = 72.4 amps
Transformer "B" = 8523va/120v = 71 amps
Transformer "C" = 10603va/120v = 88.3 amps

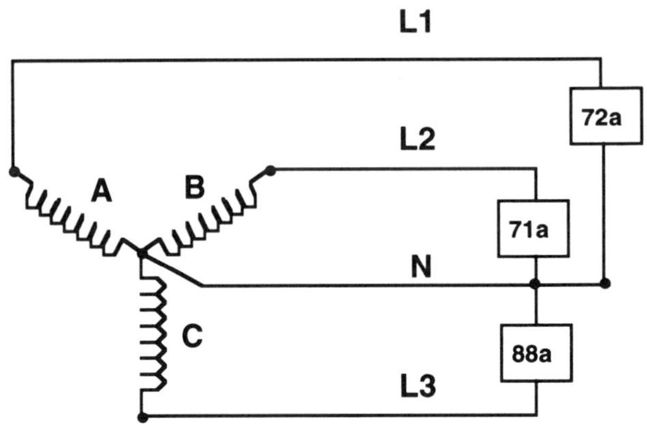

The maximum load on the neutral conductor would be *88 amperes.*

PHASE LOADS

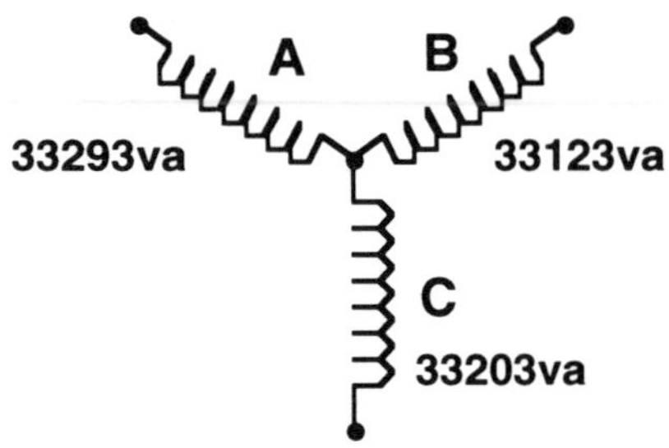

33293va/120v = 277a

33123va/120v = 276a

33203va/120v = 277a

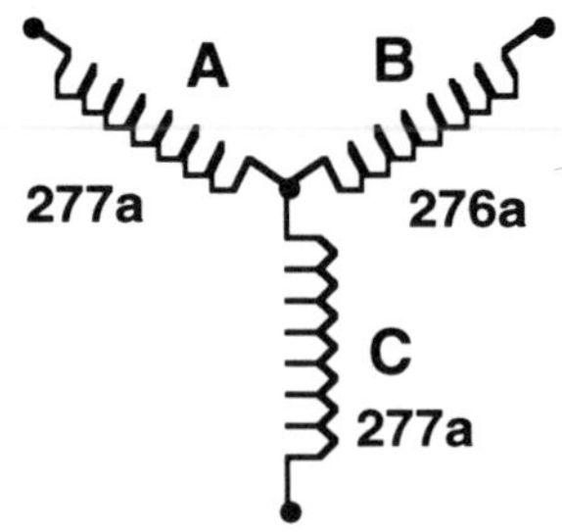

LINE LOADS

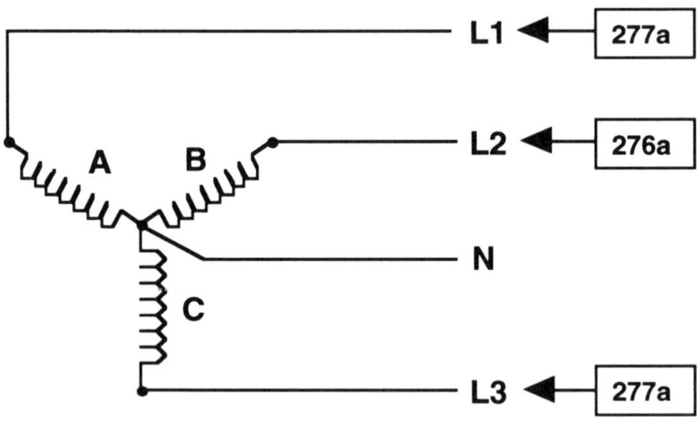

PHASE

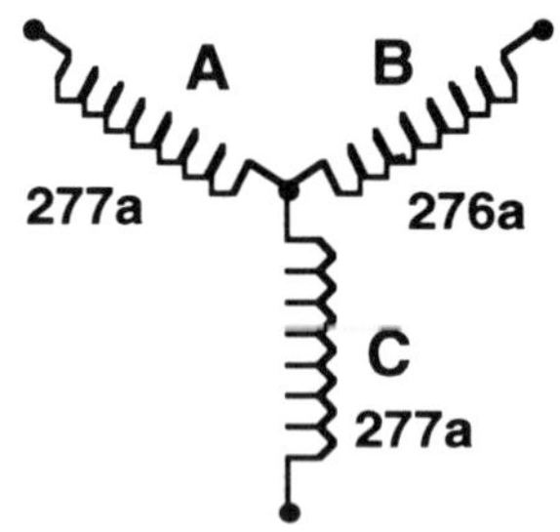

The line amps are the same as the phase amps wye-connected.

L1	L2	L3
33293va	33123va	33203va = **99619 va total**

99619va/208v/1.732 = **276.5 amps**

SIZING THE CONDUCTORS - WYE

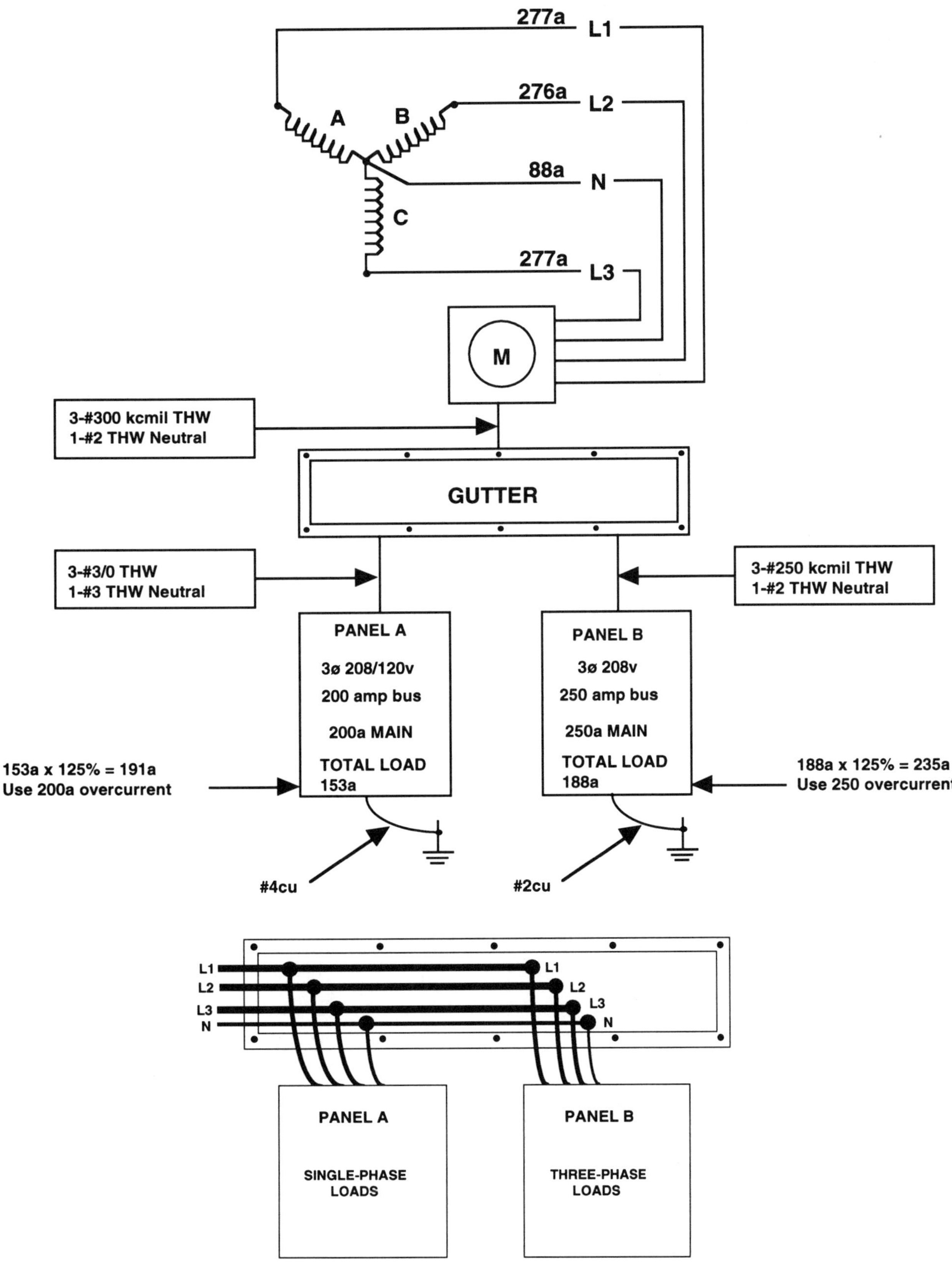

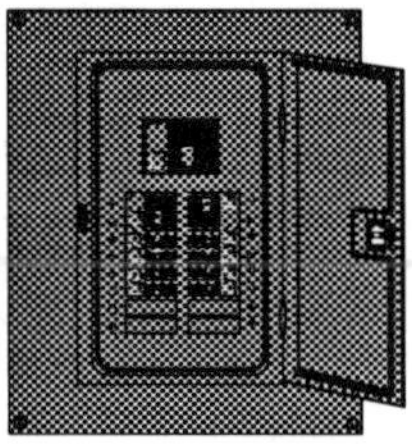

Different manufacturers of panelboards make different sizes of ampere ratings and spaces for single-phase and three-phase panelboards.

Manufacturer "X"

SINGLE-PHASE

Ampere Rating	Spaces
60a	12-18-24
100a	12-18-24
150a	30-36-42
225a	12-18-24-30-36-42

Manufacturer "Y"

SINGLE-PHASE

Ampere Rating	Spaces
100a	12-20
125a	30-42
150a	30-42
175a	30-42
200a	30-42
225a	30-42

Manufacturer "X"

THREE-PHASE

Ampere Rating	Spaces
100a	12-18-24-30-36-42
150a	12-18-24-30-36-42
225a	12-18-24-30-36-42
250a	12-18-24-30-36-42
400a	12-18-24-30-36-42
600a	12-18-24-30-36-42

Manufacturer "Y"

THREE-PHASE

Ampere Rating	Spaces
100a	12-24-30-42
110a	12-24-30-42
125a	12-24-30-42
150a	12-24-30-42
175a	12-24-30-42
200a	12-24-30-42
225a	12-24-30-42
250a	12-24-30-42
300a	12-24-30-42
350a	12-24-30-42
400a	12-24-30-42
600a	12-24-30-42

THE EQUIPMENT

It is not the task of the designer to size each piece of equipment in the branch circuit or feeder.

The electrician is the one responsible as these circuits are *field routed* and the designer would not know of each bolt, nut, washer, clamp, etc. as electricians have different methods they use, of which, must meet the *minimum* requirements of the Code.

Conduit sizing is from Code Tables 4,5 and Appendix C. Box sizing is from Article 370.

GUTTERS

There is *no* table listing the size and number of conductors permitted in a gutter. This is a calculation the designer must make.

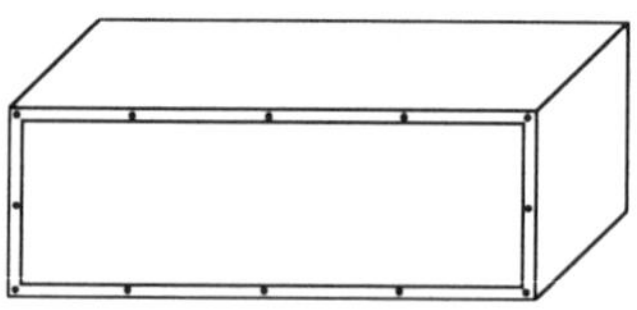

An auxiliary gutter serves the same function as a long junction box would. Sometimes it is necessary to run large conductors for a short distance but to make several *taps* to the conductors in the case of multiple services.

The Code permits gutters to supplement wiring spaces at meter centers, distribution centers, etc. The gutter shall *NOT* be used to enclose switches, overcurrent devices, appliances, or other similar equipment.

A gutter shall not extend more than 30 feet beyond the equipment it supplements.

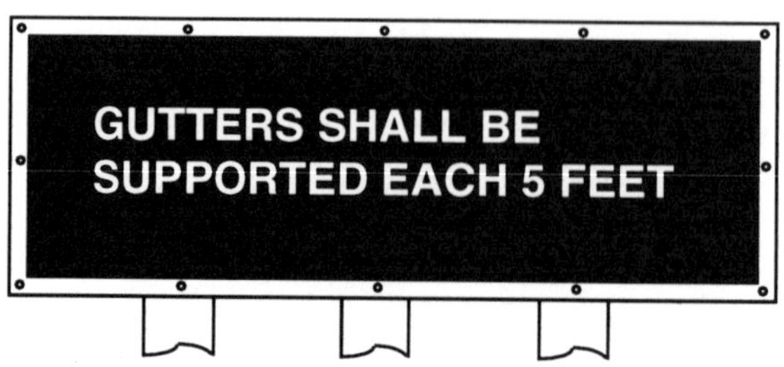

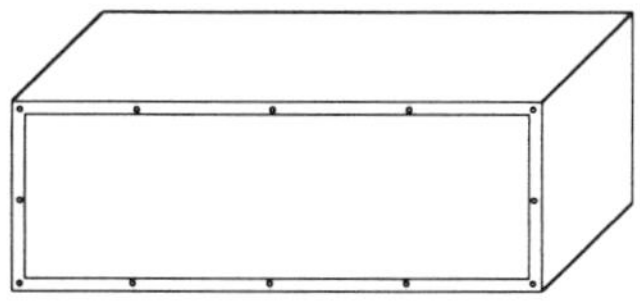

Gutters are manufactured in sizes 4" x 4", 6" x 6", 8" x 8", 10" x 10" and 12" x 12" square in lengths of one foot up to 10 feet.

A total of 30 conductors can be installed in a gutter *without* applying Note 8 derating of ampacity. Over 30 circuit conductors, Note 8 would apply.

The total cross sectional area occupied by the *conductors* at any cross section of the gutter must not exceed *20%* of the gutter interior cross sectional area at that point.

The total cross sectional area occupied by the *mass* of conductors *and splices* at any cross section of the gutter must not exceed 75% of the gutter interior cross sectional area at that point.

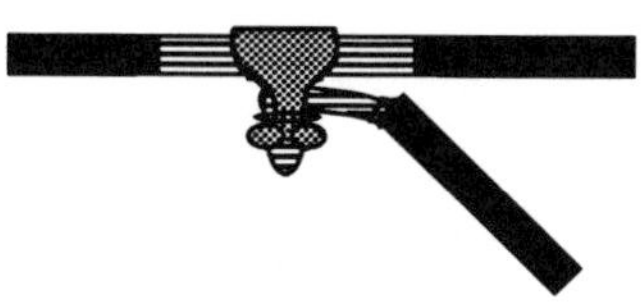

Most splices for larger conductors with split-bolt connectors are usually twice the size of the conductors being spliced.

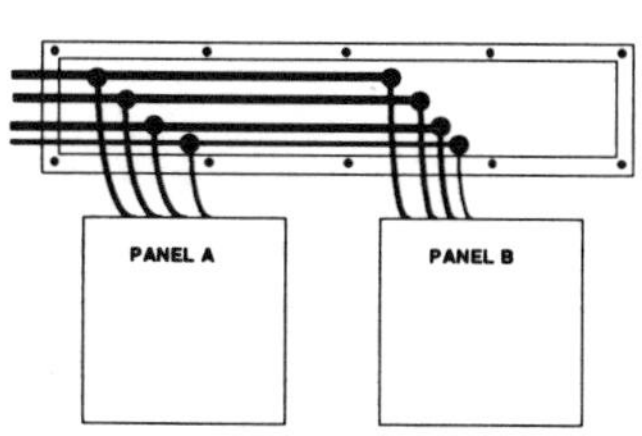

The staggering of the splices must be done to minimize the area taken up at the same cross section. This will keep the mass of splices from all adding up at the same cross section.

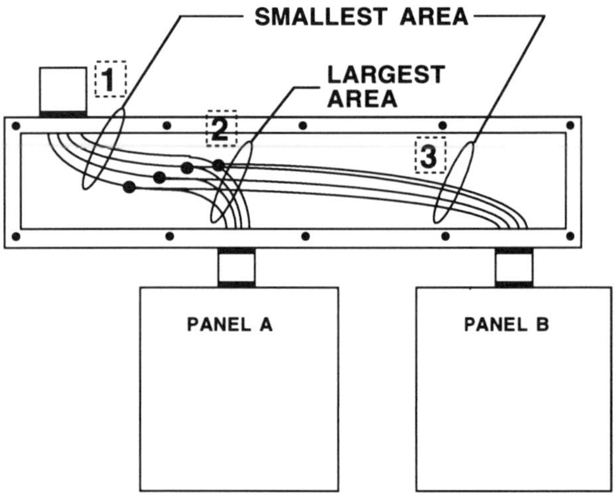

The service has 4 - #300 THHN conductors which are tapped to 8 - #3/0 THHN conductors.

To size the gutter required, the largest area of conductor concentration must be used. Code Table 5 lists the area square inch for conductors with THHN insulation.

TABLE 5 - THHN INSULATION	
AWG SIZE	AREA SQ. IN.
#1/0	.1855
#2/0	.2223
#3/0	.2679
#4/0	.3237
#250	.397
#300	.4608
#350	.5242
#400	.5863
#500	.7073

The area of concentration at 1 would be 4 - #300 THHN conductors. Table 5 lists a #300 at .4608 sq.in. x 4 conductors = 1.8432 cross sectional area.

The area of concentration at 2 would be 8 - #3/0 THHN conductors. Table 5 lists a #3/0 at .2679 sq.in. x 8 conductors = 2.1432 cross sectional area.

The area of concentration at 3 would be 4 - #3/0 THHN conductors. Table 5 lists a #3/0 at .2679 sq.in. x 4 conductors = 1.0716 cross sectional area.

The largest area of conductor concentration in the gutter would be at point 2 of 2.1432 cross sectional area.

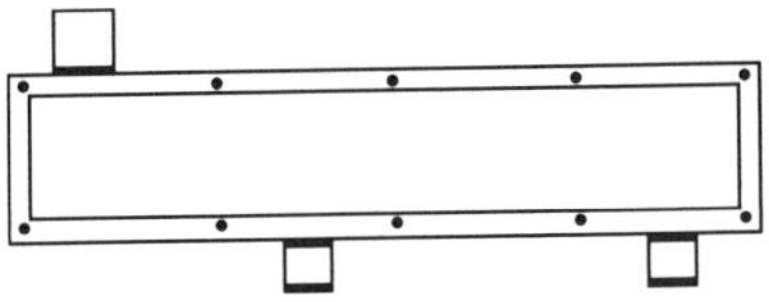

The Code states the most the gutter can be filled is 20% (the reciprocal is 5) multiply the largest conductor area of 2.1432 x 5 (reciprocal of 20%) = 10.716 cross sectional area required.

This requires a gutter of an area of at least 10.716 square inches. A *4" x 4"* gutter would provide an area of 16 sq.in. Table 373-6a must also be checked for bending space.

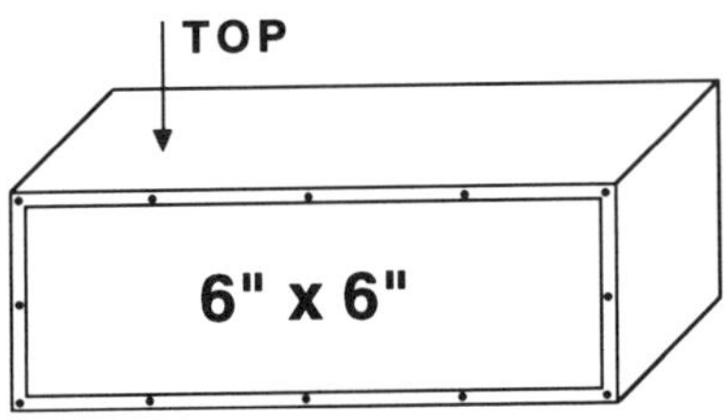

Code Table 373-6a lists the minimum wire-bending space and minimum width of gutters. Since the service enters the top of the gutter rather than the end, the table requires a minimum bending depth of 5" for a single #300 conductor. This requires a 6" x 6" minimum gutter.

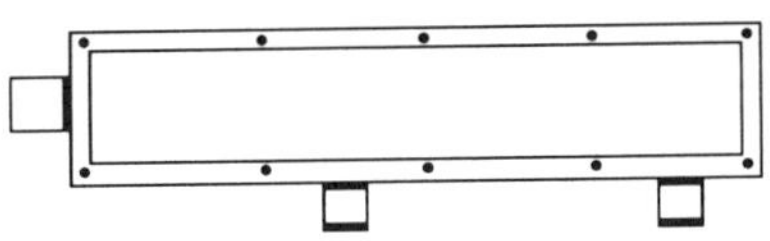

If the service conductors entered the end instead of the top, Table 373-6a would be applied to the bending space of the #3/0 conductor going out the bottom of the gutter. The table requires a bending space of only 4" for a #3/0 conductor.

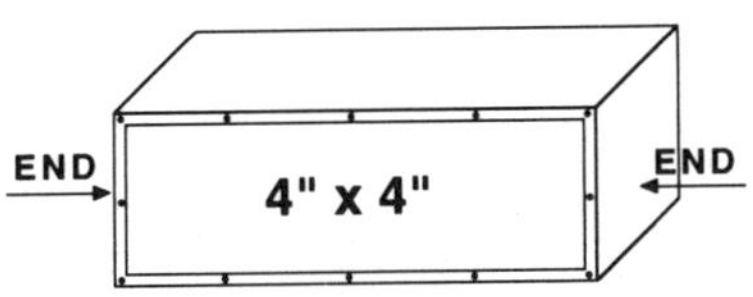

A 4" x 4" gutter would be suitable if the service conductors entered the end of the gutter instead of the top.

WIREWAYS

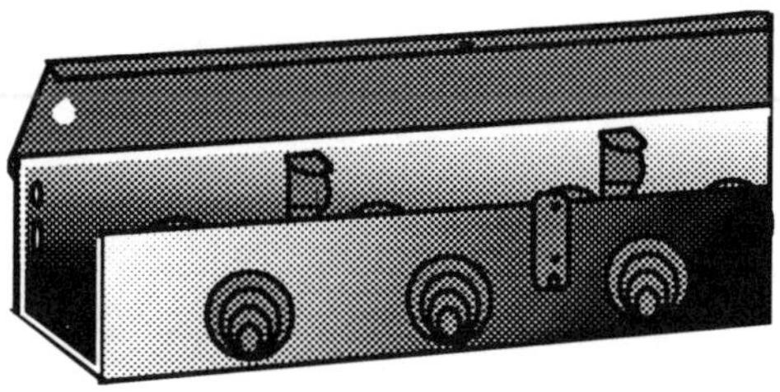

A wireway is a sheet-metal trough used instead of conduit to run large wires a considerable distance to several locations. The cover may be hinged or removable.

A gutter can only extend 30 feet from the equipment. A wireway has no limit on the distance it is run. Wireways are available in lengths of 1, 2, 3, 4, 5, and 10 feet, and in cross sections of 2 1/2" x 2 1/2" up to 8" x 8".

The conductor fill and sizing follow the same 20% and 75% fill area rules for a gutter.

SPLICER/REDUCER

When applying the variables to circuit designing often it is required to use a larger conductor for voltage drop, ampacity corrections, cost, etc. The question that comes up is how can you properly install a #4 conductor to a 20 amp circuit breaker? The terminal lug isn't big enough.

WIRE RANGE	
Maximum	**Minimum**
#2	#14
#1/0	#14
#4/0	#6
#250	#6
#350	#6
#500	#3/0
#750	#250
#1000	#500

UL listed splicer/reducer lugs are available for this purpose.

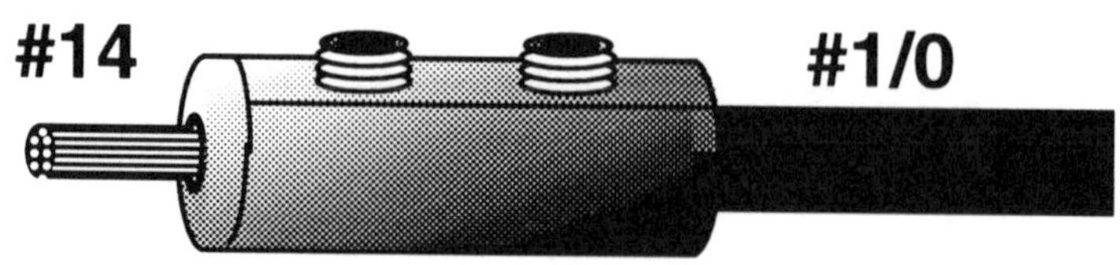

FORMULAS

$$W = I^2R$$

FORMULAS

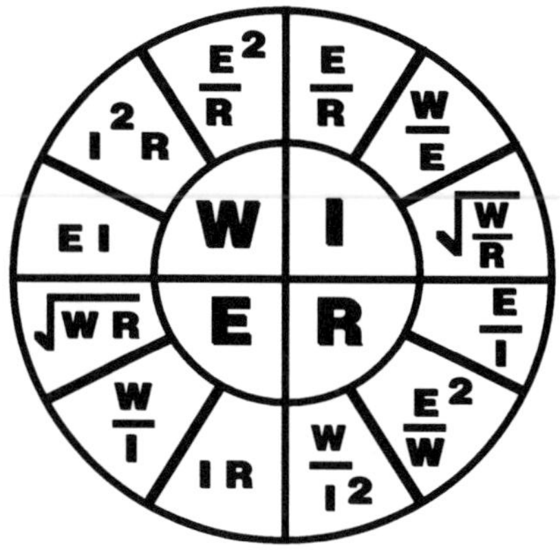

POWER

746 watts	=	1 hp (output work)
E x I	=	watt (true power)
E x I	=	va (volt-amp, apparent power)
E x I x 1.732	=	3 ø va

POWER FACTOR of an AC circuit is the ratio between the power in watts and the apparent power in volt-amps. PF is normally expressed in percent. Unity being 1.0. **PF = W/VA**

EFFICIENCY

$$\text{EFFICIENCY} = \frac{\text{OUTPUT}}{\text{INPUT}} \qquad \text{INPUT} = \frac{\text{OUTPUT}}{\text{EFFICIENCY}} \qquad \text{OUTPUT} = \text{INPUT} \times \text{EFF}$$

•Output is work (hp) secondary

•INPUT is the primary

COST

$$\text{COST} = \frac{\text{WATTS} \times \text{HOURS USED} \times \text{RATE PER KWH}}{1000}$$

TO FIND	DC	1 ø AC	3 ø AC
Amperes when horsepower is known	$\frac{\text{hp} \times 746}{\text{E} \times \%\text{ Eff}}$	$\frac{\text{hp} \times 746}{\text{E} \times \%\text{ Eff} \times \text{PF}}$	$\frac{\text{hp} \times 746}{1.732 \times \text{E} \times \%\text{ Eff} \times \text{PF}}$
Amperes when kilowatts is known	$\frac{\text{KW} \times 1000}{\text{E}}$	$\frac{\text{KW} \times 1000}{\text{E} \times \text{PF}}$	$\frac{\text{KW} \times 1000}{1.732 \times \text{E} \times \text{PF}}$
Amperes when KVA is known		$\frac{\text{KVA} \times 1000}{\text{E}}$	$\frac{\text{KVA} \times 1000}{\text{E} \times 1.732}$
Kilowatts	$\frac{\text{E} \times \text{I}}{1000}$	$\frac{\text{E} \times \text{I} \times \text{PF}}{1000}$	$\frac{\text{E} \times \text{I} \times 1.732 \times \text{PF}}{1000}$
KVA		$\frac{\text{E} \times \text{I}}{1000}$	$\frac{\text{E} \times \text{I} \times 1.732}{1000}$
Horsepower	$\frac{\text{E} \times \text{I} \times \%\text{ Eff}}{746}$	$\frac{\text{E} \times \text{I} \times \%\text{ Eff} \times \text{PF}}{746}$	$\frac{\text{E} \times \text{I} \times \%\text{ Eff} \times \text{PF} \times 1.732}{746}$

VOLTAGE DROP

The Code suggests in sections 210-19a FPN, 215-2b FPN and 310-15 FPN keeping voltage drop at *3% for branch circuits* and *5% total* will provide reasonable efficiency of operation.

Voltage drop is *wasted electricity* due to heating the conductors. Power Loss = VD x I

VOLTAGE DROP FORMULAS

To find:

VOLTAGE DROP ...... $VD = \frac{2 \times K \times D \times I}{CM}$ (or) $VD = I \times R$

WIRE SIZE .... $CM = \frac{2 \times (K) \times D \times I}{VD\ permitted}$

DISTANCE .... $D = \frac{CM \times VD\ permitted}{2 \times K \times I}$

LOAD .... $I = \frac{CM \times VD\ permitted}{2 \times K \times D}$

* The "2" in the formulas is for single-phase circuits. This is the conductor to and from the load.

* For 3 phase voltage drop calculations, change the "2" in the formula to **1.732**

* "K" is the resistance of a circular mil-foot. Exact $K = \frac{R \times CM}{1000'}$

* When using the formula to find "WIRE SIZE" use the **approximate (K) factor** of 12.9 for copper and 21.2 for aluminum.

* "D" is the distance *one way* in a circuit.

* "I" is the load in amperes. For motors use the Full Load Current from the motor tables.

* "CM" is the size of conductor in circular mils, found only in Table 8.

* "VD permitted" is the *percentage of the applied source voltage*, 3% of the source for a branch circuit and 5% of the total applied to the system. Example: On a branch circuit that has a source voltage of 120, the voltage drop permitted would be 120v x 3% = 3.6 volts. Example : On a branch circuit that has a source voltage of 208, the voltage drop permitted would be 208v x 3% = 6.24v. Example: A system with an applied voltage of 240, the *total* voltage drop permitted on the entire service would be 240v x 5% = 12 volts.

EXACT K @ 75°C

DESIGNED BY TOM HENRY
CODE ELECTRICAL CLASSES

AWG	Area Circular Mils	Copper Un-coated Resistance	**Copper Un-coated EXACT K**	Copper Coated Resistance	**Copper Coated EXACT K**	Aluminum Resistance	**Aluminum EXACT K**
#14 SOLID	4110	3.07	**12.6177**	3.19	**13.1109**	5.06	**20.7966**
#14 STRANDED	4110	3.14	**12.9054**	3.26	**13.3986**	5.17	**21.2487**
#12 SOLID	6530	1.93	**12.6029**	2.01	**13.1253**	3.18	**20.7654**
#12 STRANDED	6530	1.98	**12.9294**	2.05	**13.3865**	3.25	**21.2225**
#10 SOLID	10 380	1.21	**12.5598**	1.26	**13.0788**	2.00	**20.76**
#10 STRANDED	10 380	1.24	**12.8712**	1.29	**13.3902**	2.04	**21.1752**
#8 SOLID	16 510	0.764	**12.61364**	0.786	**12.97686**	1.26	**20.8026**
#8 STRANDED	16 510	0.778	**12.84478**	0.809	**13.35659**	1.28	**21.1328**
#6	26 240	0.491	**12.88384**	0.510	**13.3824**	0.808	**21.20192**
#4	41 740	0.308	**12.85592**	0.321	**13.39854**	0.508	**21.20392**
#3	52 620	0.245	**12.8919**	0.254	**13.36548**	0.403	**21.20586**
#2	66 360	0.194	**12.87384**	0.201	**13.33836**	0.319	**21.16884**
#1	83 690	0.154	**12.88826**	0.160	**13.3904**	0.253	**21.17357**
#1/0	105 600	0.122	**12.8832**	0.127	**13.4112**	0.201	**21.2256**
#2/0	133 100	0.0967	**12.87077**	0.101	**13.4431**	0.159	**21.1629**
#3/0	167 800	0.0766	**12.85348**	0.0797	**13.37366**	0.126	**21.1428**
#4/0	211 600	0.0608	**12.86528**	0.0626	**13.24616**	0.100	**21.16**
250 kcmil	250 000	0.0515	**12.875**	0.0535	**13.375**	0.0847	**21.175**
500 kcmil	500 000	0.0258	**12.9**	0.0265	**13.25**	0.0424	**21.2**
1000 kcmil	1 000 000	0.0129	**12.9**	0.0132	**13.2**	0.0212	**21.2**

MOTOR FORMAT Designed by *Tom Henry* ©

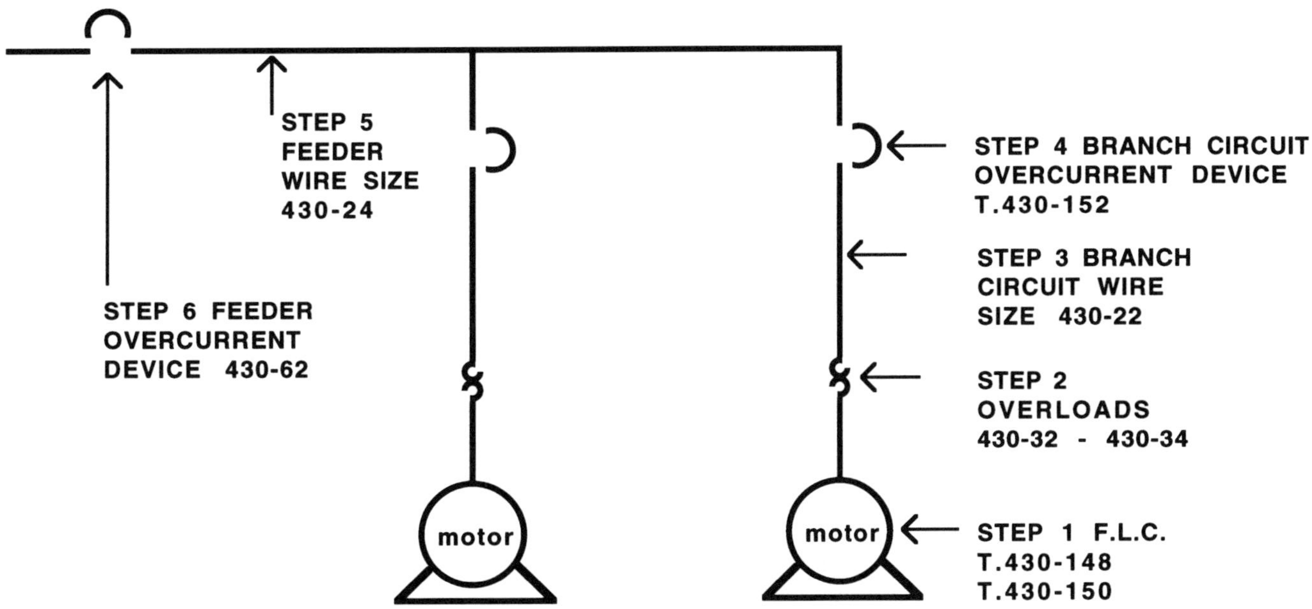

STEP 1 FLC (full load current) Table 430-148 1 ø current Table 430-150 3 ø current

STEP 2 Motor running over**load** protection. Thermal protector called a "heater".
- 430-32 Minimum size • 430-34 Maximum size
- If motor **nameplate** is given, use it, if not use FLC from Tables

STEP 3 Branch circuit conductor sizing 430-22 FLC x 125% = Required ampacity
- Turn to Table 310-16 and select the conductor with the required ampacity

STEP 4 Branch circuit over**current** device shall be selected from Table 430-152. First select the **type** of motor (1 ø, 3 ø, AC, DC, wound rotor, etc.) next select the **type** of protective device (non-time delay fuse, dual element fuse, inverse time breaker) now select the **percentage** from the proper column and multiply it times the FLC of the motor. Use 240-6 to select the **standard size** overcurrent device. • When the value found does **not** match a standard size the Code permits the **next higher STANDARD size** for a branch circuit overcurrent device (430-52 ex.1).

STEP 5 Feeder conductor size 430-24. Multiply the **largest** rated motor in FLC by 125% and add the FLC of all other motors connected to the **SAME** feeder conductor for the total amps.

STEP 6 Feeder overcurrent protection 430-62. Select the **largest** branch circuit **overcurrent device** and add all the other motor FLC connected on the **same** feeder conductor to find the amperage required to select the feeder fuse or CB. • The Code does **NOT** permit going to a higher size if the value does not match a standard size from 240-6. Drop **DOWN** to the next smaller standard size. Code permits going up only on a branch circuit.

CONDUCTOR SIZING - HEAT DERATINGS

WIRE SIZE : $\dfrac{\text{LOAD}}{\text{CORRECTION FACTOR X NOTE 8}}$ = REQUIRED AMPACITY FROM TABLE 310-16

SINGLE DWELLING

Sq. Ft. x 3 + 1500 x .35 + 3000 =

NET GENERAL LIGHTING & SMALL APPLIANCE LOAD

SINGLE DWELLING

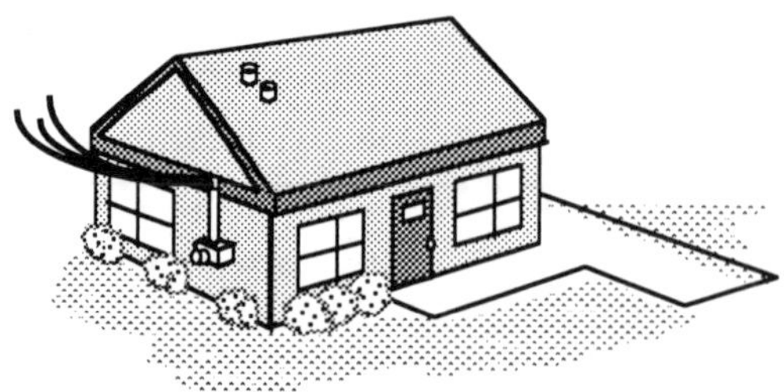

MINIMUM NUMBER OF BRANCH CIRCUITS

$\dfrac{\text{Square footage x 3va}}{\text{120v}} = \dfrac{\text{amperage}}{\text{15 or 20}} = \text{number of circuits}$

RECEPTACLES

THERE IS NO LIMIT ON A CIRCUIT FOR A DWELLING

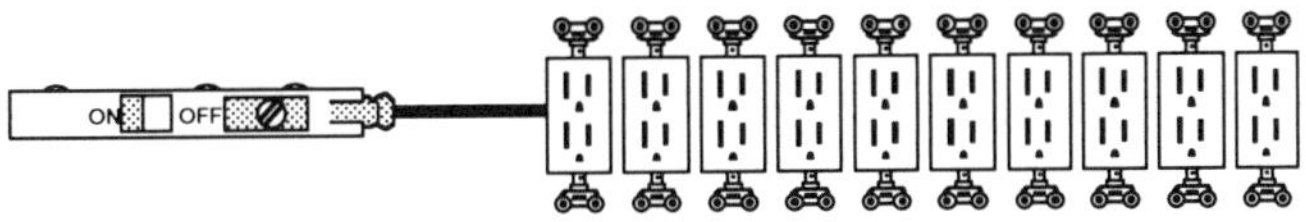

OFFICE BUILDING

FOR NON-DWELLING THE 180 VA APPLIES
Example: 20a circuit x 120v = 2400va/180va = 13

Table 210-22 A 20 AMP CIRCUIT CAN BE LOADED TO 20 AMPS

Table 210-21b2 A SINGLE RECEPTACLE CAN ONLY BE LOADED 80%

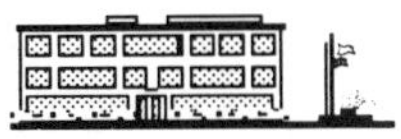

OFFICE BUILDING

Table 220-13
NON-DWELLING OVER 10 KVA DEMAND

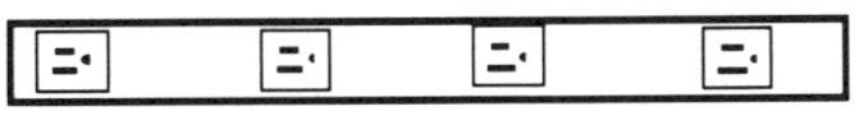

220-3c ex.1
MULTIOUTLET ASSEMBLIES 180 VA PER 5 EACH FEET

TYPICAL AMBIENT TEMPERATURES

LOCATION	TEMPERATURE	Minimum Rating of Required Conductor Insulation
Well ventilated, normally heated buildings	30°C (86°F)	* (see note below)
Buildings with such major heat sources as power stations or industrial processes	40°C (104°F)	75°C (167°F)
Poorly ventilated spaces such as attics	45°C (113°F)	75°C (167°F)
Furnaces and boiler rooms (minimum) (maximum)	 40°C (104°F) 60°C (140°F)	 75°C (167°F) 90°C (194°F)
Outdoors in the shade	40°C (104°F)	75°C (167°F)
In thermal insulation	45°C (113°F)	75°C (167°F)
Direct solar exposure	45°C (113°F)	75°C (167°F)
Places above 60°C (140°F)	-----	110°C (230°F)

*Note: 60°C for up to and including #8 copper and up to and including #6 aluminum.
75°C for over #8 copper and #6 aluminum.

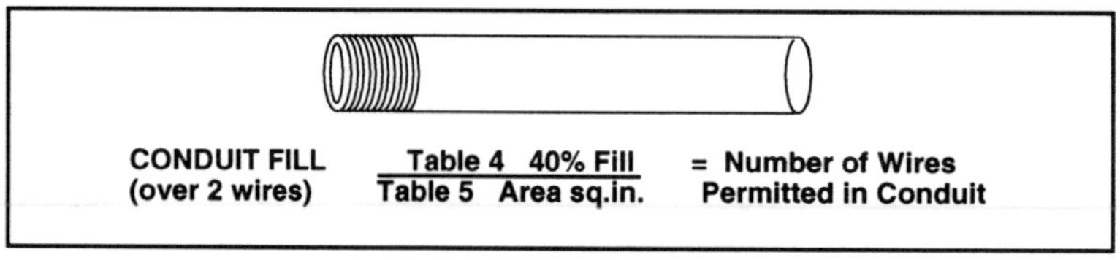
CONDUIT FILL
(over 2 wires)
Table 4 40% Fill
Table 5 Area sq.in.
= Number of Wires
Permitted in Conduit

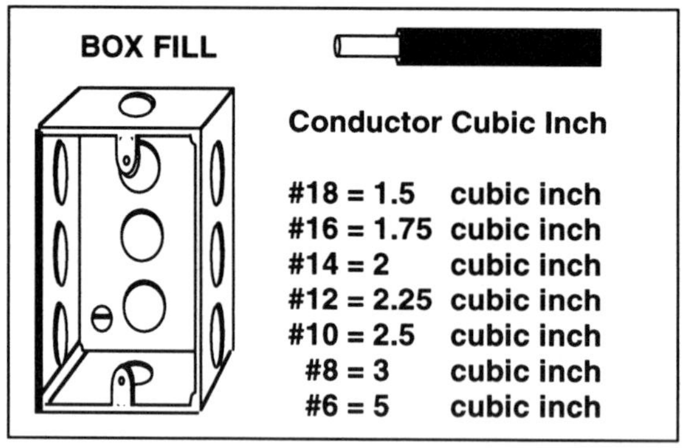
BOX FILL
Conductor Cubic Inch
#18 = 1.5 cubic inch
#16 = 1.75 cubic inch
#14 = 2 cubic inch
#12 = 2.25 cubic inch
#10 = 2.5 cubic inch
#8 = 3 cubic inch
#6 = 5 cubic inch

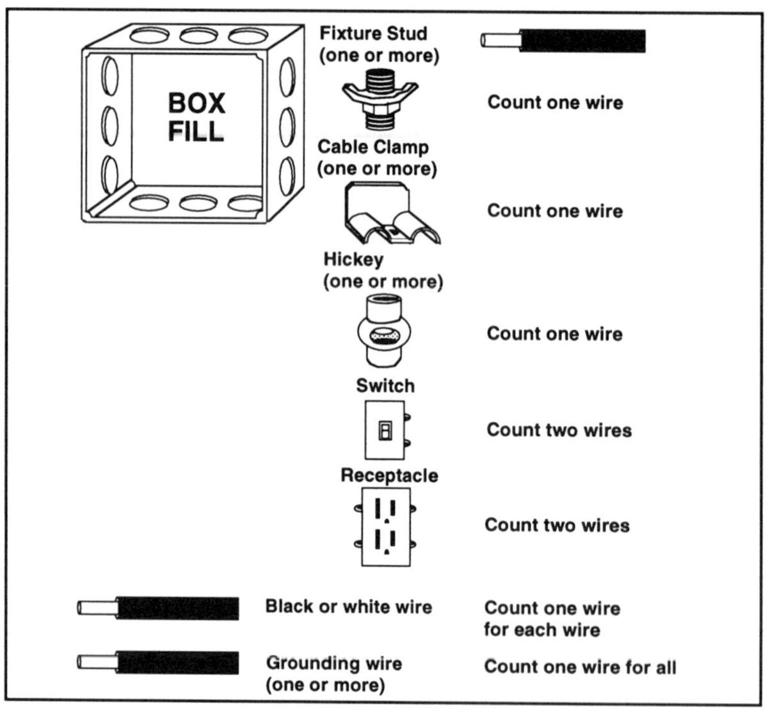
BOX
FILL
Fixture Stud
(one or more)
Count one wire
Cable Clamp
(one or more)
Count one wire
Hickey
(one or more)
Count one wire
Switch
Count two wires
Receptacle
Count two wires
Black or white wire
Count one wire
for each wire
Grounding wire
(one or more)
Count one wire for all

EXAMPLES

i.e.

BRANCH CIRCUIT TO RANGE

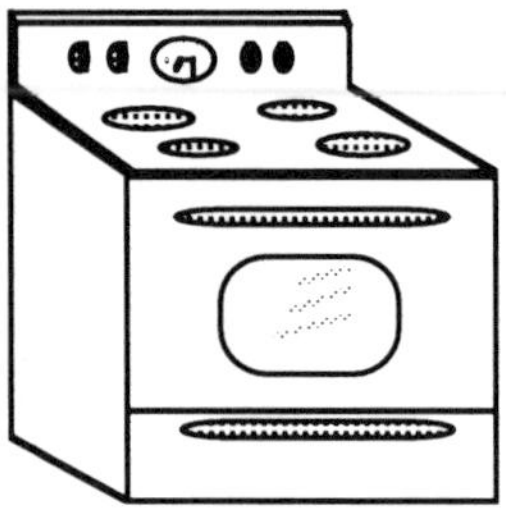

Design the branch circuit to a 12 kw electric range 240/120v single-phase. The conductors are NMB installed through the attic which reaches a maximum ambient 122°F.

The connected load must be determined first. 12,000w/240v = *50 amperes* maximum load.

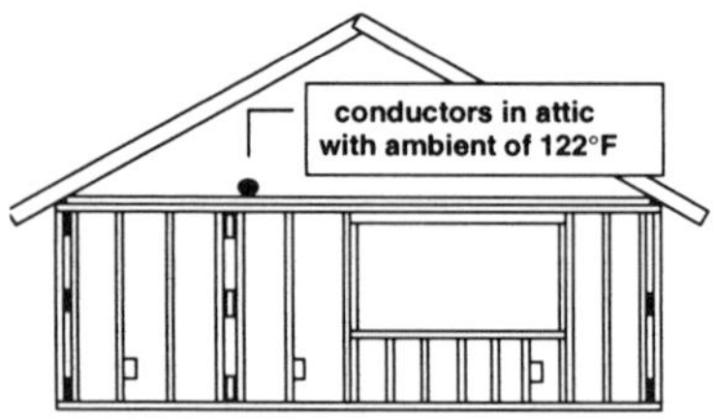

Table 310-16 correction factor
for 60°C ampacity conductors at 122°F = .58

The nonmetallic sheathed cable NMB (romex) has a 90°C insulation, but, Code section 336-26 states it has an *ampacity of 60°C*. The .58 correction factor is from the 60°C column.

Wire Size = 50 amps/.58 correction factor = *86.2* amps conductor ampacity required.

Table 310-16 would require a *minimum* #2 NMB cable for this branch circuit.

The circuit breaker size would be 50 amps.

•Note: In general, the Code does not consider cooking equipment to be a continuous load, even though, at certain times the oven may be on for 3 hours or more.

The Code permits in Note 4 of Table 220-19 a demand for a branch circuit to household cooking equipment. This is the only place in the Code that permits a demand on a *branch circuit*. Demands are for *feeders* where all the loads are not on at the same time. I have always disagreed with Note 4. It's insane to think that the oven and all the burners would *never* be turned on simultaneously.

Section 210-19b ex.2 states the branch circuit neutral for cooking equipment appliances shall have an ampacity of not less than 70% of the branch circuit rating and shall not be smaller than #10.

The Code violation occurs when the electrician (designer) takes the nameplate load of 50 amps and goes to Table 310-16 and selects a #6 which has an ampacity of 55 amps.

It would be complete insanity to assume the attic temperature would *never exceed 86°F*!

If the attic ambient temperature reached 96°F the correction factor is .82.

Wire Size = 50amps/.82 = 60.9 required ampacity. This would require a #4 romex.

BRANCH CIRCUIT TO CLOTHES DRYER

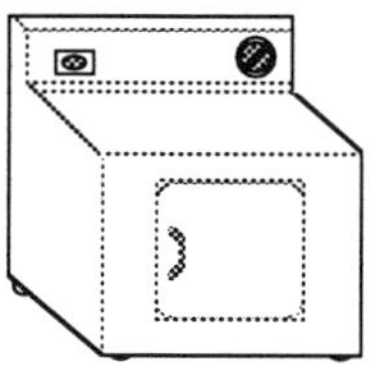

Design the branch circuit to a 5 kw electric clothes dryer 240/120v single-phase. The conductors are NMB installed through the attic which reaches a maximum ambient 114°F.

The connected load must be determined first. 5,000w/240v = *20.8 amperes* maximum load.

Wire Size = 20.8 amps/.58 correction factor = *35.8* amps conductor ampacity required.

Table 310-16 would require a *minimum* #8 NMB cable for this branch circuit.

The circuit breaker size would be 30 amps.

•Note: A 25 amp breaker would not be selected as Section 210-23 states the rating of cord and plug connected appliances shall not exceed 80% the rating of the circuit.

The neutral conductor would also be a minimum #8, Section 210-19b ex.2 which is the 70% neutral for a branch circuit, applies *only* to cooking equipment, not to a *branch circuit* of a clothes dryer.

Again, the Code violation occurs when a #10 conductor which has an ampacity of 30 is selected with no consideration what so ever to the ambient heat where the conductors are to be installed!

BRANCH CIRCUIT TO MOTOR

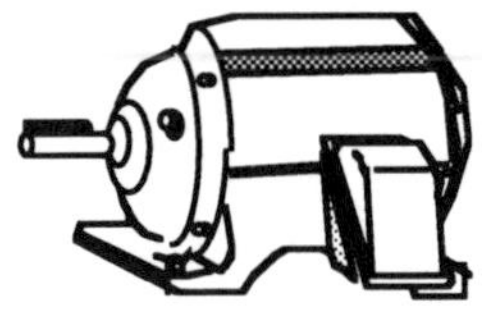

Design the branch circuit to a 10 hp, 208v three-phase, motor. The conductors are THHN installed in metal conduit through the high bay area of the industry which reaches a maximum ambient 105°F. The length of the branch circuit is 86 feet.

The full-load current must be determined first. Table 430-150 lists the F.L.C. at 30.8 amps.

CEILING AMBIENT REACHES 105°F

Table 310-16 correction factor
for 90°C ampacity conductors at 105°F = .87

Wire Size = 30.8 amps/.87 correction factor = *35.4* amps x 125% (430-22) = 44.2 ampacity required.

Table 310-16 would require a *minimum* #8 THHN which has an ampacity of 55, but can only be loaded to the 60°C ampacity of 40 amps.

The inverse circuit breaker size 30.8a x 250% (T.430-152) = 77a = 80 amp breaker.

Now check to see if the #8 THHN will meet the voltage drop requirements of the Code for 86 feet.

Wire Size: $CM = \frac{1.732 \times 12.9 \times 86' \times 30.8a}{6.24 \text{ vd}}$ = 9484 cm required minimum.

From Code Table 8, a #8 has 16,510 circular mils, so a #8 THHN would meet the Code.

PARKING LOT LIGHTS - VOLTAGE DROP

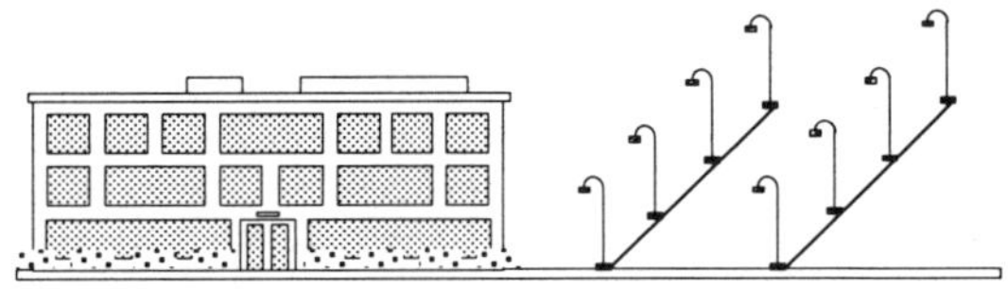

The following diagram shows eight parking lot lights at 2 amps each. Four of the lights are wired with a #10 stranded uncoated conductor; three of the lights are wired with a #8 stranded uncoated conductor. What size conductor is required from the 240v single-phase source to the first light 200 feet in distance?

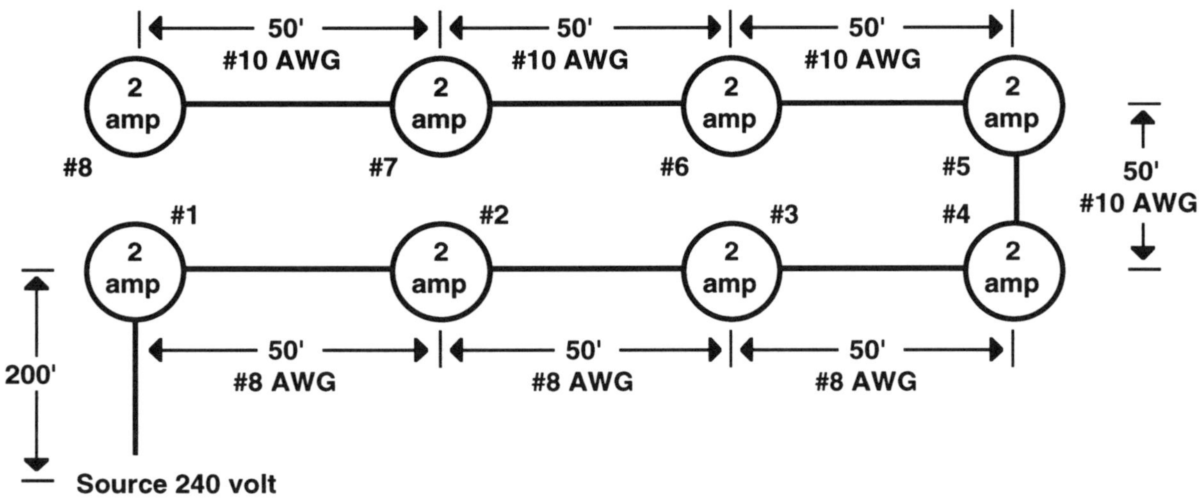

Solution: The Code allows a VD of 3% of 240v = 7.2 volts. First you must find the voltage drop between the lights to determine how many volts *are left* to drop in the 200 feet to the source. There is 100 feet of conductor between each of the lights, use Code Table 8.

VOLTAGE DROP BETWEEN LIGHTS

#8 - #7 #10 resistance = .124 ohm x 2 amp = .248 volts dropped
#7 - #6 #10 resistance = .124 ohm x 4 amp = .496 volts dropped
#6 - #5 #10 resistance = .124 ohm x 6 amp = .744 volts dropped
#5 - #4 #10 resistance = .124 ohm x 8 amp = .992 volts dropped
#4 - #3 #8 resistance = .0778 ohm x 10 amp = .778 volts dropped
#3 - #2 #8 resistance = .0778 ohm x 12 amp = .9336 volts dropped
#2 - #1 #8 resistance = .0778 ohm x 14 amp = *1.0892 volts dropped*

5.2808 volts dropped

7.2 total voltage drop permitted
-5.2808 volts dropped between the lights
1.9192 volts *left to drop* in this branch circuit

Now you can size the conductor required between the source and the first light.

$$CM = \frac{2 \times 12.9 \times 200 \text{ feet} \times 16 \text{ amp}}{1.9192 \text{ volts left to drop}} = 43{,}017.9 \text{ CM required}$$

Table 8: #3 conductor has a cma of 52,620

Answer: *#3 conductor*

Summary: Excessive voltage drop causes problems such as the overheating of motors, appliances not operating properly, insufficient lighting, etc. Low voltage on a circular handsaw can cause dangerous "kickback" of the blade.

It is impossible to eliminate voltage drop *completely* because of the resistance in the conductors to the load.

Voltage drop is *wasted* power (watts). The Code recommendation to limit voltage drop to 3% on branch circuits and 5% total is very inefficient.

CONTINUOUS LOAD

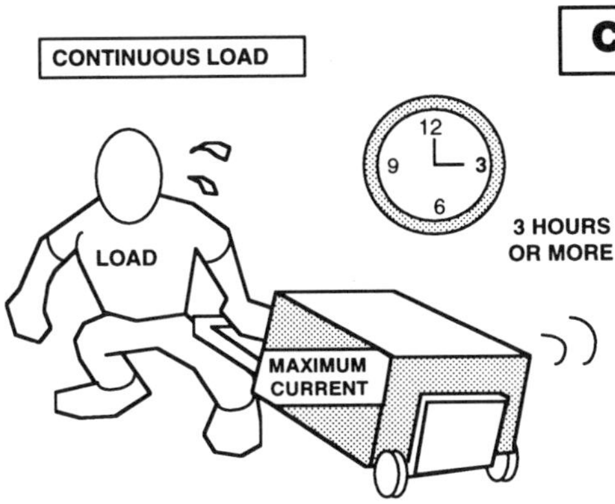

The first step in design is start with the load, either computed load or connected load. The computed load for feeders and services contains the demand factors which actually reduces the connected load.

As we have seen, the ambient temperature which the conductors are installed affects the sizing as well as the distance the conductors are run.

The next variable to consider is the load continuous or noncontinuous. In general, the Code recognizes everything to be continuous *except,* a dwelling, hotel or motel guest room, and receptacles.

If the load is *noncontinuous*, a receptacle circuit, a 20 amp rated receptacle circuit could be loaded to 20 amps.

If the circuit was continuous, a 20 amp rated circuit could only be loaded to 16 amps (80% per 384-16c).

A continuous load can really affect conductor sizing.

Example: The service load is calculated to be 130 amperes of *noncontinuous* load.

A 150 amp circuit breaker would be the next higher standard size permitted. Using THW 75°C conductors, #1 THW has an ampacity of 130. Where the ampacity of the conductors (130) does not correspond to a standard ampere rating, the next higher standard size (150) is permitted per section 240-3b.

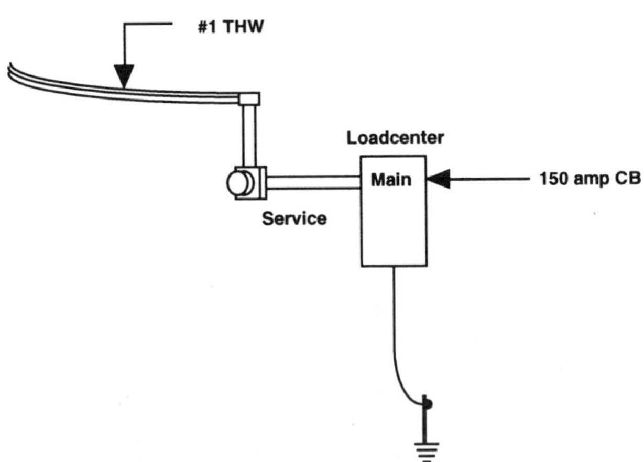

If THHN conductors were used instead of THW, it would still require a #1 THHN. A #2 THHN has an ampacity of 130 amps and the noncontinuous load is 130 amps. But a #2 THHN can only be loaded to the 75°C ampacity of 115 amps and the calculated load is 130 amps.

Example: The service load is calculated to be 130 amperes of which 84 amps is noncontinuous and 46 amps is continuous. 75°C equipment.

Noncontinuous load	=	84a
Continuous load 46a x 125%	=	57.5a
		141.5 amps

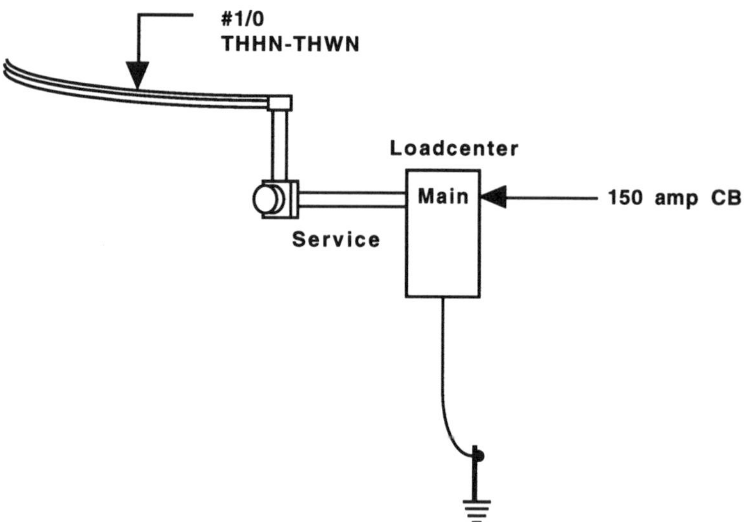

Section 240-6 lists a 150 amp overcurrent device as the next higher standard size. A #1/0 THHN has a ampacity of 170 and can be loaded to the 75°C ampacity of 150. The load is 141.5 amps.
The 150a breaker can be loaded 80% for *continuous load*, 150a x 80% = 120 amps. The continuous load is 46 amps.

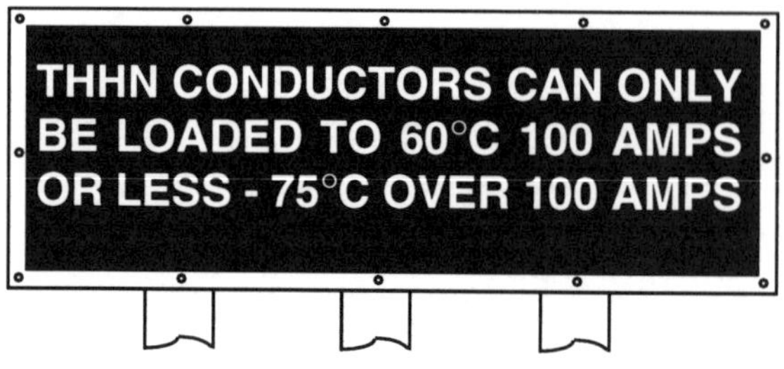

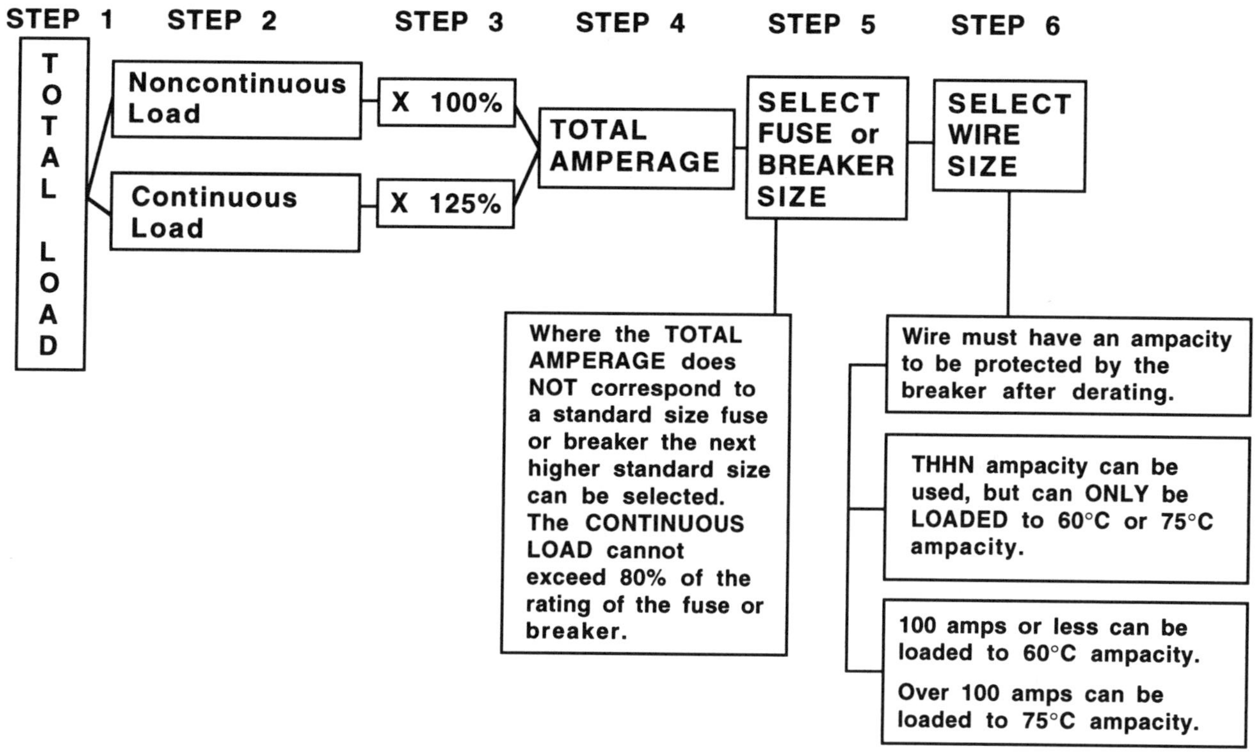

Example: The service load is calculated to be 125 amps of which 75 amps is continuous and 50 amps is noncontinuous. 60°C equipment.

75a x 125% = 93.75a
50a x 100% = 50a
143.75a

Section 240-6 lists the next higher size breaker at 150 amps. A #2/0 THHN has an ampacity of 195 and can be loaded to the 60°C ampacity of 145. The load is 143.75 amps.

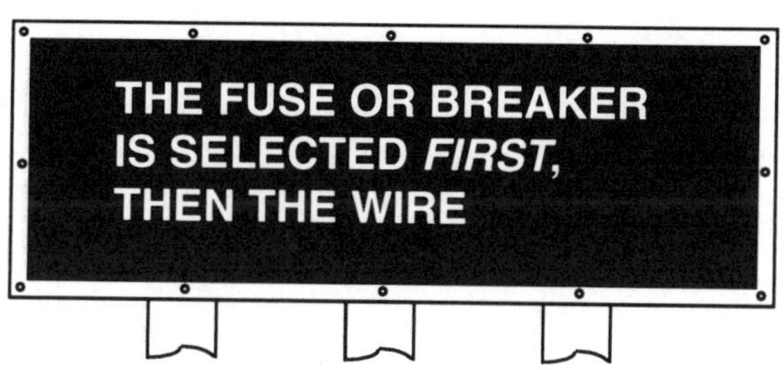

COST OF POWER LOSS

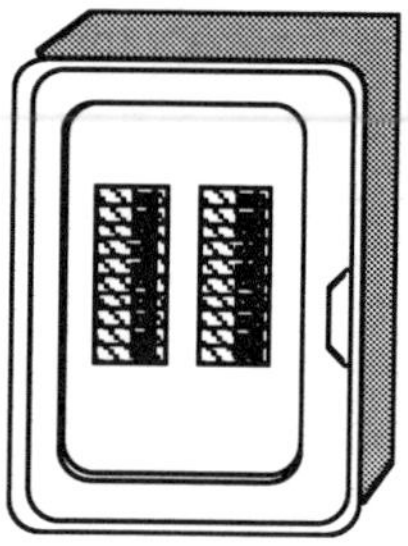

A feeder to a 200 amp single-phase 240/120v panel is run with #3/0 THHN conductors which have an ampacity of 225 amps, and can be loaded to the 75°C ampacity of 200 amps. The feeder is 100 feet in length.

The distribution panel supplies industrial machinery which has a recorded load of approximately 160 amperes for 12 hours a day.

The voltage drop of the feeder would be: $VD = \frac{2 \times 12.85348 \times 100' \times 160a}{167800cm} = 2.45v$

Power Loss = VD x I   2.45vd x 160a = 392 watts wasted

$\text{Cost} = \frac{392w \times 12 \text{ hrs used per day} \times .08 \text{ cents per kwh}}{1000} = .38 \text{ cents per day}$

Based on 22 working days per month = 22 x .38 cents = *$8.36* wasted electricity

3 - #3/0 THHN conductors would require a minimum 1 1/2" EMT conduit. By installing a larger conductor, a #250 kcmil THHN, it would require a larger EMT conduit of 2", but the savings would be:

$VD = \frac{2 \times 12.875 \times 100' \times 160a}{250{,}000cm} = 1.65v$

Power Loss = 1.65vd x 160a = 264 watts wasted

$\text{Cost} = \frac{264w \times 12 \text{ hrs used per day} \times .08 \text{ cents per kwh}}{1000} = .25 \text{ cents per day}$

22 working days x .25 cents per day = $5.50 wasted electricity

A savings of $2.86 per month or *$34.32* per year for *one* panelboard.

COMPUTED LOAD

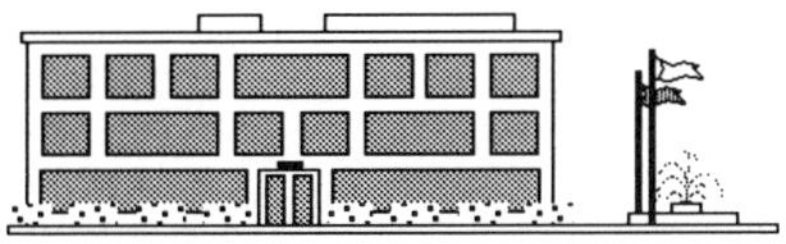

Where the load is computed on a volt-amps per square foot basis, the wiring system up to and including the branch-circuit panelboard(s) shall be provided to serve not less than the calculated load.

The computed load for general lighting of a 30,000 sq.ft. office building would be:

30,000 sq.ft. x 3.5va (Table 220-3b) x 125% = 131,250 va minimum required for general lighting.

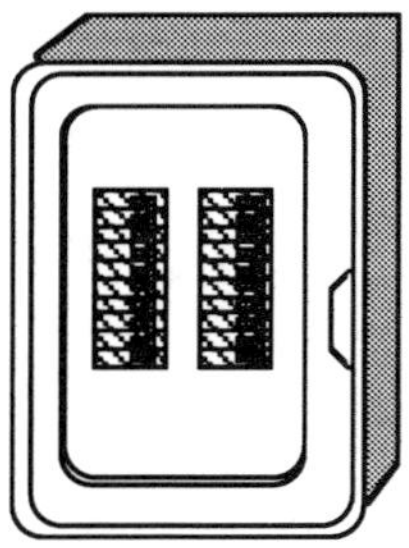

The *panelboard busbars* must have a minimum capacity of 131,250 va for general lighting.

The last sentence of Code section 220-4d makes it clear that the number of branch circuits required is determined by the actual *connected load*, not the computed load.

If the office had an *actual connected load of 60,000 va* for light fixtures, the number of branch circuits required would be:

Each 20 amp circuit would calculate 20a x 120v = 2400va x 80% (continuous load) = 1920 va per branch circuit.

60,000va/1920va = 31.25 or *32* branch circuits would be required for the general lighting.

The conductor and panelboard sizing is based on the computed load and the number of branch circuits required is based on the connected load.

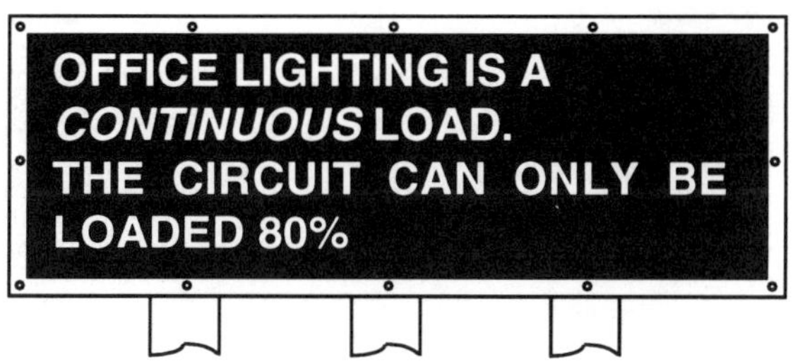

RECEPTACLE BRANCH CIRCUITS

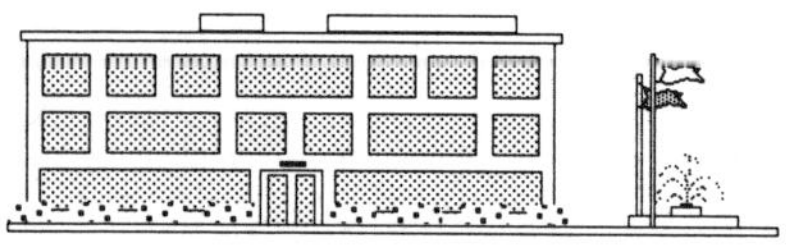

For the same 30,000 sq.ft. office the branch circuits required for receptacles would calculate as follows:

Table 220-3b requires 1 volt-amp per square foot for *unknown* receptacle loads.

30,000 sq.ft. x 1va (Table 220-3b) = 30,000va.

Each 20 amp circuit would calculate 20a x 120v = 2400va.

In general, the Code does *not* consider receptacles as a continuous load.

30,000va/2400va = 12.5 or *13 branch circuits* for receptacles.

The office building requires a minimum of 45 branch circuits. Thirty-two lighting circuits and thirteen receptacle circuits.

32

13

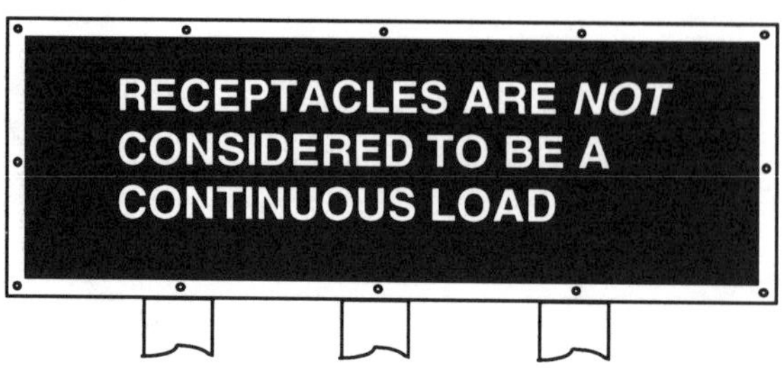

THE QUIZZES

?

DESIGN QUIZ #1

•The following questions are to be answered after reviewing pages 1 - 20.

1. When designing the proper size of a branch circuit, the first thing that must be determined is the _____.

(a) voltage (b) phase (c) load (d) distance

2. The Code recommends the voltage drop on a branch circuit should not exceed _____.

(a) 1% (b) 2% (c) 3% (d) 5%

3. Homeruns should be limited to a maximum length of _____.

(a) 100' (b) 200' (c) 50' (d) 150'

4. Good designing would provide one spare circuit for each _____ circuits in use.

(a) 2 (b) 3 (c) 4 (d) 5

5. Good designing would limit voltage drop to no more than _____.

(a) 1% (b) 2% (c) 3% (d) 5%

6. In general, the Code permits going up to the next higher size overcurrent device up to _____ amps.

(a) 200 (b) 500 (c) 600 (d) 800

7. A #2 THW is considered protected by a _____ amp breaker.

(a) 115 (b) 125 (c) 130 (d) 150

8. A #250 kcmil THW is considered protected by a _____ amp breaker.

(a) 255 (b) 260 (c) 275 (d) 300

9. A #6 THHN is considered protected by a ____ amp breaker.

(a) 75 (b) 80 (c) 85 (d) 90

10. A #1/0 THHN is considered protected by a ____ amp breaker.

(a) 170 (b) 175 (c) 185 (d) 200

11. If the branch circuit conductors are going to be installed in an ambient temperature that exceeds ____ the derating of conductor ampacity is required.

(a) 70°C (b) 86°C (c) 122°F (d) 86°F

12. Ambient temperature is the ____.

(a) average (b) maximum (c) surrounding temperature (d) temperature of the conduit

13. Note 8 is a reduction of ampacity when there are more than ___ current carrying conductors.

(a) 4 (b) 5 (c) 6 (d) 3

14. Ampacity is defined as the current in amperes that a conductor can carry ____.

(a) intermittently (b) continuously (c) periodically (d) instantaneously

15. A conductor can be ____.

(a) rectangular (b) aluminum (c) solid (d) all of these

16. The current flowing through a conductor produces ____.

(a) emf (b) heat (c) resistance (d) voltage

17. A conductor in free air will carry ____ current than a conductor in a conduit.

(a) more (b) less (c) same (d) none of these

18. The greatest hazard a conductor encounters is ____.

(a) mechanical damage (b) heat (c) water (d) cold temperatures

19. The type letter on a conductor that indicates "moisture resistant" is ____.

(a) MR (b) m (c) W (d) R

20. Which conductor has the highest insulation rating?

(a) RHW (b) THW (c) RHH (d) THWN

21. Heat = _____.

(a) I^2R (b) E/I (c) W/I^2 (d) IR

22. Conductors in parallel are required to be derated in ampacity when Note 8 applies.

(a) true (b) false

23. Note 8 would not apply if the conduit size was increased.

(a) true (b) false

24. It is better to design for parallel conductors rather than one large conductor.

(a) true (b) false

25. Conductors with higher temperature insulation ratings will have a ____ ampacity than conductors with lower temperature insulation ratings.

(a) same (b) higher (c) lower (d) average

DESIGN QUIZ #2

•The following questions are to be answered after reviewing pages 21 - 49.

1. If equipment is not marked, it is 60°C rated for 100 amps and less. Over 100 amps is 75°C rated. If the equipment is rated higher, it would have to be marked to indicate the higher temperature.

(a) true (b) false

2. A 90°C conductor such as THHN connected to a 60°C rated receptacle can be loaded to the ___ ampacity.

(a) 90°C (b) 60°C (c) 75°C (d) all of these

3. When a conductor is chosen for the circuit, the designer or installer must know the termination ratings for the _____.

(a) breaker (b) switch (c) receptacle (d) entire circuit

4. When two metallic parts are tightly connected having different operating temperatures (60°C-90°C), the higher-temperature part (90°C) will give heat to the lower temperature part (60°C) and raise its temperature to above 60°C.

(a) true (b) false

5. THW has a thinner insulation than THHN. This is why more THW's can be installed in a conduit.

(a) true (b) false

6. A #8 THHN has a ____ than a #6 TW.

(a) smaller cma (b) greater voltage drop (c) higher resistance (d) all of these

7. A #1 THHN feeding a 150 amp panel would have an ampacity of ____.

(a) 110 (b) 130 (c) 150 (d) 170

8. A conduit has four current-carrying #250 kcmil THHN conductors with an ampacity of ____. The equipment is 75°C rated.

(a) 255 amps (b) 290 amps (c) 204 amps (d) 232 amps

9. A #2 THHN connected to 75°C equipment in an ambient of 95°F would have an ampacity of ____ amps.

(a) 115 (b) 124.8 (c) 130 (d) 95

10. A #4/0 THHN connected to 75°C equipment in an ambient of 105°F would have an ampacity of____ amps.

(a) 260 (b) 230 (c) 226.2 (d) 236.6

11. A #3/0 THHN connected to 75°C equipment in an ambient of 104°F and four current carrying conductors in the conduit would have an ampacity of ____ amps.

(a) 163.8 (b) 165 (c) 200 (d) 180

12. A #6 THHN connected to 60°C equipment in an ambient of 90°F would have an ampacity of ____ amps.

(a) 75 (b) 72 (c) 65 (d) 55

13. A #4 THHN connected to 60°C equipment with four current carrying conductors would have an ampacity of ____ amps.

(a) 95 (b) 85 (c) 76 (d) 70

14. A #8 THHN connected to 60°C equipment in an ambient of 122°F would have an ampacity of ____ amps.

(a) 45.1 (b) 40 (c) 50 (d) 55

15. A #8 THHN connected to 60°C equipment in an ambient of 123°F and four current carrying conductors would have an ampacity of ____ amps.

(a) 33.44 (b) 40 (c) 50 (d) 41.8

16. A #14 THHN connected to 60°C equipment in an ambient of 104°F can be loaded to ___ amps.

(a) 25 (b) 22.75 (c) 20 (d) 15

17. A #12 THHN connected to 60°C equipment with four current carrying conductors can be loaded to ____ amps.

(a) 20 (b) 24 (c) 25 (d) 30

18. A #1/0 THW connected to 60°C equipment in an ambient of 114°F would have an ampacity of ____ amps.

(a) 125 (b) 150 (c) 112.5 (d) 93.75

19. A #500 THW connected to 75°C equipment in an ambient of 96°F and with eight current carrying conductors would have an ampacity of ____ amps.

(a) 364.8 (b) 334.4 (c) 234.08 (d) 250

20. A #4/0 THW connected to 75°C equipment in an ambient of 123°F would have an ampacity of ____ amps.

(a) 154.1 (b) 195 (c) 230 (d) 172.5

21. A #6 TW connected to 60°C equipment in an ambient of 122°F would have an ampacity of ____ amps.

(a) 39.05 (b) 31.9 (c) 45.1 (d) 50.05

22. A #4 TW connected to 60°C equipment in an ambient of 140°F would have an ampacity of ____ amps.

(a) 70 (b) 28.7 (c) 40.6 (d) there is no correction factor, 60°C insulation is not to be used

23. A #4 TW connected to 60°C equipment in an ambient of 70°F would have an ampacity of ____ amps.

(a) 70 (b) 75 (c) 75.6 (d) 80

24. A #350 THHN connected to 75°C equipment in an ambient of 45°C would have an ampacity of ____ amps.

(a) 285 (b) 310 (c) 304.5 (d) 300

25. A #750 THHN connected to 75°C equipment in an ambient of 104°F would have an ampacity of ____ amps.

(a) 535 (b) 499.2 (c) 475 (d) 450

DESIGN QUIZ #3

•The following questions are to be answered after reviewing pages 50 - 66.

1. The total load 75 amps is continuous. What size breaker is required?

(a) 90 (b) 95 (c) 100 (d) 125

2. What size THHN conductors are required in Question #1 using 60°C equipment?

(a) #2 (b) #3 (c) #4 (d) #6

3. What size THHN conductors are required in Question #1 using 75°C equipment?

(a) #2 (b) #3 (c) #4 (d) #6

4. The total load 20 amps is continuous. What size breaker is required?

(a) 15 (b) 20 (c) 25 (d) 30

5. What size THHN conductors are required in Question #4 using 60°C equipment?

(a) #14 (b) #12 (c) #10 (d) #8

6. A feeder supplies a panel which has a continuous load of 140 amps. What size breaker is required?

(a) 150 (b) 175 (c) 200 (d) 225

7. What size THW conductors are required in Question #6 using 75°C equipment?

(a) #4/0 (b) #3/0 (c) #2/0 (d) #1/0

8. What size THHN conductors are required in Question #6 using 75°C equipment?

(a) #4/0 (b) #3/0 (c) #2/0 (d) #1/0

9. What is the maximum size breaker the Code permits for a *noncontinuous* load of 375 amps using #500 kcmil THW conductors?

(a) 300 amp (b) 350 amp (c) 375 amp (d) 400 amp

10. What size breaker is required for a 375 amp *continuous* load?

(a) 400 (b) 450 (c) 475 (d) 500

11. What size THHN conductors are required for Question #10 using 75°C equipment?

(a) #600 kcmil (b) #750 kcmil (c) #500 kcmil (d) #700 kcmil

12. What size THW conductors are required for Question #10 using 75°C equipment?

(a) #600 kcmil (b) #750 kcmil (c) #500 kcmil (d) #700 kcmil

13. What size breaker is required for a 480 amp *continuous* load?

(a) 700 (b) 600 (c) 500 (d) 450

14. What size THHN conductors are required for Question #13 using 75°C equipment?

(a) #1500 kcmil (b) #1250 kcmil (c) #1000 kcmil (d) #900 kcmil

15. What size THW conductors are required for Question #13 using 75°C equipment?

(a) #1500 kcmil (b) #1250 kcmil (c) #1000 kcmil (d) #900 kcmil

16. What size breaker is required for a 525 amp *continuous* load?

(a) 600 (b) 700 (c) 800 (d) 1000

17. What size THHN conductors are required for Question #16 using 75°C equipment?

(a) #2000 kcmil (b) #1750 kcmil (c) #1500 kcmil (d) #1250 kcmil

18. What size THW conductors are required for Question #16 using 75°C equipment?

(a) #2000 kcmil (b) #1500 kcmil (c) #1250 kcmil (d) #1000 kcmil

19. What size breaker is required for a 630 amp *continuous* load?

(a) 600 (b) 700 (c) 800 (d) 1000

20. What size THHN conductors are required for Question #19 using 75°C equipment?

(a) #2000 kcmil (b) #1750 kcmil (c) parallel two #500 kcmil (d) parallel two #600 kcmil

21. What size THW conductors are required for Question #19 using 75°C equipment?

(a) #2000 kcmil (b) #1750 kcmil (c) parallel two #500 kcmil (d) parallel two #600 kcmil

22. What size breaker is required for a 900 amp *continuous* load?

(a) 2000 (b) 1600 (c) 1200 (d) 1000

23. What size THHN conductors are required for Question #22 using 75°C equipment? Parallel two conductors.

(a) #900 kcmil (b) #1000 kcmil (c) #1250 kcmil (d) #1500 kcmil

24. What size THW conductors are required for Question #22 using 75°C equipment? Parallel two conductors.

(a) #750 kcmil (b) #1000 kcmil (c) #1250 kcmil (d) #1500 kcmil

DESIGN QUIZ #4

•The following questions are to be answered after reviewing pages 67 - 75.

1. A continuous load is a load that continues for ___ hours or more.

(a) 2 (b) 3 (c) 4 (d) 6

2. A breaker supplying a continuous load cannot exceed ___ of its rating.

(a) 70% (b) 75% (c) 80% (d) 100%

3. The definition of ampacity is "the current in amperes that a conductor can carry ____ under the conditions without exceeding its temperature rating".

(a) normally (b) continuously (c) indefinitely (d) periodically

4. The † indicates the maximum ____.

(a) ampacity (b) voltage (c) overcurrent device (d) distance

5. If the distance of a single-phase branch circuit is 75 feet, it would require ____ feet of wire.

(a) 75 (b) 100 (c) 150 (d) 225

6. The formula for Power Loss is ____.

(a) VD x W (b) E x I (c) R/I (d) VD x I

7. The 120 volt branch circuit is supplying receptacles. The equipment is 60°C. There are four current carrying conductors and the ambient temperature is 122°F. What size THHN is required if you wanted to load the circuit to 20 amps?

(a) #14 (b) #12 (c) #10 (d) #8

8. The 120 volt branch circuit is supplying receptacles. The equipment is 60°C. What size breaker is required?

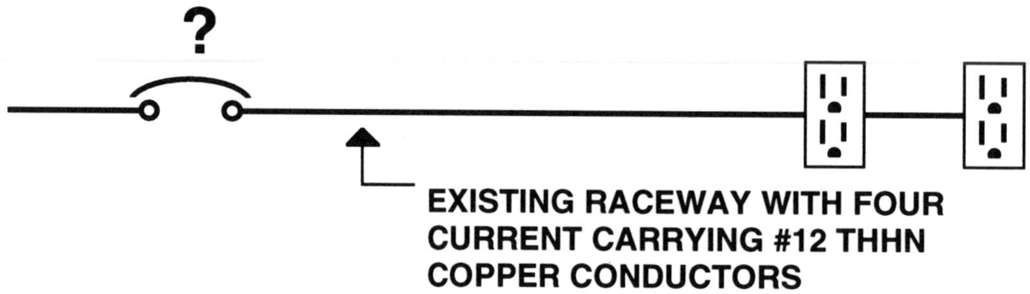

(a) 15 amp (b) 20 amp (c) 25 amp (d) 30 amp

9. The 120 volt branch circuit is supplying 8 lights at 2 amps each which are continuously loaded. What size breaker is required?

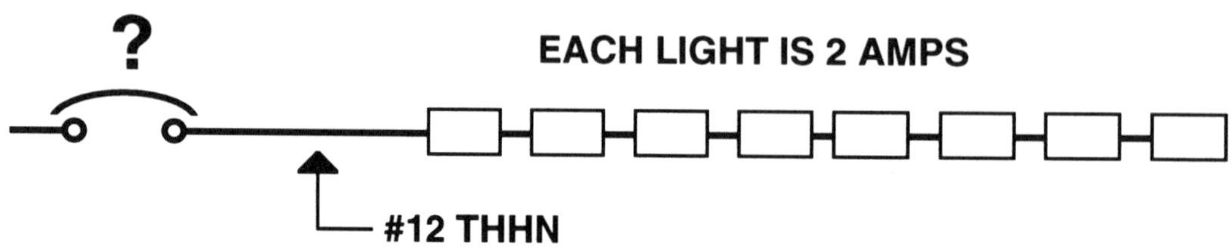

(a) 15 amp (b) 20 amp (c) 25 amp (d) 30 amp

10. The 120 volt branch circuit is supplying receptacles. The equipment is 60°C. There are 8 current carrying wires in the conduit. What size THHN conductor is required if you wanted to load the circuit 16 amps?

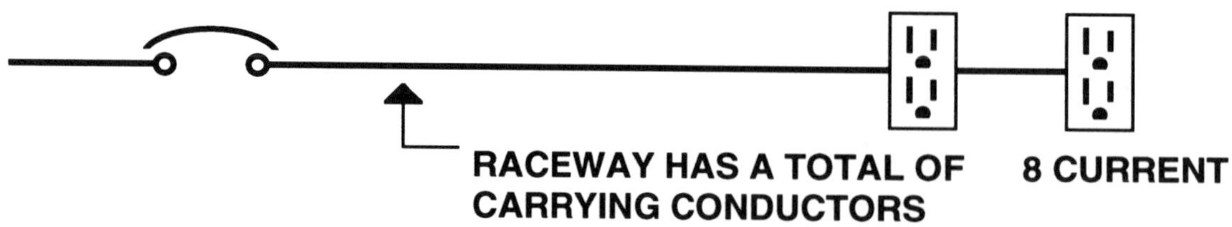

(a) #14 (b) #12 (c) #10 (d) #8

11. The 120 volt branch circuit is supplying receptacles. The equipment is 60°C. There are 8 current carrying wires in the conduit. What size THHN conductor is required if you wanted to load the circuit *20 amps*?

(a) #14 (b) #12 (c) #10 (d) #8

12. The 120 volt branch circuit is supplying 8 lights at 2 amps each. The ambient temperature is 114°F. What THHN conductors are required?

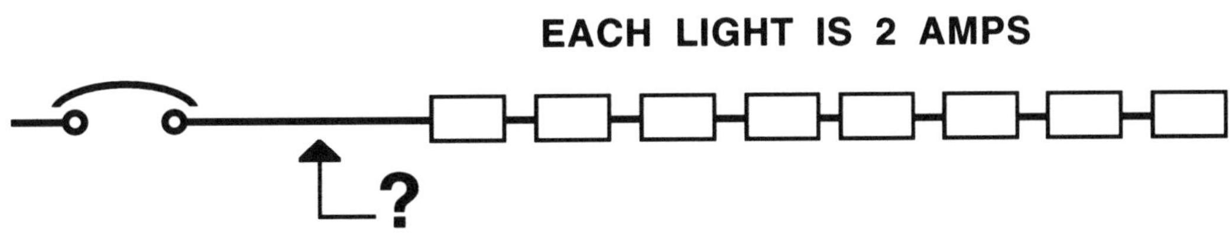

(a) #14 (b) #12 (c) #10 (d) #8

13. The 120 volt branch circuit is supplying 8 lights at 2 amps. The ambient temperature is 114°F with eight current carrying conductors in the conduit. What THHN conductors are required?

(a) #14 (b) #12 (c) #10 (d) #8

14. A dimmer panel has 24 - 20 amp 120 volt circuits continuously loaded. Using 60°C rated equipment what size THHN does the Code require?

(a) #6 (b) #8 (c) #10 (d) #12

15. A 1" conduit has 29 - #12 THHN conductors carrying current. What is the maximum load permitted on these conductors?

(a) 30 amps (b) 25 amps (c) 20 amps (d) 13.5 amps

16. The 120 volt branch circuit is supplying receptacles. The equipment is 60°C. There are 12 current carrying conductors and the ambient temperature is 122°F. What size THHN is required if you wanted to load the circuit to 20 amps?

(a) #14 (b) #12 (c) #10 (d) #8

DESIGN QUIZ #5

•The following questions are to be answered after reviewing pages 76 - 106.

1. Which of the following receptacle(s) in a commercial building would calculate at 180 volt-amps?

I.  II.  III. 

(a) I only (b) II only (c) I and II only (d) I, II and III

2. In an office building the maximum number of receptacles permitted on a 15 amp circuit is ____.

(a) 6 (b) 10 (c) 12 (d) 13

3. What is the maximum distance the Code recommends for a 15 amp load on a 15 amp, 120v branch circuit using #14 THHN stranded conductors?

(a) 38 feet (b) 51 feet (c) 76 feet (d) 100 feet

4. If a #12 THHN stranded conductor was installed in Question #3, how far could the branch be run?

(a) 38 feet (b) 47 feet (c) 60 feet (d) 100 feet

5. If a #10 THHN stranded conductor was installed in Question #3, what would be the voltage drop of the branch circuit if it was installed 35 feet in distance?

(a) 3.6 volts (b) 3.45 volts (c) 2 volts (d) 1.3 volts

6. What is the voltage drop on the branch circuit in Question #3 for 35 feet?

(a) 3.297 (b) 3.45 (c) 3.6 (d) 4.14

7. The power loss for Question #5 is ____ watts.

(a) 19.5 (b) 30 (c) 51.75 (d) 54

8. The power loss for Question #6 is ____ watts.

(a) 62.1 (b) 51.75 (c) 49.4 (d) 54

9. The cost per day for power loss for the #10 wire would be ____ for an 8 hour day @ .08 cents per kwh.

(a) 1.2¢ (b) 1.9¢ (c) 3.3¢ (d) 3.45¢

10. The cost per day for power loss for the #14 wire would be ____ for an 8 hour day @ .08 cents per kwh.

(a) 3.16¢ (b) 3.31¢ (c) 3.45¢ (d) 3.97¢

11. What size wire is required for the 15 amp, 120 volt load if it was run 100 feet in distance and limited to 2% voltage drop?

(a) #14 (b) #12 (c) #10 (d) #8

12. What size wire is required for the 15 amp, 120 volt load if it was run 100 feet in distance and to the Code recommendation *3%* voltage drop?

(a) #14 (b) #12 (c) #10 (d) #8

13. What is the maximum amperage the Code permits on a 120 volt receptacle branch circuit using #12 THHN stranded wire and the total distance is 100 feet?

(a) 25 amps (b) 20 amps (c) 15 amps (d) 9 amps

14. What is the maximum amperage the Code permits on a 120 volt receptacle branch circuit using *#10* THHN stranded wire and the total distance is 100 feet?

(a) 12 amps (b) 14.5 amps (c) 16 amps (d) 24 amps

15. If the electric heat unit was rated 10kw @240 volt single-phase and was connected to a 208 volt single-phase system, the kw output would be approximately ____.

(a) 10 (b) 7.5 (c) 12 (d) 11.5

16. What is the ampacity of a #12 romex (NMB)?

(a) 20 (b) 25 (c) 30 (d) 40

17. What is the maximum size circuit breaker the Code allows for a #12 romex supplying receptacles?

(a) 15 (b) 20 (c) 25 (d) 30

18. At least one ____ shall be installed for the laundry in a dwelling.

(a) duplex receptacle (b) 240/120v receptacle (c) single receptacle (d) 30 amp receptacle

19. The laundry outlet required is for the ____.

(a) clothes dryer (b) dishwasher (c) washing machine (d) iron

20. The *rating* of a branch circuit is determined by the ____ size.

I. fuse II. breaker III. wire

(a) I only (b) I and II only (c) III only (d) II only

DESIGN QUIZ #6

•The following questions are to be answered after reviewing pages 76 - 106.

1. If the fastened in place air conditioner is plugged into a 20 amp rated circuit, the maximum current rating of the air conditioner is ____ amps.

(a) 8 (b) 10 (c) 12 (d) 16

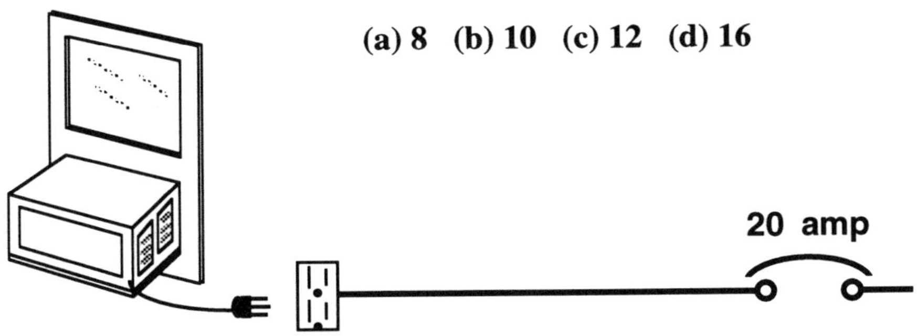

2. What minimum size romex (NMB) is required by the Code to supply a branch circuit to the 3 ton air conditioner shown below?

(a) #10 (b) #8 (c) #6 (d) #4

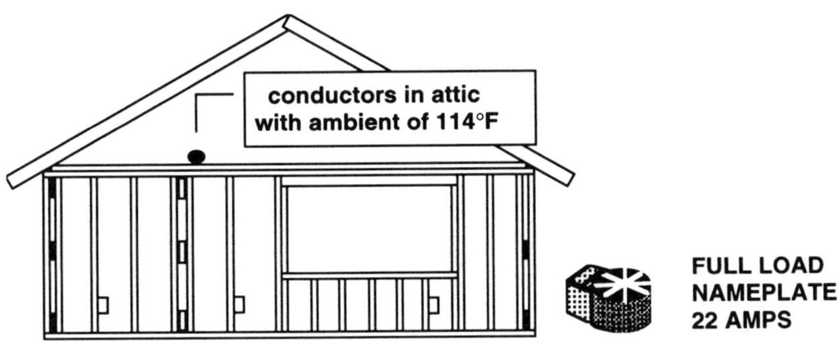

3. The romex conductors supplying the branch circuit to a clothes dryer which has a nameplate rating of 21 amps are installed through the attic which reaches an ambient temperature of 115°F. What is the minimum size romex conductors the Code would allow?

(a) #10 (b) #8 (c) #6 (d) #4

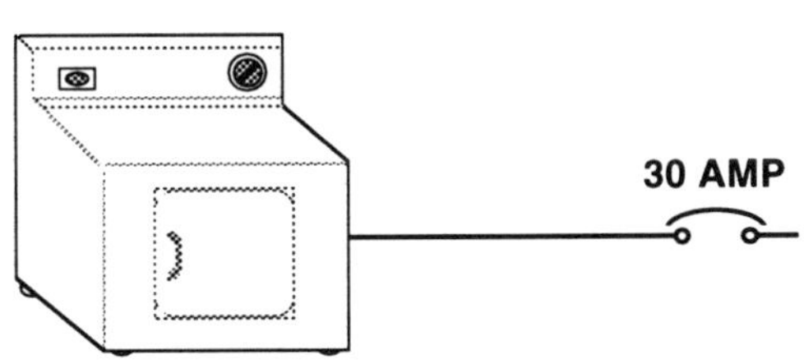

4. What is the voltage drop on a branch circuit using #10 solid conductors that supplies a 240 volt air conditioner with a 22 amp full current rating and the circuit is 100' in distance?

(a) 3% (b) 7.2 volts (c) 3.6 volts (d) 5.324 volts

5. The power loss in Question #4 would be ____ watts.

(a) 66 (b) 158 (c) 79 (d) 117

6. The cost of power loss for operating the air conditioner for a 30 day month in Question #4 for eight hours each day and using #10 romex would be ____. Use 8¢ per kwh.

(a) $2.25 (b) $1.52 (c) $3.03 (d) $1.27

7. If a #8 Romex was installed in Question #4 instead of a #10, the cost of power loss would be ____.

(a) $1.42 (b) $2.17 (c) $2.46 (d) $3.18

8. What is the maximum size breaker for a branch circuit supplying a hermetic refrigerant motor-compressor that draws a full-load current of 25 amps?

(a) 40 (b) 45 (c) 50 (d) 60

9. What size breaker would be installed first to see if it would start the motor-compressor?

(a) 40 (b) 45 (c) 50 (d) 60

10. Each motor-compressor shall be protected against overload and failures by ____.

I. an integral (built-in) thermal protector
II. a fuse or circuit breaker rated at not more than 125% of the motor-compressor load
III. a separate overload relay selected to trip at not more than 140% of the motor-compressor load

(a) I only (b) II only (c) III only (d) I, II and III

11. Which of the following statements is *false*?

(a) The breaker provides ground fault protection.
(b) The overload relay can withstand repeated trip cycles without having to be replaced.
(c) The overload relay provides ground fault protection.
(d) The overload relay holds during the starting period.

12. "Within sight from" is defined as equipment that is visible and not more than ___ feet from the other equipment.

(a) 50 (b) 75 (c) 100 (d) 150

13. Circuit breakers used as switches shall be installed so the center of the grip of the operating handle, when in its highest position, will not be more than ____ above the floor.

(a) 8' (b) 8' 6" (c) 6' 6" (d) 7'

14. HACR breakers are ____.

I. suitable for refrigeration equipment comprised of multimotor loads
II. of the instantaneous type and listed for group installation on branch circuits
III. limited to installations where equipment is marked

(a) I and III only (b) I only (c) I and II only (d) I, II and III

15. The wiring of an outdoor sign that is portable or mobile and is readily accessible shall be provided with factory installed GFCI.

(a) true (b) false

DESIGN QUIZ #7

•The following questions are to be answered after reviewing pages 107 - 147.

1. Which of the following statements is *false*?

(a) A higher service factor provides temporary extra power when needed by the motor.
(b) Selecting a motor that is too large for the load is good designing.
(c) When a motor is underloaded it is not efficient.
(d) It's best not to use a motor above its rated horsepower.

2. An elevator motor is an example of a continuous duty rated motor.

(a) true (b) false

3. A "code letter" on a motor indicates ____.

(a) the wiring must be installed to the present Code as marked
(b) the motor is continuous rated
(c) the approximate kva consumed per hp in a locked rotor condition
(d) the motor is equipped with overcurrent protection

4. Which of the following statements is *false*?

(a) The smaller the motor, the higher the locked-rotor current per horsepower.
(b) The ratio of starting current to running current could be 600% or more.
(c) Motors are designed to operate as much as 10% over their rated voltage.
(d) Any motor over 1/4 hp or larger has on its nameplate a "code letter".

5. The motor shown below will draw approximately ____ amps on start.

(a) 27 (b) 35 (c) 44 (d) 61

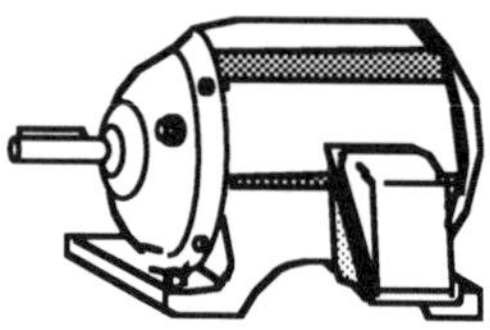

MOTOR NAMEPLATE
1 hp 1ø 240v
CODE LETTER M

6. Which of the following statements is *false*?

(a) Autotransformer starting is used generally in the larger hp motors.
(b) The instantaneous trip breaker is the regular circuit breaker.
(c) A wound rotor motor has brushes.
(d) Most everyday motors are full voltage start.

7. Which of the following does the Code recognize as a standard size fuse?

(a) 601 amps (b) 85 amps (c) 75 amps (d) 650 amps

8. A 230v, 5hp, 3ø motor has a full-load current rating of 15.2 amperes. Which of the following is the *maximum* allowable standard rating of a dual element (time-delay) fuse used to provide short-circuit and ground-fault protection?

(a) 15 amps (b) 20 amps (c) 25 amps (d) 30 amps

9. What size inverse time circuit breaker would be selected for the motor shown below?

(a) 20 amp (b) 30 amp (c) 40 amp (d) 50 amp

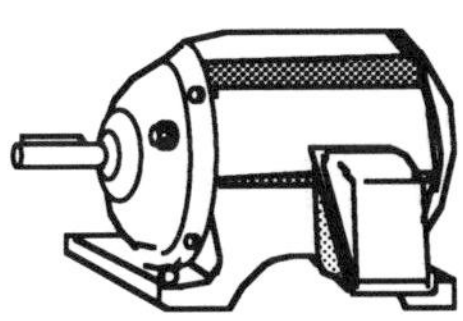

MOTOR NAMEPLATE
1 1/2 hp 1ø 115v

10. Which of the following statements is *false*?

(a) The code letter marked on the motor nameplate is an important part of designing.
(b) A knife blade fuseholder is for 60 amps and over.
(c) A cartridge fuseholder is for up to 60 amps.
(d) Fuseholders must be sized to accommodate the largest size fuse permitted.

11. Overloads can ____.

I. come from a bearing on a shaft freezing up
II. be electrical as well as mechanical
III. come from low voltage

(a) I only (b) II only (c) I and II only (d) I, II and III

12. Which of the following statements is *false*?

(a) Thermal overloads can be subdivided into two types; solder pot and bi-metallic.
(b) Overload relays can be classified as being either thermal or magnetic.
(c) The normal rating for most motors is 30°C.
(d) All overloads shorten the life of a motor by deteriorating the motor insulation.

13. Which of the following statements is *false*?

(a) The difference between winding and ambient temperature is called the service factor.
(b) With an increase of 10°C above the rated temperature the motor life is cut in half.
(c) Temperature rise is over and above the ambient while the motor is delivering its rated hp.
(d) Motors and motor controllers are subject to a 104°F ambient temperature limit.

14. What size overload is required for the motor shown below?

(a) 10 amps (b) 11.5 amps (c) 12.5 amps (d) 13 amps

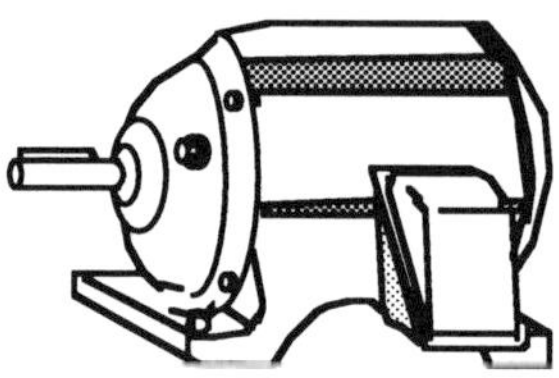

MOTOR NAMEPLATE
1 1/2 hp 1ø 230v
SF 1.15

15. Which of the following statements is *false*?

(a) The Code requires an overload to be installed in each phase of a three phase motor.
(b) The difference between a relay and a motor starter is the relay provides thermal protection.
(c) Manual starters cannot provide automatic operation of motors.
(d) Manual starters usually provide across the line starting.

16. What size standard disconnect would be required for the motor shown below?

(a) 30 amp (b) 50 amp (c) 60 amp (d) 75 amp

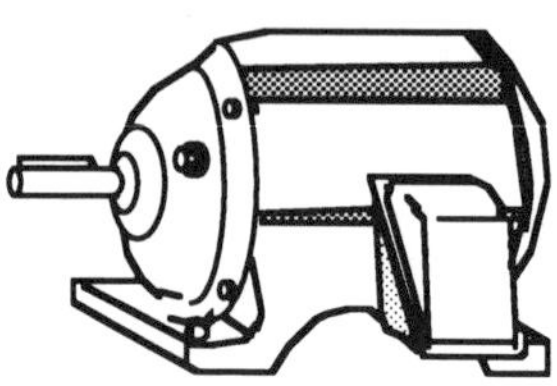

MOTOR NAMEPLATE
10 hp 1ø 230v
F.L.C. 50 amps

DESIGN QUIZ #8

•The following questions are to be answered after reviewing pages 107 - 147.

1. Which of the following statements is *false*?

(a) When starting larger motors voltage drop can cause the motor to stall out.
(b) Sizing the branch circuit conductors 125% larger is for voltage drop.
(c) The † does not apply to the breaker sizing for a motor.
(d) The overload device generally has a rating 25% higher than the F.L.C. of the motor.

2. What is the minimum size THHN conductor permitted for the branch circuit below?

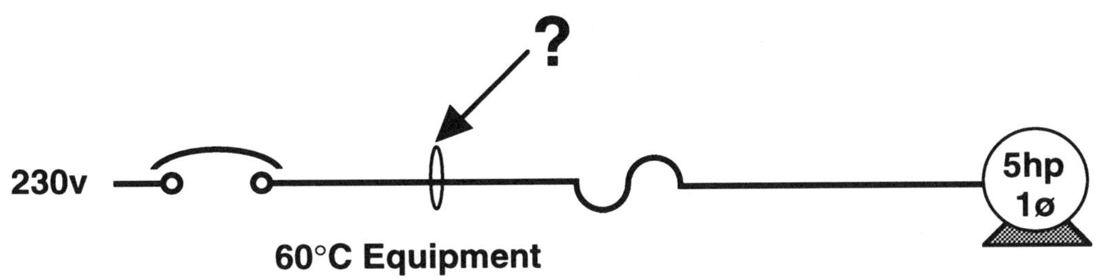

(a) #12 (b) #10 (c) #8 (d) #6

3. The 3ø motor shown below has a F.L.C. of 46.2 amps. How far can this branch circuit run from the source?

(a) 253 feet (b) 219 feet (c) 121 feet (d) 187 feet

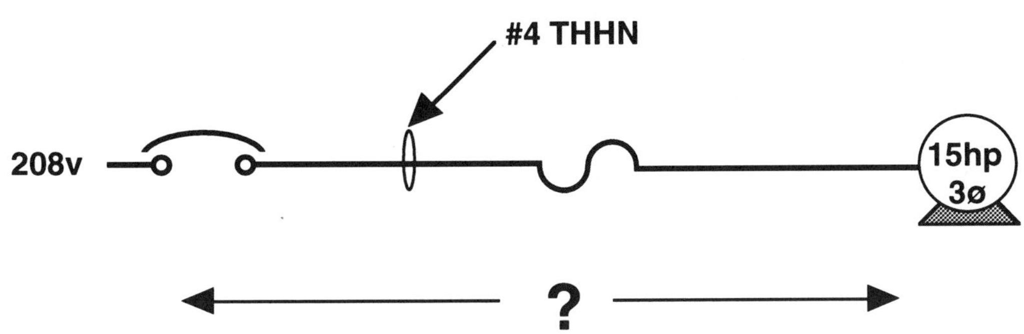

4. The lift station pump shown below would require ___ minimum size THHN conductors.

(a) #14 (b) #12 (c) #10 (d) #8

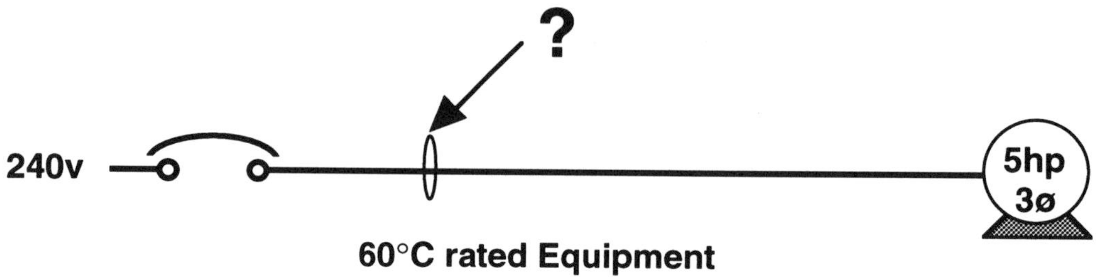

5. Which of the following statements is *false*?

(a) The Code recommendation is a maximum of 3% voltage drop on a branch circuit.
(b) Three-wire well pumps have the advantage of high starting torque motors.
(c) Two-wire well pumps require an above ground control box between switch and motor.
(d) Two-wire well pumps have a higher running torque and are more efficient.

6. Which of the following statements is *false*?

(a) A machine is considered to be an individual load, even if it has several motors.
(b) An industrial crane generally has a motor to move the hook from side to side.
(c) The disconnect to the machine is required to have overcurrent protection.
(d) A machine shall have a disconnecting means.

7. The welder shown below would require ____ THHN conductors.

(a) #10 (b) #8 (c) #6 (d) #4

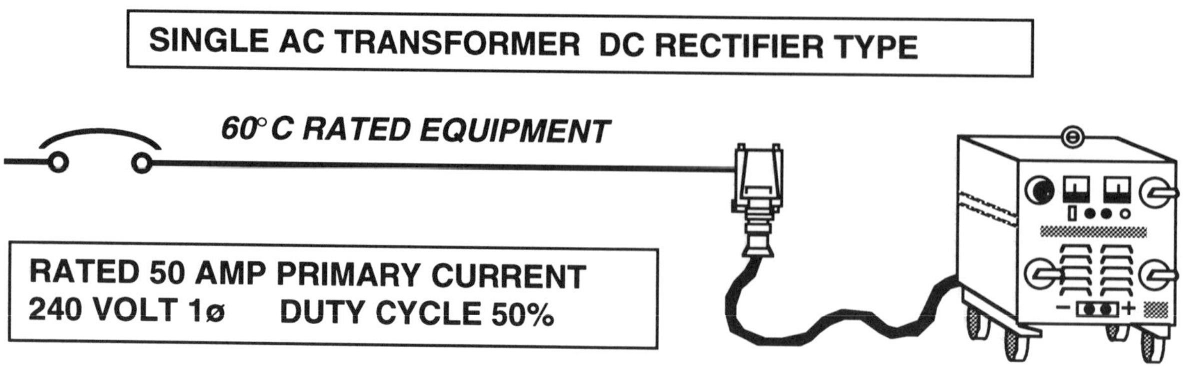

8. The welder in Question #7 would require a ____ circuit breaker.

(a) 50 amp (b) 60 amp (c) 70 amp (d) 100 amp

9. The welder shown below would require ____ THHN conductors.

(a) #10 (b) #8 (c) #6 (d) #4

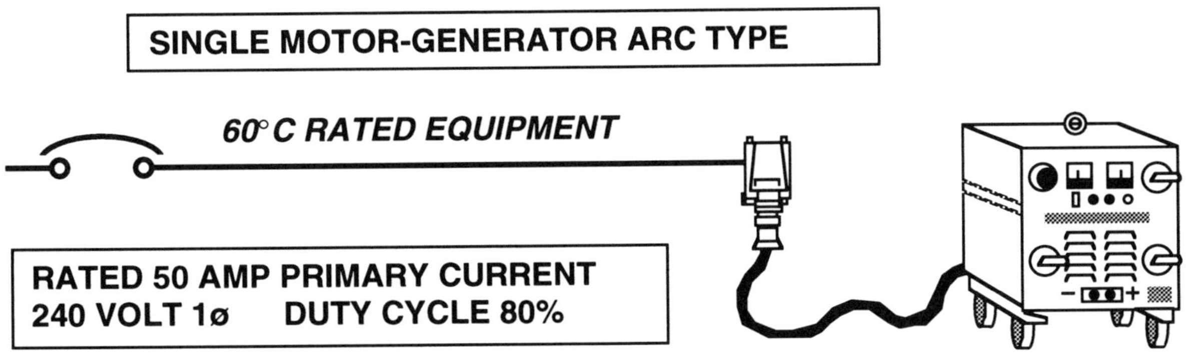

10. The welder in Question #9 would require a ____ circuit breaker.

(a) 50 amp (b) 60 amp (c) 70 amp (d) 100 amp

11. The welder shown below would require ____ THHN conductors.

(a) #10 (b) #8 (c) #6 (d) #4

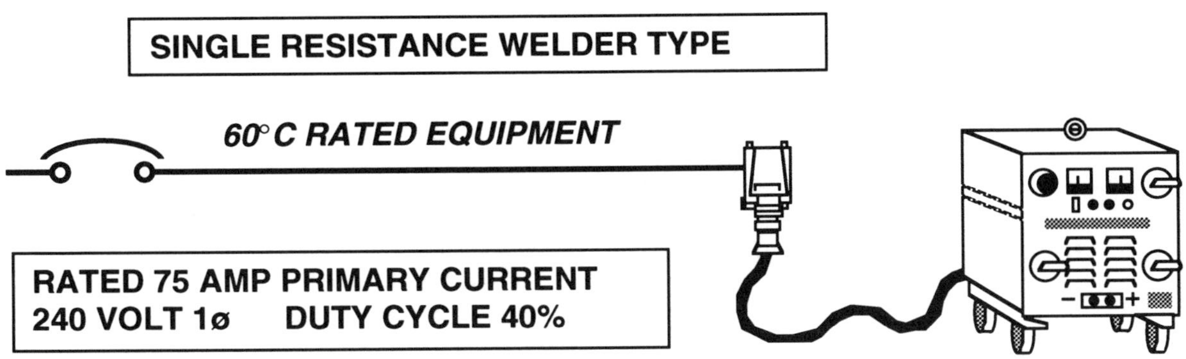

12. The welder in Question #11 would require a ____ circuit breaker.

(a) 750 amp (b) 100 amp (c) 225 amp (d) 250 amp

DESIGN QUIZ #9

•The following questions are to be answered after reviewing pages 148 - 162.

1. A 480 volt rated primary has 24 volts applied which causes secondary full load current to flow through the shorted secondary. The transformer impedance would be ____.

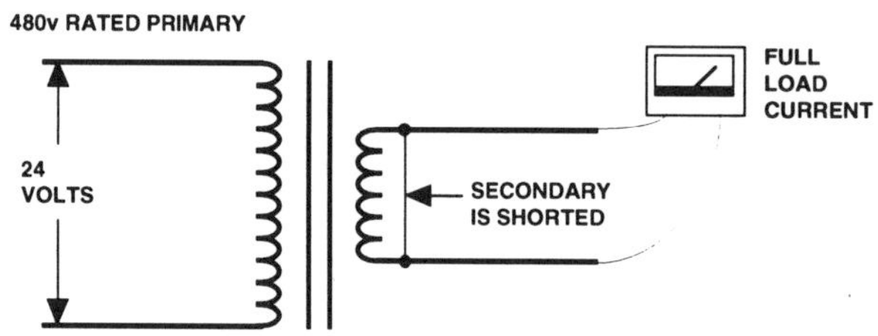

(a) 20% (b) 2% (c) 5% (d) 95%

2. What is the approximate available short-circuit current from the transformer shown below?

(a) 6950 amps (b) 20833 amps (c) 12019 amps (d) 28460 amps

3. A #12 copper wire with 75°C insulation can withstand ____ amperes for 2 cycles.

(a) 1200 (b) 2700 (c) 1900 (d) 3800

4. A #10 copper wire with THHN insulation can withstand ____ amperes for 1/2 cycle.

(a) 5147 (b) 5785 (c) 11570 (d) 8181

5. A bare copper conductor melts at ____ degrees fahrenheit.

(a) 1981 (b) 746 (c) 212 (d) 86

6. A ground fault occurs more often than a short circuit.

(a) true (b) false

7. A fault current may exist for _____.

(a) minutes (b) hours (c) days (d) all of these

8. An overcurrent device should open the circuit within _____ cycle(s) if it has an effective grounding path.

(a) 1/4 (b) 1/2 (c) 1 (d) 3

9. The conductor withstand classification(s) is/are _____.

I. melting II. terminal damage III. insulation damage

(a) I only (b) II only (c) III only (d) I, II and III

DESIGN QUIZ #10

•The following questions are to be answered after reviewing pages 163 - 176.

1. Which of the following statements is *false*?

(a) Many variables must be considered when designing feeder circuits.
(b) The Code permits demand factors to be applied to feeders.
(c) The feeder is the conductors beyond the final overcurrent device.
(d) Table 220-13 permits a demand on receptacle loads over 10 kva.

2. What is the demand on the feeder for 200 receptacles in an office building?

(a) 10 kva (b) 36 kva (c) 18 kva (d) 23 kva

3. Which of the following statements is *false*?

(a) Smaller conductors in cma will produce a greater voltage drop.
(b) More space may be required at equipment terminals when conductors are paralleled.
(c) Larger conductors can carry more amps per cma than smaller conductors.
(d) Larger conductors in cma will produce less voltage drop.

4. What size THHN conductors does the Code require for the motor feeder shown below? 60°C equipment.

(a) #1 (b) #2 (c) #3 (d) #4

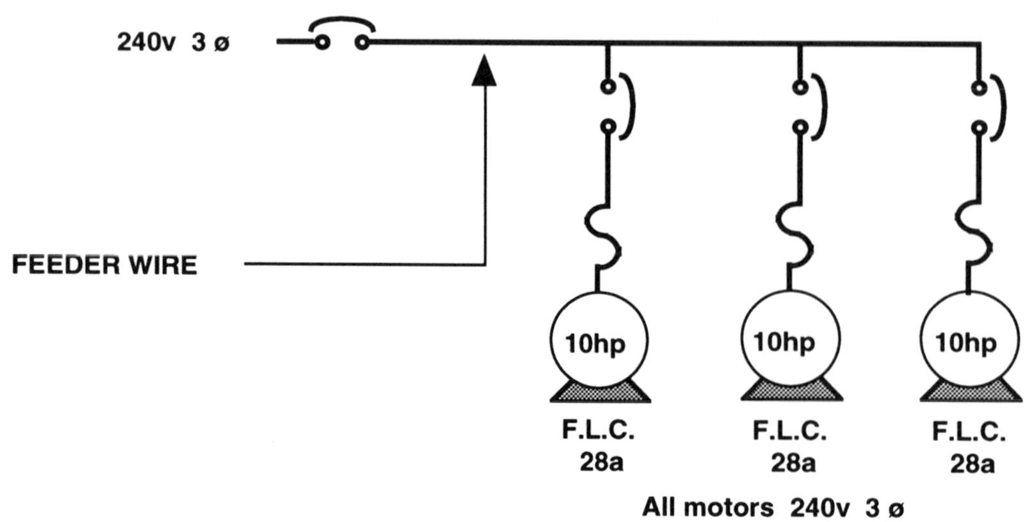

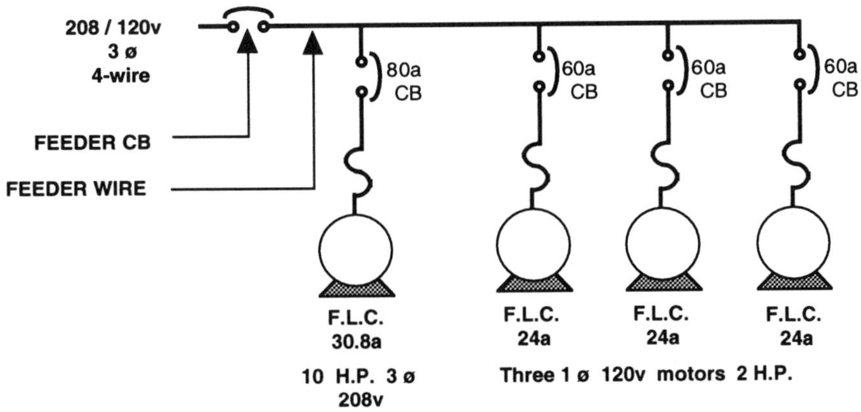

5. What is the minimum size THHN feeder conductors for the motor circuit shown above? 60°C equipment.

(a) #1/0 (b) #1 (c) #4 (d) #6

6. What size inverse-time circuit breaker for the feeder protection?

(a) 100 amp (b) 125 amp (c) 150 amp (d) 175 amp

7. What is the minimum size THHN feeder conductors for the motor circuit shown below? 60°C equipment.

(a) #4/0 (b) #3/0 (c) #2/0 (d) #1/0

8. What size inverse-time circuit breaker for the feeder protection?

(a) 225 amp (b) 250 amp (c) 275 amp (d) 300 amp

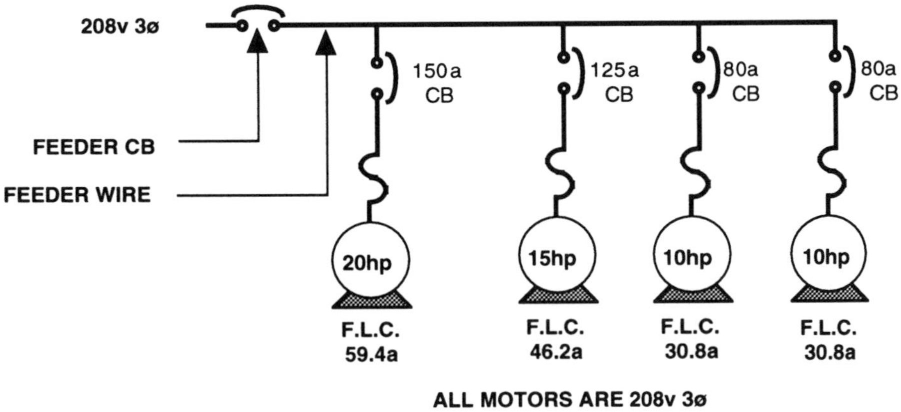

DESIGN QUIZ #11

208/120v three-phase 4-wire
Dual-element time delay fuses
THW insulation
Discharge lighting load

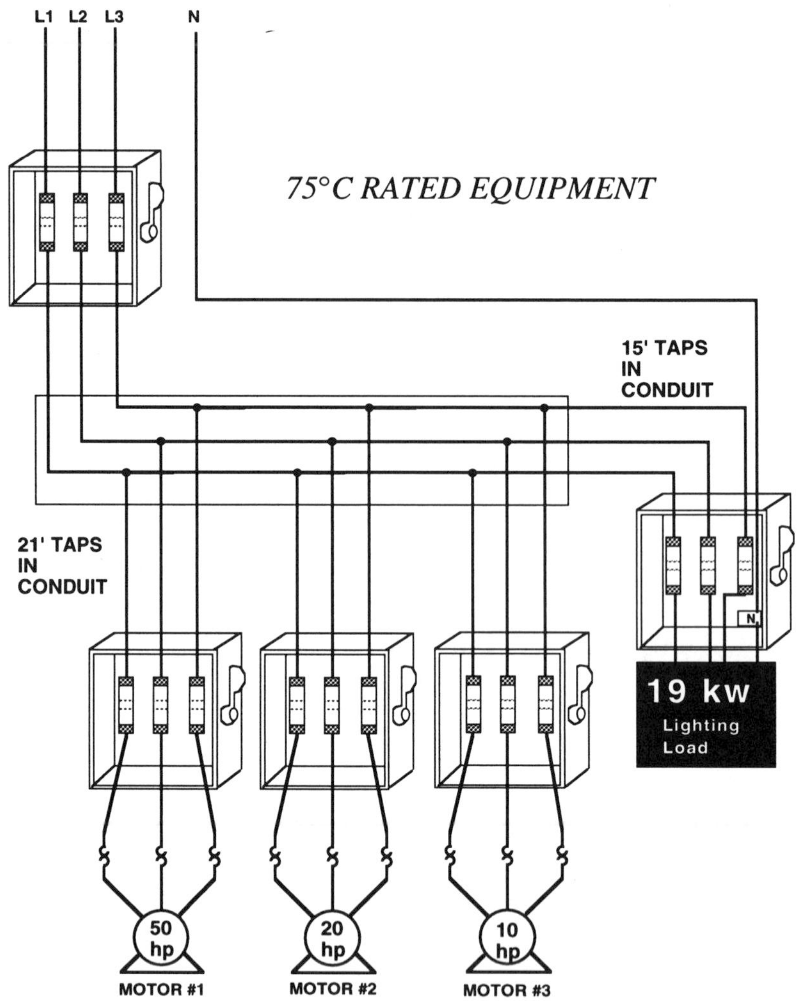

1. **Full load current**:
Motor #1 __________

Motor #2 __________

Motor #3 __________

2. **Size overloads**:
Motor #1 __________

Motor #2 __________

Motor #3 __________

3. **Branch circuit wire size**:
Motor #1 __________

Motor #2 __________

Motor #3 __________

4. **Branch fuse size**:
Motor #1 __________

Motor #2 __________

Motor #3 __________

5. **Disconnect size**:
Motor #1 __________

Motor #2 __________

Motor #3 __________

6. **Tap conductor size**:
Motor #1 __________

Motor #2 __________

Motor #3 __________

7. **Feeder conductor size**: ____________

8. **Feeder fuse size**: ___________

9. **Feeder disconnect size**: __________

10. **Tap conductors size (to lighting load)** : __________

DESIGN QUIZ #12

•The following questions are to be answered after reviewing pages 177 - 219.

1. Which of the following statements is *false*?

(a) The load side consists of those conductors that are tapped or spliced to the service drop.
(b) The number of lighting branch circuits is computed by using the connected load.
(c) The panelboard busbars must have their load-current rating based on the computed load.
(d) A panelboard with a main breaker can change the size of the wire required.

2. Which is the high-leg conductor on the 3ø delta 4-wire shown below?

(a) L1 (b) L2 (c) L3 (d) N

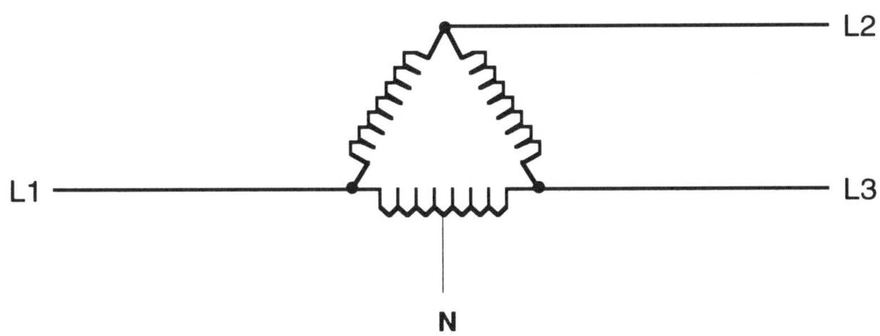

3. Which of the following statements is *false*?

(a) A panelboard marked "suitable only for service equipment" has the neutral permanently bonded to the metal enclosure.
(b) A panelboard marked "suitable for use as service equipment" has the neutral permanently bonded to the metal enclosure.
(c) A loadcenter is a light-duty type panelboard.
(d) The panelboard size is determined by the calculated load it will supply.

4. Which of the following statements is *false*?

(a) A lighting and appliance panelboard shall be protected by not more than two mains.
(b) A power panel could be classified as a lighting and appliance panelboard.
(c) A switchboard is a smaller panelboard.
(d) The overcurrent protection cannot be larger than the rating of the panelboard.

5. The UL designation for circuit limiting is ____.

(a) CTL (b) ckt (c) AIC (d) CLC

6. Each service is permitted up to ____ disconnects.

(a) 2 (b) 4 (c) 6 (d) 8

7. The definition of a lighting and appliance panelboard is one that has more than 10% of its overcurrent devices rated ____ amps or less with neutral connections.

(a) 15 (b) 20 (c) 25 (d) 30

8. Which of the following statements is *false*?

(a) The Code permits more than one set of service entrance conductors to be tapped from a single service drop or lateral.
(b) The general rule is that a building shall be supplied by one service.
(c) The best situation is a main disconnect.
(d) The Code allows eight movements of the hand to disconnect the service.

9. Which of the following statements is *false*?

(a) A separate garage with a panel can be disconnected from a panel at the house.
(b) A billboard is an example of "other structure".
(c) The service equipment is the only place where all the grounding is connected together.
(d) A ground rod is a made electrode.

10. Which of the following statements is *false*?

(a) The feeder requires an equipment grounding conductor.
(b) Only at the service equipment is the neutral bus bonded to the metal enclosure.
(c) Livestock buildings require special grounding in some cases.
(d) The main disconnect must be on the outside of a building.

11. In a 2-wire, 120 volt circuit the white wire is called the ____ conductor.

(a) grounded (b) grounding (c) ungrounded (d) neutral

12. The maximum load the neutral would be required to carry is ____ amps.

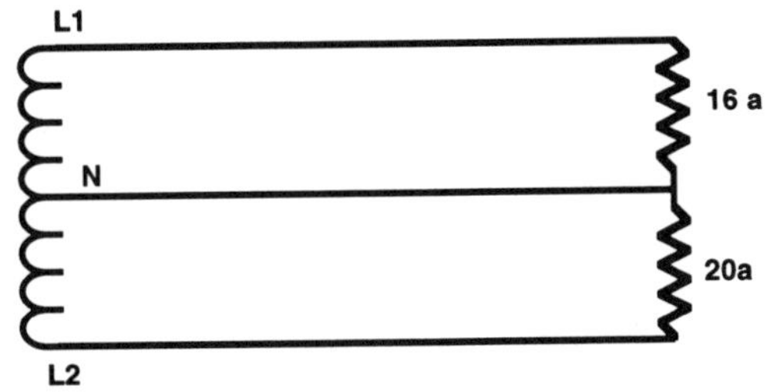

(a) zero (b) 4 (c) 16 (d) 20

13. The neutral on a branch circuit to a household clothes dryer is permitted a demand of ____%.

(a) 70% (b) 75% (c) 80% (d) there is no demand

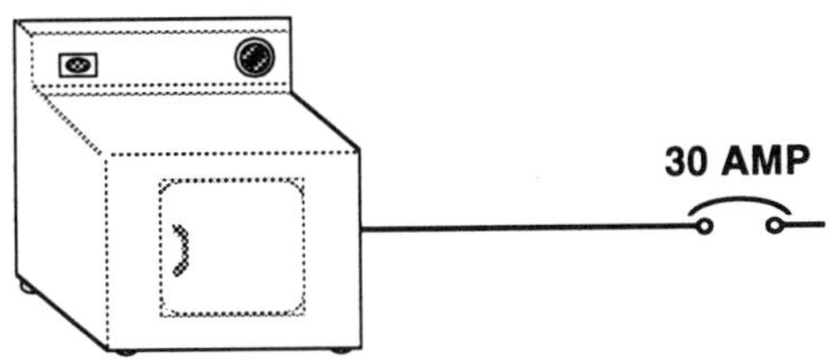

14. Examples of non-linear loads which contain harmonics are ____.

I. ballasts II. incandescent bulbs III. computers

(a) I only (b) II only (c) I and III only (d) I, II and III

15. Which of the following statements is *false*?

(a) An oversized neutral is not necessary for ballasts with 3rd harmonics.
(b) 400 Hz is very nearly the 7th harmonic of a 60 Hz system.
(c) An ammeter that reads average-rms is a reliable instrument for evaluating harmonics.
(d) Heat is increased by an increase in current frequency.

DESIGN QUIZ #13

•The following questions are to be answered after reviewing pages 220 - 247.

1. Which of the following statements is *false*?

(a) The bare conductor in SE cable is called the grounded conductor.
(b) The bare conductor in NM cable is called the grounded conductor.
(c) A screw-type bond is not adequate to pass high currents.
(d) An equipment grounding conductor must be installed in PVC raceway.

2. Which of the following statements is *false*?

(a) The equipment grounding conductor is designed to carry current continuously.
(b) The Code places no restriction on the size or length of metal conduit used for grounding.
(c) The effective grounding path must be permanent and continuous.
(d) The equipment grounding conductor must provide a low impedance path.

3. The minimum size equipment grounding conductor for a piece of electrical equipment protected by a 150 amp breaker would be a copper conductor size ____ AWG.

(a) 8 (b) 6 (c) 4 (d) 2

4. What size copper equipment grounding conductor is required for the motor circuit shown below?

(a) #8 (b) #6 (c) #4 (d) #2

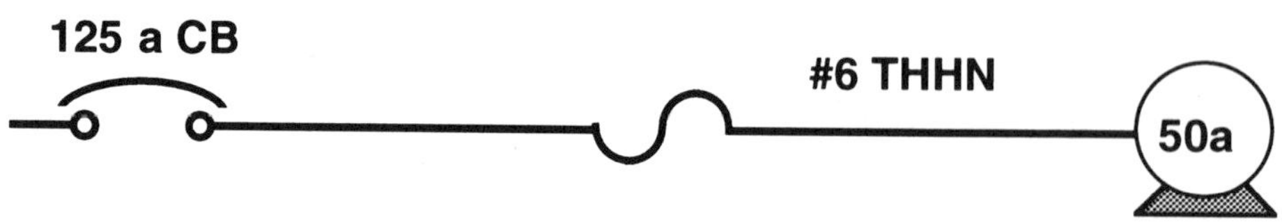

5. Which of the following statements is *false*?

(a) A ground rod cannot replace an equipment ground.
(b) Earth cannot be used as a return path for fault currents.
(c) An equipment grounding conductor can be bare or insulated.
(d) The #1/0 minimum for paralleling conductors applies to the grounding conductor.

6. Which of the following is a made electrode?

I. UNDERGROUND TANK

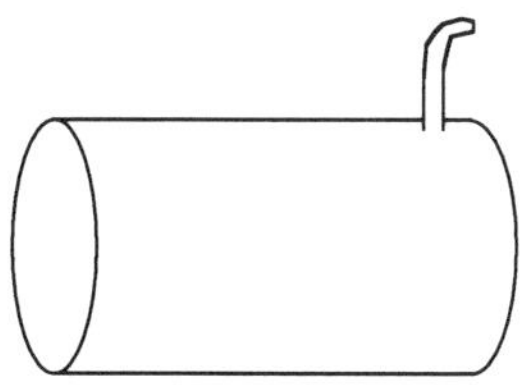

II. GROUND RING

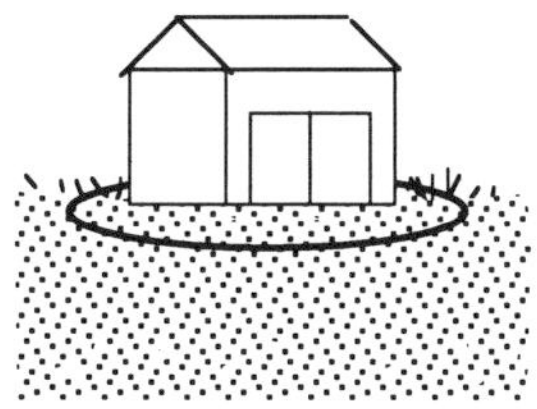

III. PLATE

(a) III only (b) II and III only (c) I and II only (d) I and III only

7. A main bonding jumper may be a ____.

I. bus II. screw III. wire

(a) I only (b) II only (c) III only (d) I, II and III

8. Which of the following statements is *false*?

(a) Load side bonding jumpers are sized from Table 250-95
(b) On the load side standard locknuts can be used on both sides of an enclosure.
(c) Line side bonding jumpers are sized from Table 250-94.
(d) The jumper shall not be less than 12 1/2% of the area of one phase when larger than 1000 kcmil copper.

9. Which of the following is an electrode?

I. ROD OR PIPE

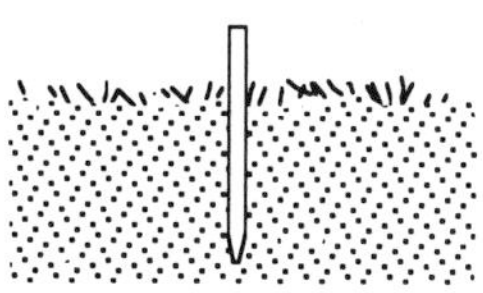

II. OTHER PIPING SYSTEMS

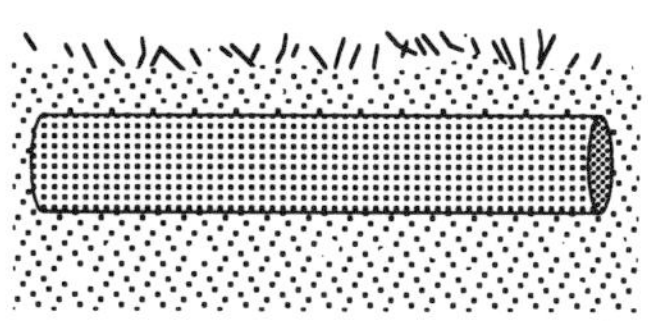

III. GROUND RING

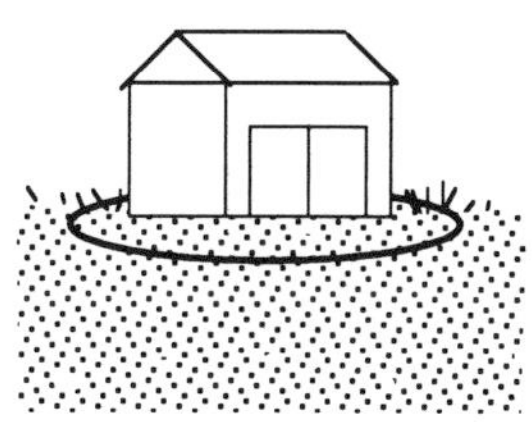

(a) I only (b) II only (c) III only (d) I and III only

10. #600 kcmil copper service conductors would require a # ____ copper grounding electrode conductor.

(a) #1/0 (b) #2/0 (c) #3/0 (d) #4/0

11. The portion of the grounding electrode conductor to a concrete-encased electrode, which is the sole connection, is not required to be larger than ____ copper.

CONCRETE-ENCASED ELECTRODE

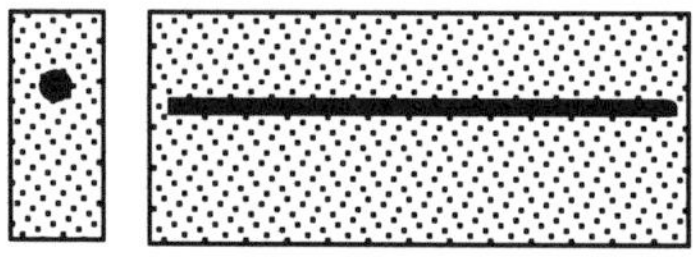

(a) #6 (b) #4 (c) #3 (d) #2

12. A ground ring consists of a minimum #2 bare copper conductor at least 20 feet in length and buried at least _____.

(a) 12" (b) 18" (c) 2' (d) 30"

13. Which of the following is the only electrode that must be supplemented with another electrode?

I. UNDERGROUND WATER PIPE

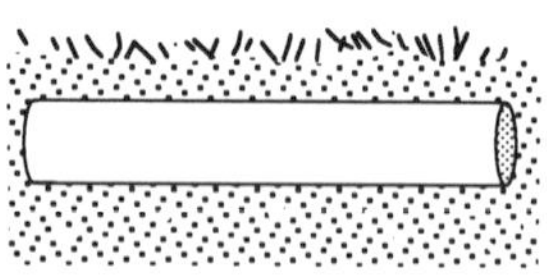

II. CONCRETE-ENCASED ELECTRODE

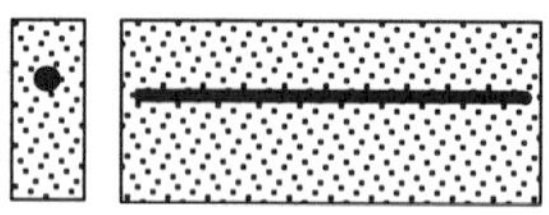

III. METAL FRAME OF A BUILDING

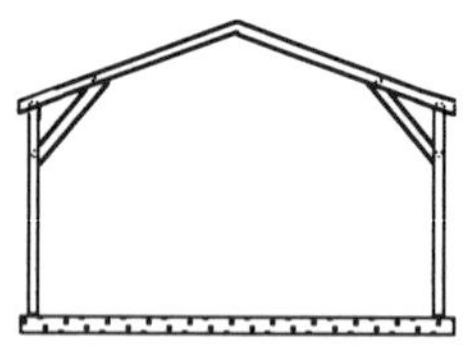

IV. GROUND RING

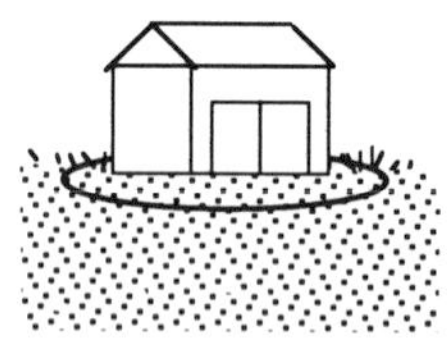

(a) I only (b) II only (c) III only (d) IV only

14. An industrial machine is fed from a 60 amp fuse using a copper equipment grounding conductor. What size copper equipment grounding conductor is required?

(a) #12 (b) #10 (c) #8 (d) #6

15. A service with #2 THW copper phase conductors would require what size copper equipment bonding jumper for bonding around concentric rings?

(a) #10 (b) #8 (c) #6 (d) #4

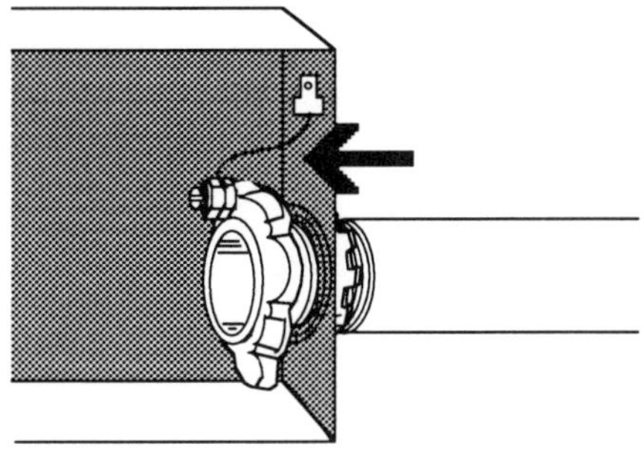

16. Two - #400 kcmil THW copper conductors are paralleled per phase. The overcurrent device protecting the circuit is 500 amperes. What size copper equipment grounding conductor is required for each PVC conduit?

(a) #1/0 (b) #1 (c) #2 (d) #4

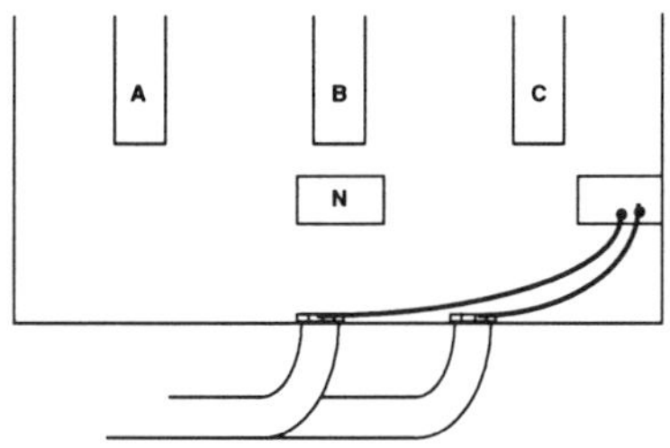

17. The service entrance contains 3 - #500 kcmil copper conductors paralleled per phase. What size copper main bonding jumper is required?

(a) #4/0 (b) #3/0 (c) #250 kcmil (d) #300 kcmil

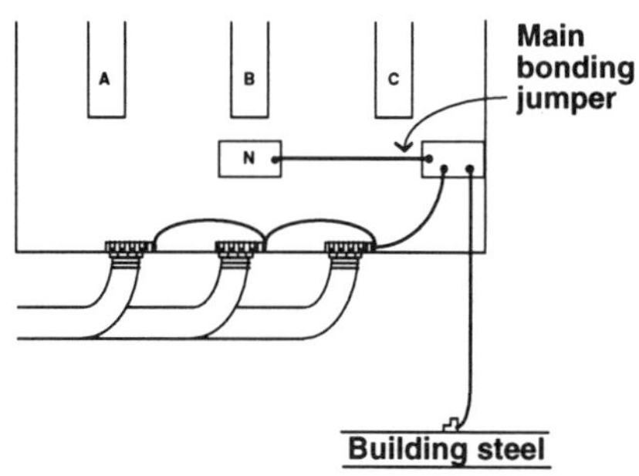

18. The service entrance contains 3 - #500 kcmil copper conductors paralleled per phase. What size copper grounding electrode conductor is required?

(a) #4/0 (b) #3/0 (c) #1/0 (d) #2/0

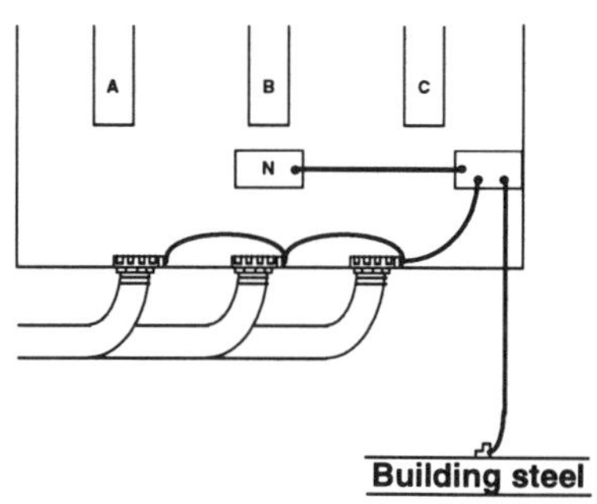

19. A 200 amp service has #3/0 THW copper service entrance conductors. What size copper bonding jumper is required to bond around the water meter?

(a) #2 (b) #4 (c) #6 (d) #8

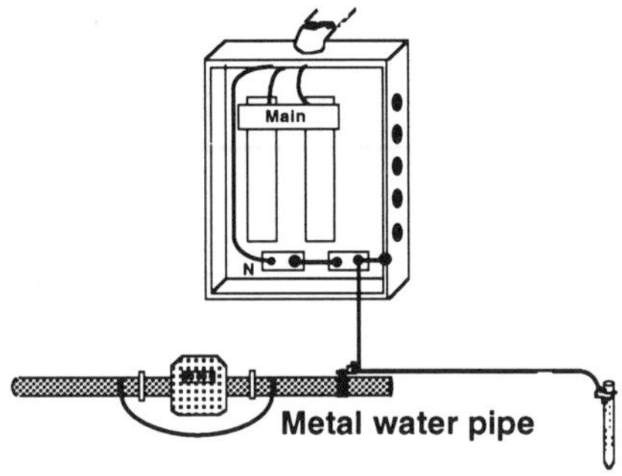

20. Each conduit contains 4 - #500 kcmil copper service conductors A-B-C-N. The largest phase conductor in each conduit is #500 kcmil. Each conduit has an individual bonding jumper connecting to the grounding bus. What size copper bonding jumper is required for each conduit?

(a) #1/0 (b) #2/0 (c) #1 (d) #2

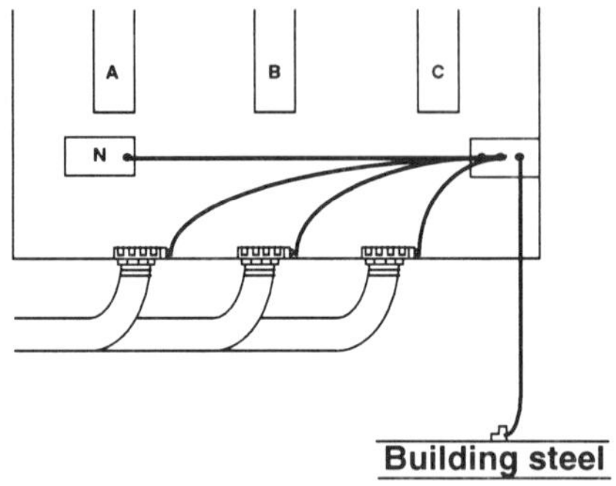

21. A service has 2 - #750 kcmil copper conductors paralleled per phase. What size copper main bonding jumper is required?

(a) #1/0 (b) #2/0 (c) #3/0 (d) #4/0

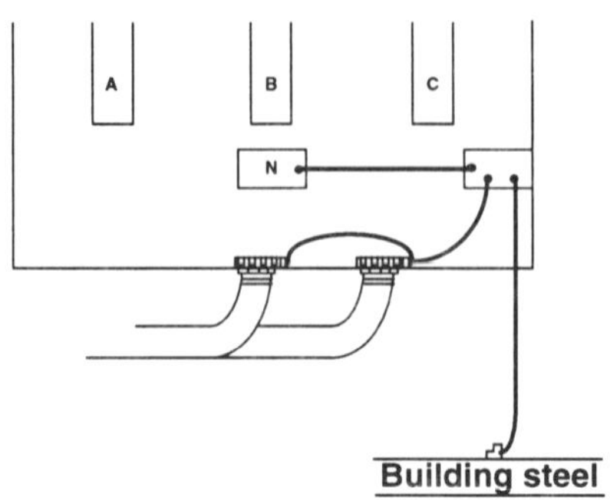

22. A 1 1/4" PVC conduit contains 10 - 20 amp lighting circuits. Each single-phase lighting circuit is protected with a 20 amp circuit breaker. What size **single** copper equipment grounding conductor can be installed in the PVC conduit?

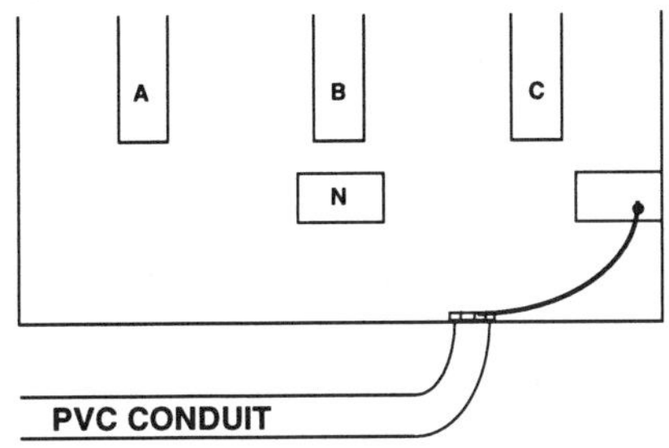

(a) #6 (b) #8 (c) #10 (d) #12

23. The service supply transformer has a grounded neutral, even though the distribution on the load side has all three-phase loads, no neutral loads, the Code still requires a neutral conductor to be installed from the transformer to the service panel. The service conductors are #500 kcmil THW copper conductors per phase. The main circuit breaker is 350 amperes. What size copper neutral conductor is required by the Code?

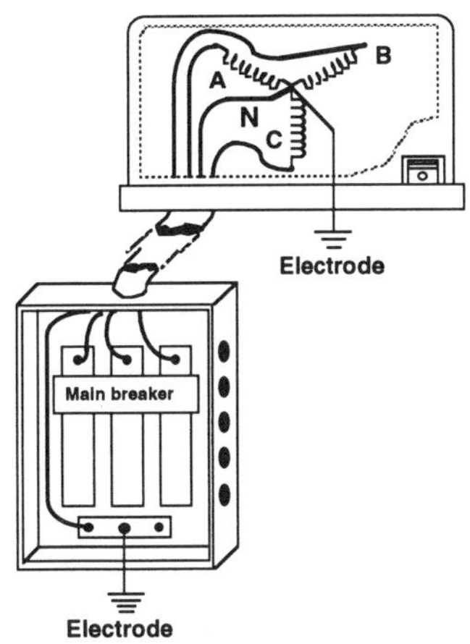

(a) #1/0 (b) #2/0 (c) #3/0 (d) #4/0

24. A feeder to a 200 amp distribution panel had the #3/0 conductors increased to #4/0 THW copper to allow for voltage drop. The larger feeder conductors are protected by a 200 amp circuit breaker. What size copper equipment grounding conductor is required for the PVC conduit?

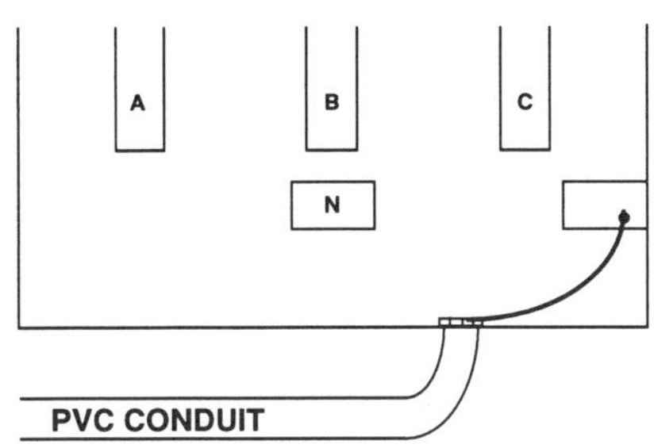

(a) #2 (b) #4 (c) #6 (d) #8

DESIGN QUIZ #14

•The following questions are to be answered after reviewing pages 248 - 265.

1. A single-phase transformer with a 240/120v secondary has the following loads:

2 - 2500 watt 1ø 240v
1 - 1200 watt 1ø 120v
1 - 3000 watt 1ø 240v
4 - 1500 watt 1ø 120v

With the loads balanced as closely as possible, what would be the maximum neutral current?

(a) 35 amps (b) 25 amps (c) 57.9 amps (d) 66.08 amps

2. In the sketch shown below, with all loads turned on, the current flowing in the neutral would be _____ amps.

(a) 162.5 (b) 85.75 (c) 76.75 (d) 9

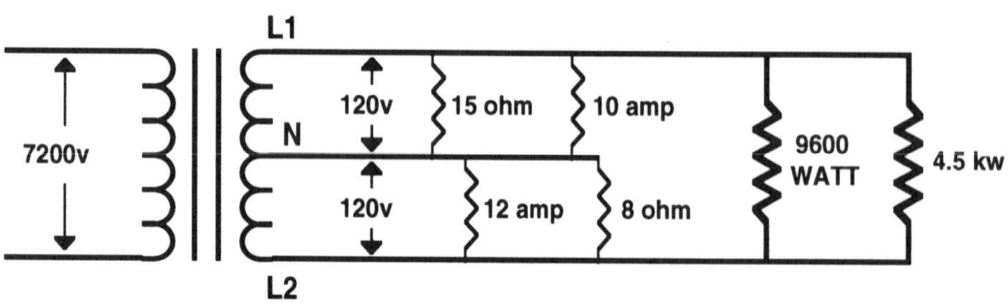

3. In the sketch shown above, the maximum neutral current would be _____ amps.

(a) 9 (b) 162.5 (c) 85.75 (d) 27

4. A single-phase transformer 480v primary, 240/120v secondary with a center tap neutral. The transformer is rated 48 kva, a 240v load is 24 kva, a balanced 120v load would need a neutral of _____ amps to carry the maximum neutral current.

(a) 50 (b) 100 (c) 150 (d) 200

5. The following loads are connected to a single-phase transformer 240/120v secondary with a center-tap neutral:

1 - 5000w water heater 240v
4 - 1200w lighting loads 120v
1 - 6000w space heat 240v
5 - 1200w appliance loads 120v

With all loads turned on and balanced as closely as possible, what is the current flowing in the neutral?

(a) 10 amps (b) 181.6 amps (c) 90.8 amps (d) 100 amps

6. If the total load on a center-tapped 240/120v transformer is 100kw, and the 120v loads were balanced and the 240v load on this phase was 20kw, the maximum neutral current, under any condition, would be approximately ____ amps.

(a) 0 (b) 192 (c) 333 (d) 667

7. A three-phase 4-wire wye connected transformer has a current of 25 amps on L1, 20 amps on L2, and 15 amps on L3. These loads are line to neutral. What is the current flowing in the neutral?

(a) 8.66 amp (b) 20 amp (c) 30 amp (d) 50 amp

8. A three-phase delta-delta high-leg transformer has a primary of 480 volts and a secondary of 240/120v. One of the loads consists of 240 kva 1 ø 120v balanced, the maximum current in the primary of the neutral transformer due to this load would be ____ amperes.

(a) 1000 (b) 2000 (c) 500 (d) 250

9. If the light loads are connected 3000w, 3000w, and 1500w per phase A-B-C, what is the maximum neutral load?

(a) 4500w (b) 3000w (c) 1500w (d) 7500w

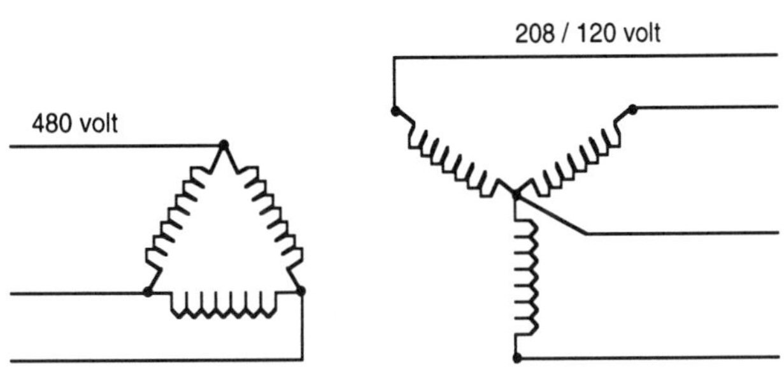

LOADS:

2 - 10 H.P. 208v 3 ø motors
4 - 4000w 208v 1 ø heaters
5 - 1500w 120v 1 ø light circuits

10. In the sketch below, with the 120v light banks connected 6000w, 3000w, and 3000w per phases A-B-C, what is the maximum neutral current in amps?

(a) 25 (b) 50 (c) 75 (d) 100

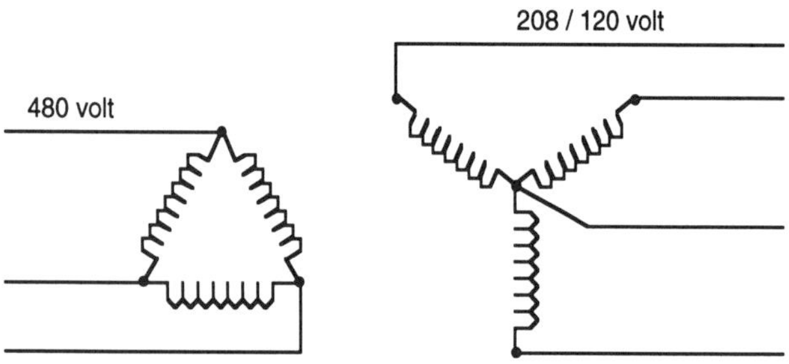

LOADS:

1 - 15 H.P. 3 ø 208v motor
4 - 3000w light banks 120v
5 - 2500w heat strips 208v 1 ø
2 - 1500w dryer 208v 1 ø
1 - 800w cook unit 208v 1 ø

11. If the secondary load of a 3 ø 480/208/120v transformer is:

9 - 2000 watts of 120v lighting
9 - 2000 watts of 208v 1 ø loads

The neutral must have a load current rating of at least ____ amps.

(a) 50 (b) 75 (c) 86.5 (d) 150

12. The system is 240/120v three-phase high-leg, the load consists of:

10 - 20 amp branch-circuit lighting loads with each circuit loaded 80%
1 - 6 kw 240v 1 ø load
1 - 9 kw 240v 3 ø load

If only the three-phase load is attached to the high-leg and all other loads are supplied by the neutral transformer, what is the load on the grounded neutral transformer?

(a) 19.2 kw (b) 25.2 kw (c) 28.2 kw (d) 34.2 kw

DESIGN QUIZ #15

•The following questions are to be answered after reviewing pages 266 - 293.

1. Which of the following statements is *false*?

(a) All loads will connect to a line wire, but not all loads connect to the neutral.
(b) Actual connected loads are shown in a load balancing schedule.
(c) The Code permits reducing the size of the connected load when determining the service size.
(d) The neutral carries the balanced current.

2. Which of the following statements is *false*?

(a) The reason for balancing loads on the neutral is to ensure that the neutral conductor never has to carry a large amount of current.
(b) The 1500va laundry calculated in the service size is for the washing machine.
(c) The kitchen is the heaviest loaded area of the house, 120v.
(d) 70% neutral demand can be applied to a household water heater.

3. Which of the following statements is *false*?

(a) The panelboard size is determined by the connected load it will supply.
(b) The main breaker cannot be larger than the rating of the panelboard.
(c) The main breaker can be rated less than the rating of the panelboard.
(d) If the panelboard is used for service equipment, it must be marked suitable for service.

4. Which of the following statements is *false*?

(a) Different manufacturers of panelboards make different ampere ratings and spaces.
(b) The three-phase wye will have one transformer larger than the other two.
(c) The wye connected transformer provides better balancing of the neutral loads.
(d) The delta has a high-leg.

5. Which of the following statements is *false*?

(a) The electrician is responsible for the field routing of the circuits.
(b) There is no table listing the numbers of conductors permitted in a gutter.
(c) A gutter shall not extend more than 20 feet beyond the equipment it supplements.
(d) It is not the task of the designer to size each piece of equipment.

6. Which of the following statements is *false*?

(a) A wireway cover may be hinged.
(b) A #2 wire can be connected to a #14 wire.
(c) A wireway can only extend 30 feet beyond the equipment.
(d) Table 373-6a applies to conductor bending space.

7. Which of the following statements is *false*?

(a) A gutter can enclose a switch.
(b) The staggering of splices will minimize the area taken up in a gutter.
(c) 30 conductors can be installed in a gutter without applying Note 8.
(d) The area of square inch for a #4/0 THHN is .3278.

8. Which of the following statements is *false*?

(a) The high-leg of a delta generally carries less current than the other two legs.
(b) A panelboard may have a busbar current rating larger than the calculated load.
(c) A panelboard may have a busbar current rating less than the connected load.
(d) A panelboard may have a busbar current rating less than the calculated load.

9. Which of the following is *not* considered a fastened in place appliance in section 220-17?

(a) water heater (b) disposal (c) clothes dryer (d) dishwasher

10. Which of the following generally has a neutral connection?

(a) air conditioner (b) water heater (c) clothes dryer (d) electric heat

FINAL EXAM

ELECTRICAL DESIGNER

FINAL EXAM

•The following questions are to be answered *OPEN BOOK* using your Code book after reviewing the entire book.

1. A branch circuit supplies a 200 amp load 150 feet from the 240v single-phase source. What is the voltage drop on the #3/0 THHN copper conductors? 75°C rated equipment.

(a) 3%   (b) 4.59 volts   (c) 7.2 volts   (d) 8.14 volts

2. What minimum size THHN aluminum conductors would the Code allow for Question #1?

(a) #3/0   (b) #4/0   (c) #250 kcmil   (d) #300 kcmil

3. What size THHN conductors are required for a branch circuit feeding 7 - 400 watt lights, 500 feet from the 240 volt single-phase source? 60°C rated equipment.

(a) #10   (b) #8   (c) #6   (d) #4

4. What is the minimum size THHN conductors for a 125 amp continuous load? 75°C equipment.

(a) #1   (b) #1/0   (c) #2/0   (d) #3/0

5. What size THHN conductors are required for a 20 amp rated branch circuit feeding receptacles? The conduit is above a lay-in ceiling where the maximum ambient temperature reaches 105°F. The equipment is 60°C.

(a) #14   (b) #12   (c) #10   (d) #8

6. What is the 30 day power loss cost of a pump that draws 20 amps and has a voltage drop of 6 volts. The kwh rate is 8¢ and the pump operates 12 hours per day.

(a) $3.46   (b) $2.17   (c) $1.54   (d) $1.12

7. What size gutter is required below?

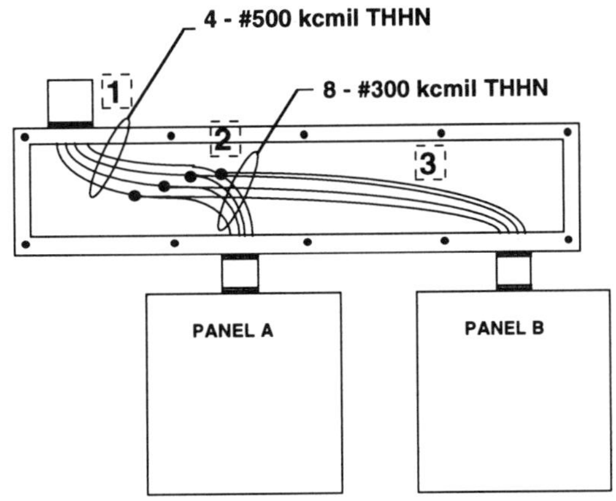

(a) 4" x 4"   (b) 6" x 6"   (c) 8" x 8"   (d) 10" x 10"

8. All conductors in raceways; in or on cars of elevators shall have flame-retardant insulation. Which of the following is a flame-retardant insulation?

(a) THHN   (b) MI   (c) PFA   (d) TFE

9. A pump has a 15-minute rated motor and a nameplate current rating of 22 amps. The required ampacity of the conductors to this pump motor would be ____ amps.

(a) 27.5   (b) 22   (c) 18.7   (d) 19.8

10. An air conditioning hermetic refrigerant motor-compressor has a rated-load of 20 amps marked on the nameplate. What size branch circuit breaker is required?

(a) 20 amp   (b) 25 amp   (c) 35 amp   (d) 50 amp

11. How far can a 20 amp, 120v branch circuit feeding receptacles in an office be run using #12 THHN solid conductors?

(a) 46.6'   (b) 86.4'   (c) 101'   (d) 93.2'

12. The service load is calculated to be 210 amps of which 140 amps is continuous and 70 amps is noncontinuous. What is the minimum size THHN conductors the Code would permit? 75°C equipment.

(a) #2/0 (b) #3/0 (c) #4/0 (d) #250 kcmil

13. What is the minimum size THHN conductors allowed feeding an 80 amp load? 60°C equipment.

(a) #1 (b) #2 (c) #3 (d) #4

14. What is the minimum size THHN conductors allowed feeding an 80 amp load with four current carrying conductors in the conduit? 60°C equipment.

(a) #1 (b) #2 (c) #3 (d) #4

15. What is the minimum size THHN conductors allowed feeding an 80 amp load with four current carrying conductors in the conduit in an ambient temperature of 105°F? 60°C equipment.

(a) #4 (b) #3 (c) #2 (d) #1

16. What is the minimum size THHN conductors allowed feeding an 80 amp *continuous* load? 60°C equipment.

(a) #1 (b) #2 (c) #3 (d) #4

17. What is the minimum size THHN conductors allowed feeding an 90 amp load with four current carrying conductors in the conduit? 60°C equipment.

(a) #1 (b) #2 (c) #3 (d) #4

18. What is the minimum size THHN conductors allowed feeding an 90 amp load with four current carrying conductors in the conduit in an ambient temperature of 105°F? 60°C equipment.

(a) #1 (b) #2 (c) #3 (d) #4

19. What is the minimum size THHN conductors allowed feeding an 80 amp *continuous* load? 75°C equipment.

(a) #1 (b) #2 (c) #3 (d) #4

20. What is the minimum size THHN conductors allowed feeding an 80 amp load with four current carrying conductors in the conduit? 75°C equipment.

(a) #1 (b) #2 (c) #3 (d) #4

21. What is the minimum size THHN conductors allowed feeding an 80 amp load with four current carrying conductors in the conduit in an ambient temperature of 105°F? 75°C equipment.

(a) #2 (b) #1 (c) #1/0 (d) #2/0

22. What is the approximate available short-circuit current from the transformer shown below?

(a) 208,333 amps (b) 120,284 amps (c) 120,192 amps (d) 69,395 amps

23. A 120 volt rated primary has 2.4 volts applied which causes secondary full load current to flow through the shorted secondary. The transformer impedance would be ____.

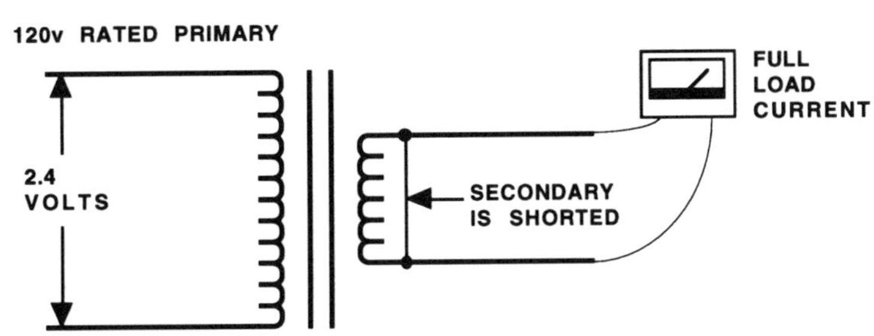

(a) 20% (b) 2% (c) 5% (d) 95%

Shown below is a 3ø, 208/120v, 4-wire feeder supplying three - 15kw loads of electric discharge lighting and a 30kw heater load. 75°C equipment.

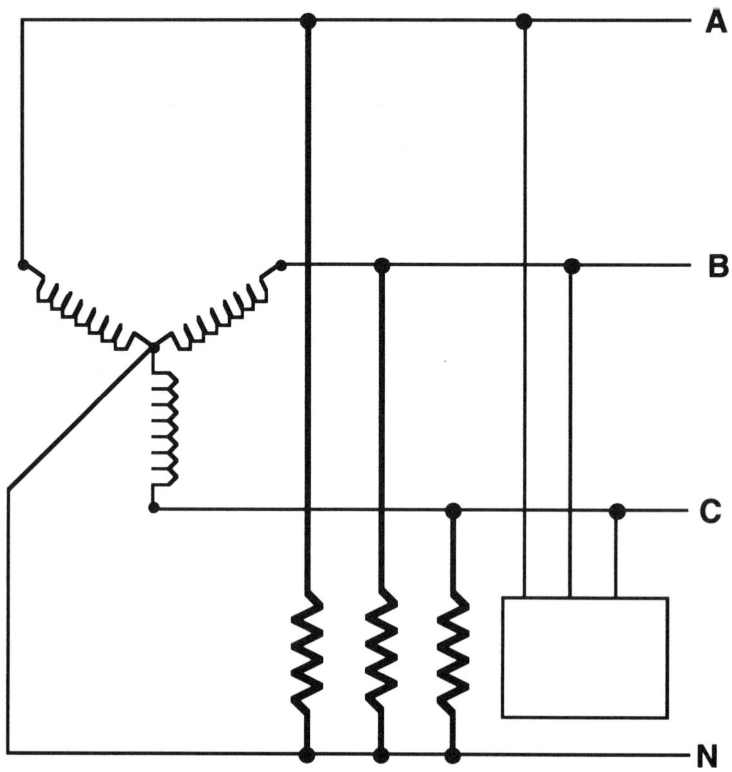

24. The minimum size breaker required to protect phase conductor "A" would be ____ amps.

(a) 200 (b) 225 (c) 250 (d) 300

25. The minimum size THHN conductors permitted would be ____ copper.

(a) #4/0 (b) #250 (c) #300 kcmil (d) #350 kcmil

THE ANSWERS

DESIGN QUIZ #1

1. (c) load
2. (c) 3%
3. (a) 100'
4. (d) 5
5. (a) 1%
6. (d) 800 amps
7. (b) 125a
8. (d) 300a
9. (b) 80a
10. (b) 175a
11. (d) 86°F
12. (c) surrounding
13. (d) more than 3
14. (b) continuously
15. (d) all of these
16. (b) heat
17. (a) more
18. (b) heat
19. (c) W
20. (c) RHH
21. (a) I^2R
22. (a) true
23. (b) false
24. (a) true
25. (b) higher

DESIGN QUIZ #2

1. (a) true
2. (b) 60°C
3. (d) entire circuit
4. (a) true
5. (b) false
6. (d) all of these
7. (b) 130a
8. (d) 290a x 80% = 232a
9. (a) 115a 75°C rating
10. (c) 260a x .87 =226.2a
11. (a) 225a x 80% =180a x .91 = 163.8a
12. (d) 55a
13. (d) 70a
14. (b) 40a
15. (a) 55a x .76 = 41.8a x 80% = 33.44a
16. (d) † 15a
17. (a) † 20a
18. (c) 150a x .75 = 112.5a
19. (c) 380a x .88 = 334.4a x 70% = 234.08a
20. (a) 230a x .67 = 154.1a
21. (b) 55a x .58 = 31.9a
22. (d)
23. (c) 70a x 1.08 = 75.6a
24. (c) 350a x .87 = 304.5a
25. (c) 475a 75°C rating

DESIGN QUIZ #3

1. (c) 100a 75a x 125% = 93.75a
2. (a) #2 THHN with an ampacity of 130a is protected by 100a CB, can be loaded to 60°C 95a
3. (b) #3 THHN with an ampacity of 110a is protected by 100a CB, can be loaded to 75°C 100a
4. (c) 25a CB 20a x 125 = 25a 240-6: 25 amp is a standard size
5. (c) #10 THHN with an ampacity of 40 is protected by 25a CB •Note: A #14 THHN has an ampacity of 20, but † 15a is largest CB, #12 THHN has an ampacity of 25, but † 20a is the largest CB. The 20a *continuous* load requires a *minimum 25a CB.*
6. (b) 175a CB 140a x 125% = 175a 240-6: 175 amp is a standard size
7. (c) #2/0 THW with an ampacity of 175a is protected by a 175a CB, can be loaded to 175a
8. (c) #2/0 THHN with an ampacity of 195 can be protected by the next higher standard size a 175a CB, can be loaded to 75°C 175a
9. (d) 400a CB 375a *noncontinuous* use next higher standard size 400a per 240-3b
10. (d) 500a CB 375a x 125% = 468.75a next higher standard size 240-6 = 500a CB
11. (b) #750 THHN with an ampacity of 520 can be protected by a 500a CB, can be loaded to 75°C 475a
12. (b) #750 THW with an ampacity of 475 can be protected by a 500a CB, can be loaded to 475a
13. (b) 600a CB 480a x 125% = 600a 240-6: 600a is a standard size
14. (a) #1500 THHN with an ampacity of 705 is protected by a 600a CB, can be loaded to 75°C 625a
15. (a) #1500 THW with an ampacity of 625 is protected by a 600a CB, can be loaded to 625a
16. (b) 700a CB 525a x 125% = 656.25a 240-6: 700a is a standard size
17. (a) #2000 THHN with an ampacity of 750 is protected by a 700a CB, can be loaded to 75°C 665a
18. (a) #2000 THW with an ampacity of 665 is protected by a 700a CB, can be loaded to 665a
19. (c) 800a CB 630a x 125% = 787.5a 240-6: 800a is a standard size
20. (d) 2-#600THHN with a total ampacity of 950 is protected by a 800a CB, can be loaded to 75°C 840a
21. (d) 2-#600 THW with a total ampacity of 840 can be protected by a 800a CB, loaded to 840a
22. (c) 1200a CB 900a x 125% = 1125a 240-6: 1200a is a standard size
23. (d) 2-#1500 THHN with a total ampacity of 1410 can be protected by a 1200a CB, can be loaded to the 75°C ampacity of 2 x 625a = 1250 amps
24. (d) 2-#1500 THW with a total ampacity of 1250 can be protected by a 1200a CB, loaded to 1250amps

DESIGN QUIZ #4

1.	(b) 3 hours	Definitions 100
2.	(c) 80%	384-16c
3.	(b) continuously	Definitions 100
4.	(c) O.C.D.	Footnotes Table 310-16
5.	(c) 150'	distance is 75' x 2 = 150' length of wire
6.	(d) VD x I	page 74
7.	(c) #10 THHN	The load is 20a/.80x.82 = 30.48 required ampacity or a #10 THHN 40 ampacity x .80 (Note 8) = 32a x .82 correction factor = 26.24 maximum load (load is 20a)
8.	(b) 20a CB	#12 THHN = 30 ampacity x 80% (Note 8) = 24a †20a CB
9.	(b) 20a CB	The total load is 16a x 125% = 20a CB
10.	(b) #12 THHN	16a load/.80 = 20 required ampacity - A #14 THHN has an ampacity of 25, but † it can only be loaded to 15a - A #12 THHN has an ampacity of 30, † can load it to 20
11.	(b) #12 THHN	20a/.70 = 28.57 required ampacity - #12 THHN = 30, † load to 20 amps
12.	(b) #12 THHN	16a load/.80 = 20 required ampacity - A #14 THHN has an ampacity of 25, but † it can only be loaded to 15a - A #12 THHN has an ampacity of 30, † can load it to 20
13.	(b) #12 THHN	16a load/.82x.70 = 27.87 required ampacity or #12 THHN = 30 ampacity x .82 x .70 = 17.22 ampacity
14.	(b) #8 THHN	Each circuit can be loaded - 20a CB x 80% = 16 amps - 16a/.35 Note 8 = 45.7 required ampacity or #8 THHN = 55 ampacity x 35% (Note 8) = 19.25 ampacity
15.	(d) 13.5 amps	30a x 45% (Note 8) = 13.5
16.	(d) #8 THHN	20a load/.50x.82 = 48.78 required ampacity #8 THHN = 55 ampacity

DESIGN QUIZ #5

1. (d) I, II and III
2. (b) 10 — 15a x 120v = 1800va/180va = 10 receptacles
3. (a) 38' — $D = \dfrac{4110\text{cm} \times 3.6\text{vd}}{2 \times 12.9054 \times 15\text{a}} = 38.21$
4. (c) 60' — $D = \dfrac{6530\text{cm} \times 3.6\text{vd}}{2 \times 12.9294 \times 15\text{a}} = 60.6$
5. (d) 1.3vd — $VD = \dfrac{2 \times 12.8712 \times 35' \times 15\text{a}}{10380\text{cm}} = 1.3$
6. (a) 3.297vd — $VD = \dfrac{2 \times 12.9054 \times 35' \times 15\text{a}}{4110\text{cm}} = 3.297$
7. (a) 19.5w — 1.3vd x 15a = 19.5
8. (c) 49.4w — 3.297vd x 15a = 49.4
9. (a) 1.2¢ — $\dfrac{19.5\text{w} \times 8 \text{ hrs.} \times .08 = 1.2}{1000}$
10. (a) 3.16¢ — $\dfrac{49.4\text{w} \times 8 \text{ hrs.} \times .08 = 3.16}{1000}$
11. (d) #8 — $CM = \dfrac{2 \times 12.9 \times 100' \times 15\text{a}}{2.4 \text{ vd}} = 16125$ cm required Table 8 = #8 @ 16510
12. (d) #8 — $CM = \dfrac{2 \times 12.9 \times 100' \times 15\text{a}}{3.6\text{vd}} = 10750$ cm required Table 8 + #8 @ 16510
13. (d) 9 amps — $I = \dfrac{6530\text{cm} \times 3.6\text{vd}}{2 \times 12.9294 \times 100'} = 9.09$
14. (b) 14.5 amps — $I = \dfrac{10380\text{cm} \times 3.6\text{vd}}{2 \times 12.8712 \times 100'} = 14.5$
15. (b) 7.5 kw — Find fixed resistance $R = E^2/W$ 240x240/10000w = 5.76Ω
    $W = E^2/R$ 208x208/5.76Ω = 7511w/1000 = 7.5 kw
16. (b) 25 ampacity — Table 310-16 60°C column
17. (b) 20a CB — † Table 310-16 footnotes
18. (c) single — 210-52f
19. (c) washing machine
20. (b) I & II only — 210-3

DESIGN QUIZ #6

1. (b) 10 amps — 210-23a 50%
2. (b) #8 — load/correction factor = 22a/.58 = 37.9 required ampacity
3. (b) #8 — load/correction factor = 21a/.58 = 36.2 required ampacity
4. (d) 5.324 volts — $VD = \frac{2 \times 12.5598 \times 100' \times 22a}{10380cm} = 5.324$
5. (d) 117 watts — Power Loss = VD x I 5.324vd x 22a = 117w
6. (a) \$2.25 — $\frac{117w \times 8 \text{ hrs.} \times .08¢}{1000} =$ .07488 x 30 days = \$2.25
7. (a) \$1.42 — $VD = \frac{2 \times 12.61364 \times 100' \times 22a}{16510cm} = 3.36 \text{ volts}$
   Power Loss = VD x I = 3.36vd x 22a = 74 watts
   $\text{Cost} = \frac{74w \times 8 \text{ hrs.} \times .08¢}{1000} =$ .04736 x 30 days = \$1.42
8. (c) 50amp — 25a x 225% (440-22a) = 56.25 240-6 = 50a
9. (b) 45 amp — 25a x 175% (440-22a) = 43.75 240-6 = 45a
10. (d) I, II and III — 440-52a
11. (c) false — from page 99
12. (a) 50' — from page 100
13. (c) 6' 6" — 380-8a
14. (a) I and III only — from page 103
15. (a) true — 600-11

DESIGN QUIZ #7

1. (b) false — from page 107
2. (b) false — from page 108
3. (c) approx. kva — from page 108
4. (d) false — 1/2 hp from page 108
5. (c) 44 amps — 1 hp x 10.6 (Table 430-7b) = 10.6kva 10,600va/240v = 44a
6. (b) false — inverse time is the regular circuit breaker
7. (a) 601 amps — 240-6
8. (d) 30 amps — 15.2a x 225% (430-52 ex.2b) = 34.2 •cannot exceed 225% = 30a
9. (c) 30 amps — 20a x 150% (T.430-152 code letter "A") = 30a
10. (b) false — over 60 amps from page 114
11. (d) I, II and III — from page 116
12. (c) false — 40°C from page 116
13. (a) false — temperature rise from page 120
14. (c) 12.5 amps — 10a F.L.C. x 125% (430-32) = 12.5a
15. (b) false — the motor starter provides thermal protection, not the relay
16. (c) 60 amps — 50a F.L.C. x 115% (430-110a) = 57.5a = 60a standard size

DESIGN QUIZ #8

1. (b) false	from page 128
2. (b) #10 THHN	28a F.L.C. x 125% (430-22) = 35 required ampacity Table 310-16 a #10 THHN has an ampacity of 40 amps, but it can only be loaded to 60°C, the load is 28 amps
3. (a) 253 feet	Distance = $\frac{41740cm \times 6.24vd}{1.732 \times 12.85592 \times 46.2a}$ = 253'
4. (a) #14 THHN	15.2a F.L.C. x 140% (T.430-22a) = 21.28 required ampacity Table 310-16 a #14 THHN has an ampacity of 25 amps, but it can only be loaded to 60°C, the load is 15.2 amps
5. (c) false	three-wire well pumps require a control box
6. (c) false	from page 138
7. (b) #8 THHN	50a x .71 (630-11a) = 35.5 required ampacity Table 310-16 #10 THHN has an ampacity of 40, but the 60°C ampacity must be selected #8
8. (d) 100 amp	50a x 200% (630-12a) = 100a
9. (c) #6 THHN	50a x .91(630-21a) = 45.5 required ampacity Table 310-16 #8 THHN has an ampacity of 55, but the 60°C ampacity must be selected #6
10. (d) 100 amp	50a x 200% (630-22a) = 100a
11. (c) #6 THHN	75a x .63 (630-31a2) = 47.25 required ampacity Table 310-16 #8 THHN has an ampacity of 55, but the 60°C ampacity must be selected #6
12. (c) 225 amp	75a x 300% (630-32a) = 225a

DESIGN QUIZ #9

1. (c) 5%	24v/480v = .05	
2. (a) 6950 amps	$\frac{50kva \times 1000}{208v \times 1.732}$ = 139a	100/2 = 50 x 139a = 6950
3. (c) 1900 amps	Table on page 155	
4. (d) 8181 amps	Table on page 157	
5. (a) 1981°F	Table on page 161	
6. (a) true		
7. (d) all of these		
8. (b) 1/2 cycle		
9. (d) I, II and III		

DESIGN QUIZ #10

1.	(c) false	*prior* to the final overcurrent device
2.	(d) 23 kva	200 recpts. x 180va = 36,000va/1000 = 36kva Table 220-13 First 10kva = 10 + 26kva x 50 = 13 10+13 = 23
3.	(c) false	smaller conductors have more ampacity per circular mil area
4.	(c) #3 THHN	28a x 125% = 35a + 28a + 28a = 91 required ampacity A #4 THHN has an ampacity of 95, but can only be loaded to 60°C. The load is 28a + 28a + 28a = 84 amps. A #3 THHN is minimum.
5.	(d) #6 THHN	30.8a x 125% = 38.5a + 24a = 62.5 required ampacity. A #6 THHN has an ampacity of 75, but can be only loaded to 60°C which is 55 amps. The load is 30.8a + 24a = 54.8a. #6 THHN is the minimum.
6.	(a) 100 amp	80a CB + 24a = 104a, *can't go up on a feeder breaker*, down to 100a.
7.	(a) #4/0 THHN	59.4a x 125% = 74.25a + 46.2a + 30.8a + 30.8a = 182 required ampacity. A #2/0 THHN has an ampacity of 195, but can only be loaded to 60°C. The load is 59.4a + 46.2a + 30.8a + 30.8a = 167.2 amps. At 60°C it would require a #4/0 which can be loaded to 195.
8.	(b) 250 amp	150a + 46.2a + 30.8a + 30.8a = 257.8 Down on a feeder = 250a

DESIGN QUIZ #11

1. Motor #1 = Table 430-150 F.L.C. **143 amps**
Motor #2 = Table 430-150 F.L.C. **59.4 amps**
Motor #3 = Table 430-150 F.L.C. **30.8 amps**

2. Motor #1 = 143 x 115% (430-32) = **164.45 amps**
Motor #2 = 59.4 x 115% (430-32) = **68.31 amps**
Motor #3 = 30.8 x 115% (430-32) = **35.42 amps**

3. Motor #1 = 143 x 125% (430-22) = 178.75 required ampacity T.310-16 = **#3/0 THW**
Motor #2 = 59.4 x 125% (430-22) = 74.25 required ampacity T.310-16 = **#4 THW**
Motor #3 = 30.8 x 125% (430-22) = 38.5 required ampacity T.310-16 = **#8 THW**

4. Motor #1 = 143 x 175% (T.430-152 dual-element) = 250.25 240-6 = **250 amp**
Motor #2 = 59.4 x 175% (T.430-152 dual-element) = 103.95 240-6 = **110 amp**
Motor #3 = 30.8 x 175% (T.430-152 dual-element) = 53.9 240-6 = **60 amp**

5. Motor #1 = 143 x 115% (430-110a) = 164.45 amps minimum = 200 amp **non-fusible** or a **400 amp fuseable disconnect** for a 250 amp branch circuit fuse
Motor #2 = 59.4 x 115% (430-110a) = 68.31 amps minimum = 100 amp **non-fusible** or a **200 amp fuseable disconnect** for a 110 amp branch circuit fuse
Motor #3 = 30.8 x 115% (430-110a) = 35.42 amps minimum = 60 amp **non-fusible** or a **60 amp fuseable disconnect** for a 60 amp branch circuit fuse

6. 21 foot taps in conduit cannot be smaller than 1/3 of the feeder conductor ampacity #400 kcmil = 335 ampacity/3 = 111.6 ampacity required for feeder taps = #2 tap conductors
Motor #1 = **#3/0 THW tap conductor**
Motor #2 = **#2 THW tap conductor**
Motor #3 = **#2 THW tap conductor**

7. 143 x 125% (430-24) = 178.75 + 59.4 + 30.8 = 268.95 required ampacity + 65.92 amps (19 kw Lighting Load) = 334.87 amps = **#400 kcmil**

8. 250 amp fuse + 59.4 + 30.8 = 340.2 amps + 65.92 (52.74 x 125%) =406.12
240-6 = **400 feeder fuse**

9. 143+59.4+30.8+52.74=285.94 x 115% (430-110c2)=328.83 amps minimum = **400 amp**

10. 19,000w/208v x 1.732 = 52.74 amps actual lighting load. The fuse size to the continuous lighting load would be 52.74 x 125 (220-10b) = 65.9 = 70 amp fuses. The ampacity of the tap conductors cannot be less than 1/3 of the rating of the feeder O.C.P. device(240-21c2). 400 amp fuse/3 = 133 ampacity required = **#1/0 THW tap conductors to lighting load.**

DESIGN QUIZ #12

1. (a) false	*line* side
2. (b) L2	high-leg
3. (b) false	neutral would be isolated
4. (c) false	a switchboard is much larger than a panelboard
5. (a) CTL	circuit limiting
6. (c) six	
7. (d) 30 amps	
8. (d) false	six
9. (a) false	
10. (d) false	inside or outside
11. (a) grounded	
12. (d) 20 amps	
13. (d)	the 70% demand applies to a feeder or service, not a branch circuit
14. (c) I and III only	
15. (c) false	need a true-rms meter

DESIGN QUIZ #13

1. (b) false — grounding
2. (a) false — *not* designed to
3. (b) #6 — Table 250-95
4. (b) #6 — Table 250-95 and 250-95 ex.2
5. (d) false — they are sized from Table 250-95
6. (d) I and III only — 250-83b,c
7. (d) I, II and III
8. (d) false — #1100 kcmil copper
9. (c) III only — 250-81 electrodes
10. (a) #1/0 — Table 250-94
11. (b) #4 — 250-94 ex.1b
12. (d) 30" — 30" = 2 1/2' 250-81d
13. (a) I only — 250-81a
14. (b) #10 — Table 250-95
15. (b) #8 — Table 250-94
16. (c) #2 — Table 250-95 310-4 does not apply
17. (a) #4/0 — 3x500 = 1500 kcmil x 12 1/2% = 187,500cm Table 8 requires a #4/0
18. (b) #3/0 — Table 250-94 over 1100 kcmil requires a #3/0. The 12 1/2% does not apply to grounding electrode conductor sizing. The 12 1/2% applies to the sizing of bonding jumpers and grounded conductors.
19. (b) #4 — Table 250-94
20. (a) #1/0 — Table 250-94 #1/0 for each conduit
21. (d) #4/0 — 2x750= 1500 kcmil x 12 1/2% = 187,500 Table 8 = #4/0
22. (d) #12 — Table 250-95
23. (a) #1/0 — Table 250-94 250-23b
24. (b) #4 — 250-95 when conductors are adjusted in size to compensate for voltage drop, equipment grounding conductors shall be adjusted proportionately to circular mil area. Table 8 #3/0 = 167,800cm #4/0 = 211,600/167,800 = 1.26 increase in cma. Table 250-95 shows a #6 copper is required for a 200 amp breaker. Table 8 #6 = 26,240cm x 1.26 increase = 33,062 cm required. Table 8 = #4

DESIGN QUIZ #14

1. (a) 35 amps

L1 = 1500w + 1500w + 1200w = 4200w
L2 = 1500w + 1500w = 3000w
The maximum neutral would be L1 load 4200w/120v = 35 amperes

2. (d) 9 amps

To find the neutral current, the unbalanced on the neutral would be the L2 neutral load of 27 amps minus the L1 neutral load of 18 amps = 9 amps neutral current. •Remember, the 9600w and the 4.5kw are 240v connected and are *not* neutral loads.

3. (d) 27 amp

The maximum neutral current would be the L1 loads shut off, the neutral would have to carry the L2 to neutral load of 27 amperes.

4. (b) 100 amps

48 kva - 24 kva = 24 kva balanced on neutral, 12kv L1 to neutral and 12 kva L2 to neutral. Maximum unbalance = 12 kva. 12,000va/120v = 100 amperes.

5. (a) 10 amp

With 9 - 1200w 120v loads L1 would have 5 loads to neutral and L2 would have 4 loads to neutral, the unbalance would be one 1200w load/120v = 10 ampere. •Remember the water heater at 5000w and the heater at 6000w are 240v connected. They do *not* connect to the neutral.

6. (c) 333 amps

Total 100kw, 20kw is 240v connected. This leaves a total of 80kw left which is 120v balanced. Balanced would mean that you have 40kw L1 to neutral and 40kw L2 to neutral. The maximum unbalanced would occur if either line was shut off, leaving a maximum neutral load of 40kw, 40,000w/120v = 333 amperes.

7. (a) 8.66 amps

25a x 25a = 625a	25a x 20a = 500a
20a x 20a = 400a	20a x 15a = 300a
15a x 15a = 225a	15a x 25a = 375a
1250a	1175a

1250a - 1175a = 75a = $\sqrt{75a}$ = 8.66a

DESIGN QUIZ #14

8. (c) 500 amperes

Maximum neutral = 120,000va/120v = 1000a in secondary ratio 2/1, primary neutral transformer "C" would have to carry 1000a/2 = 500 amperes

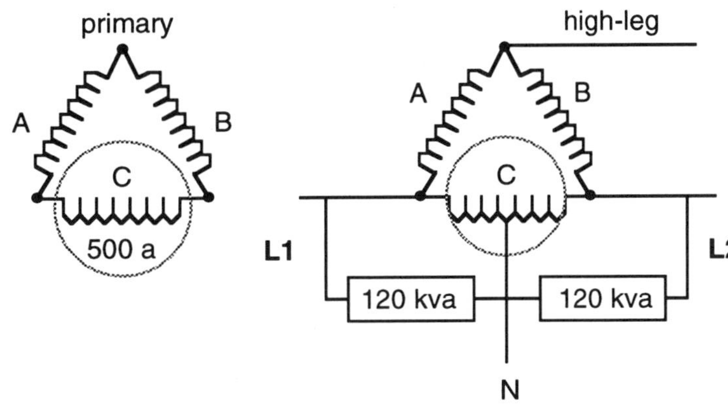

9. (b) 3000 watts

The maximum neutral load would be 3000w with the loss of the 1500w load and either 3000w load.

10. (b) 50 amp

PHASE A	PHASE B	PHASE C
3000w	3000w	3000w
3000w	-------	--------
6000w	3000w	3000w

With the loads on phases B and C off, the maximum neutral current would be 6000w/120v = 50a

11. (a) 50 amps

PHASE A	PHASE B	PHASE C	
2000w	2000w	2000w	
2000w	2000w	2000w	
2000w	2000w	2000w	
6000w	~~6000w~~	~~6000w~~	6000/120v = 50 amps

12. (c) 28.2 kw

PHASE A	PHASE B	PHASE C
3kw	3kw	3kw
		6kw
		19.2kw
		28.2kw

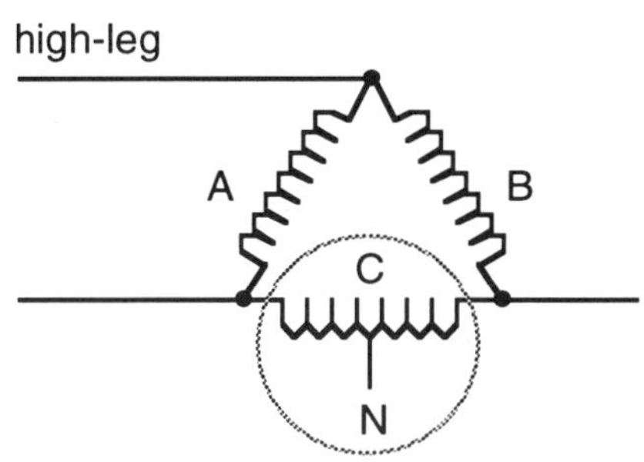

lighting circuits 10 x 20a x 120v x 80% (continuous) = 19.2kw. The *grounded neutral transformer* is "C".

DESIGN QUIZ #15

1. (d) false — the neutral carries the *unbalanced* current
2. (d) false — the 70% demand is for household cooking equipment and clothes dryers
3. (a) false — *calculated* load
4. (b) false — the delta has one transformer ("C") which is larger
5. (c) false — 30 feet
6. (c) false — no limit on wireway, the 30' limit is on a gutter
7. (a) false — *not* permitted in a gutter
8. (d) false — never less than the calculated load
9. (c) clothes dryer — 220-17ex.
10. (c) clothes dryer — 240/120v

FINAL EXAM

1. (b) 4.59vd — $VD = \frac{2 \times 12.85384 \times 200a \times 150'}{167800cm} = 4.59v$
2. (c) #250 kcmil — $CM = \frac{2 \times 21.2 \times 200a \times 150'}{7.2\ vd} = 176{,}666$ cm required

   Table 8 = #4/0 = 211,600cm, but, Table 310-16 @75°C a #4/0 can only be loaded to 180 ampacity. #250 kcmil is required.
3. (c) #6 THHN — 400w/240v = 1.7 amps x 7 lights = 11.9 amp load

   $CM = \frac{2 \times 12.9 \times 500' \times 11.9\ amps}{7.2\ vd} = 21{,}320$ cm required Table 8 = #6
4. (c) #2/0 THHN — 125a x 125% = 156.25a Use 175a breaker. A #2/0 THHN has an ampacity of 195 which is considered protected at 175a and can be loaded to the 75°C ampacity of 175. The load is 156.25 amps.
5. (b) #12 THHN — A #12 THHN has an ampacity of 30 x .87 = 26.1a, but it can only be loaded to the 60°C *ampacity,* which is 25. The load is 20 amps.
6. (a) $3.46 — Power Loss = VD x I 6vd x 20a = 120w

   $\frac{120w \times 12\ hours \times .08}{1000} = .1152 \times 30\ days = \$3.46$
7. (b) 6" x 6" — #300 kcmil THHN = .4608 x 8 = 3.6864 x 5 = 18.432 sq.in. 6" x 6" = 36 sq.in.
8. (a) THHN — Table 310-13
9. (c) 18.7 amps — Table 430-22a ex.
10. (c) 35 amp — 20a x 175% (440-22a) = 35a
11. (a) 46.6' — $\frac{6530cm \times 3.6vd}{2 \times 12.6029 \times 20a} = 46.6'$
12. (d) #250 kcmil — 70a x 100% = 70a

    140a x 125% = 175a

    245a = 250a breaker A #250 kcmil THHN has an ampacity of 290 and can be loaded to 75°C 255 amps. The continuous load is 245 amps.

FINAL EXAM

13. (c) #3 THHN — A #3 THHN has an ampacity of 110 amps and can be loaded to the 60°C rating of 85 amps.

14. (c) #3 THHN — A #3 THHN has an ampacity of 110 x 80% (note 8) = 88a. A #3 @ 60°C is 85 amps. The load is 80 amps.

15. (c) #2 THHN — A #2 THHN has an ampacity of 130 x 80% (note 8) = 104a x .87 = 90.48 a. A #2 @ 60°C is 95 amps. The load is 80 amps.

16. (a) #1 THHN — 80a x 125% = 100a breaker. A #1 THHN has an ampacity of 150 which is protected by the 100a breaker. A #1 @ 60°C is 110 amps. The load is 100 amps.

17. (b) #2 THHN — A #2 THHN has an ampacity of 130 x 80% (note 8) = 104a. A #2 @ 60°C is 95 amps. The load is 90 amps.

18. (b) #2 THHN — A #2 THHN has an ampacity of 130 x 80% (note 8) = 104a x .87 = 90.48a. A #2 @ 60°C is 95 amps. The load is 90 amps.

19. (c) #3 THHN — 80a x 125% = 100a breaker. A #3 THHN with an ampacity of 110 is protected by the 100a breaker. A #3 @ 75°C is 100 amps. The load is 100 amps.

20. (c) #3 THHN — A #3 THHN has an ampacity of 110 x 80% (note 8) = 88a. A #3 @ 75°C is 100 amps. The load is 80 amps.

21. (a) #2 THHN — A #2 THHN has an ampacity of 130 x 80% (note 8) = 104a x .87 = 90.48. A #2 @ 75°C is 115 amps. The load is 80 amps.

22. (d) 69,395 amps — 500,000va/208v/1.732 x 50 multiplier = 69,395

23. (b) 2% — 2.4v/120v = .02 or 2%

24. (b) 225a — 15kw+15kw+15kw+30kw = 75kw or 75,000w/208/1.732 = 208 amps. Next higher size = 225a.

25. (a) #4/0 THHN — A #4/0 THHN has an ampacity of 260 x 80% (note 8) = 208a. It would be considered protected by the next higher size 225 amp. The load is 208 amps.

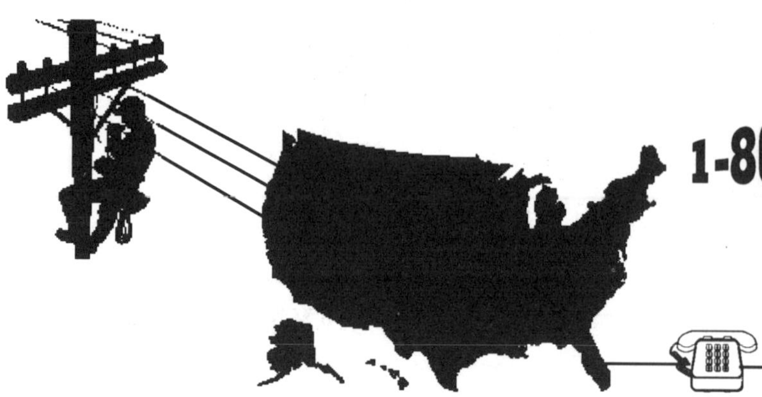